SCLERODERRIS CANKER OF CONIFERS

FORESTRY SCIENCES

Related titles previously published:

1. Baas P, ed: New Perspectives in Wood Anatomy. 1982. ISBN 90-247-2526-7
2. Prins CFL, ed: Production, Marketing and Use of Finger-Jointed Sawnwood. 1982. ISBN 90-247-2569-0
3. Oldeman RAA, et al. eds: Tropical Hardwood Utilization: Practice and Prospects. 1982. ISBN 90-247-2581-X
4. Den Ouden P and Boom BK, eds: Manual of Cultivated Conifers: Hardy in Cold and Warm-Temperate Zone. 1982. ISBN 90-247-2148-2 paperback; ISBN 90-247-2644-1 hardbound.
5. Bonga JM and Durzan DJ, eds: Tissue Culture in Forestry. 1982. ISBN 90-247-2660-3
6. Satoo T and Magwick HAI: Forest Biomass. 1982. ISBN 90-247-2710-3
7. Van Nao T, ed: Forest Fire Prevention and Control. 1982. ISBN 90-247-3050-3
8. Douglas J, ed: A Re-appraisal of Forestry Development in Developing Countries. ISBN 90-247-2830-4
9. Gordon JC and Wheeler CT, eds: Biological Nitrogen Fixation in Forest Ecosystems: Foundations and Applications. 1983. ISBN 90-247-2849-5
10. Németh MV, ed: The Virus-Mycoplasma and Rickettsia Disease of Fruit Trees. 1984. ISBN 90-247-2868-1
11. Hummel FC, ed: Forest Policy: A Contribution to Resource Development. ISBN 90-247-2883-5
12. Duryea ML and Landis TD, eds: Forest Nursery Manual: Production of Bareroot Seedlings. 1984. ISBN 90-247-2913-0

Scleroderris canker of conifers

Proceedings of an international symposium on scleroderris canker of conifers, held in Syracuse, USA, June 21–24, 1983

edited by

PAUL D. MANION
State University of New York
Syracuse, New York, USA

1984 **MARTINUS NIJHOFF/DR W. JUNK PUBLISHERS**
a member of the KLUWER ACADEMIC PUBLISHERS GROUP
THE HAGUE / BOSTON / LANCASTER

Distributors

for the United States and Canada: Kluwer Boston, Inc., 190 Old Derby Street, Hingham, MA 02043, USA
for all other countries: Kluwer Academic Publishers Group, Distribution Center, P.O.Box 322, 3300 AH Dordrecht, The Netherlands

Library of Congress Cataloging in Publication Data

International Symposium on Scleroderris Canker of Conifers (1983 : Syracuse, N.Y.)
Scleroderris canker of conifers.

(Forestry sciences ; v. 13)
Sponsored by State University of New York, College of Environmental Science and Forestry ... et al.
1. Scleroderris canker--Congresses. 2. Gremmeniella abietina--Congresses. 3. Conifers--Diseases and pests--Congresses. I. Manion, Paul D. II. College of Environmental Science and Forestry. III. Series.
SB741.S38I58 1983 634.9'75 83-23799

ISBN-13: 978-94-009-6109-8 e-ISBN-13: 978-94-009-6107-4
DOI: 10.1007/978-94-009-6107-4

Book information

This research sponsored by State University of New York, College of Environmental Science and Forestry; USDA Forest Service, North Central Forest Experiment Station; Canadian Forestry Service; IUFRO Scleroderris Canker Work Party

Copyright

TABLE OF CONTENTS

PREFACE

The "International Symposium on Scleroderris Canker of Conifers" held June 21-24, 1983 in Syracuse, New York was the result of mutual interest expressed by the USDA Forest Service, the Canadian Forestry Service, and the State University of New York College of Environmental Science and Forestry (SUNY ESF). The Symposium was an official function of the working party canker diseases (Scleroderris), Subject group S.2.06.02 Division II IUFRO.

We wish to express our thanks to the USDA Forest Service, Research and the Canadian Forestry Service for financial support of the program, and SUNY ESF for hosting the Symposium.

A major research program was launched in 1977 to study the biology of the European strain as found in North America. Support for this research resulted from special funding by the Congress of the United States to the USDA Forest Service, North Central Forest Experiment Station. Most of these funds were released as cooperative research grants to universities in the Lake States and the northeast. Many of the research papers delivered at this Symposium are the direct result of these research grants.

While the Symposium deliberately emphasized the North American experience with *Gremmeniella abietina* in order to provide a focus for the meeting, we were anxious that this Symposium would also include studies from other countries. Scleroderris canker has been a serious forest disease problem in Europe for much longer than in North America. As a result workers in Europe have been studying the causal fungus and the resulting disease for a longer time than their North American counterparts. Thus, we are especially grateful to the speakers from both Europe and Asia for sharing with the participants the benefits of their knowledge and experience. The papers presented at this Symposium represent the current world state of knowledge on Scleroderris canker. This,of course, was the primary objective of this Symposium. To bring together the majority of the scientists now working on Scleroderris canker and to review the information on this disease in Europe, Asia, and North America. To our knowledge this is the first International Symposium held on Scleroderris canker and represents the largest group of workers ever assembled to discuss this very important forest disease problem.

It is a measure of the interest in Scleroderris canker that this Symposium attracted delegates from ten countries. This is especially gratifying considering the current world wide travel restrictions.

It is our hope that this Symposium will provide: (1) a review and update of the current and historical situation with regards to this disease on a world-wide basis (2) a detailed look at the fungus causing this disease and a better understanding of the disease cycle and population dynamics of the disease (3) last and most important we hope that this Symposium will provide new ideas on how to live with this disease considering the mangerial potentials available in light of the economic liabilities and assets of the forest system.

A special thanks is due to all those people who helped with the organization of this Symposium, the field trip, and the Symposium proceedings. Phyllis Moline spent a great deal of time organizing the mailing lists and handling the correspondence to the speakers, Bruce Schneider and Terrance Podolski handled most of the logistics of the field tour. The special assistance of the Continuing Education staff of the College of Environmental Science and Forestry is graciously acknowledged. The assistance of Frank Hoerner and Bruce Kuntz and Barbara Van de Brake in running the meeting was much appreciated. Frank Hoerner and Bruce Kuntz also assisted editing the manuscript. Linda Boshart of Arlington Computer Service did the final layout and typing of the manuscript for publication.

We wish to express our appreciation for the cooperation we received from scientists and practitioners alike who provided major inputs into the success of the program, including the invited speakers who graciously accepted assignments and deadlines and the attendees of the Symposium.

As a postscript to the preface for Scleroderris canker of Conifers Symposium proceedings we have included a page of photographs illustrating various aspects of the Scleroderris disease and the *Gremmeniella abietina* pathogen. These photographs are from the files of the Co-chairmen of the Symposium.

Paul D. Manion and Darroll D. Skilling
General Program Co-Chairmen

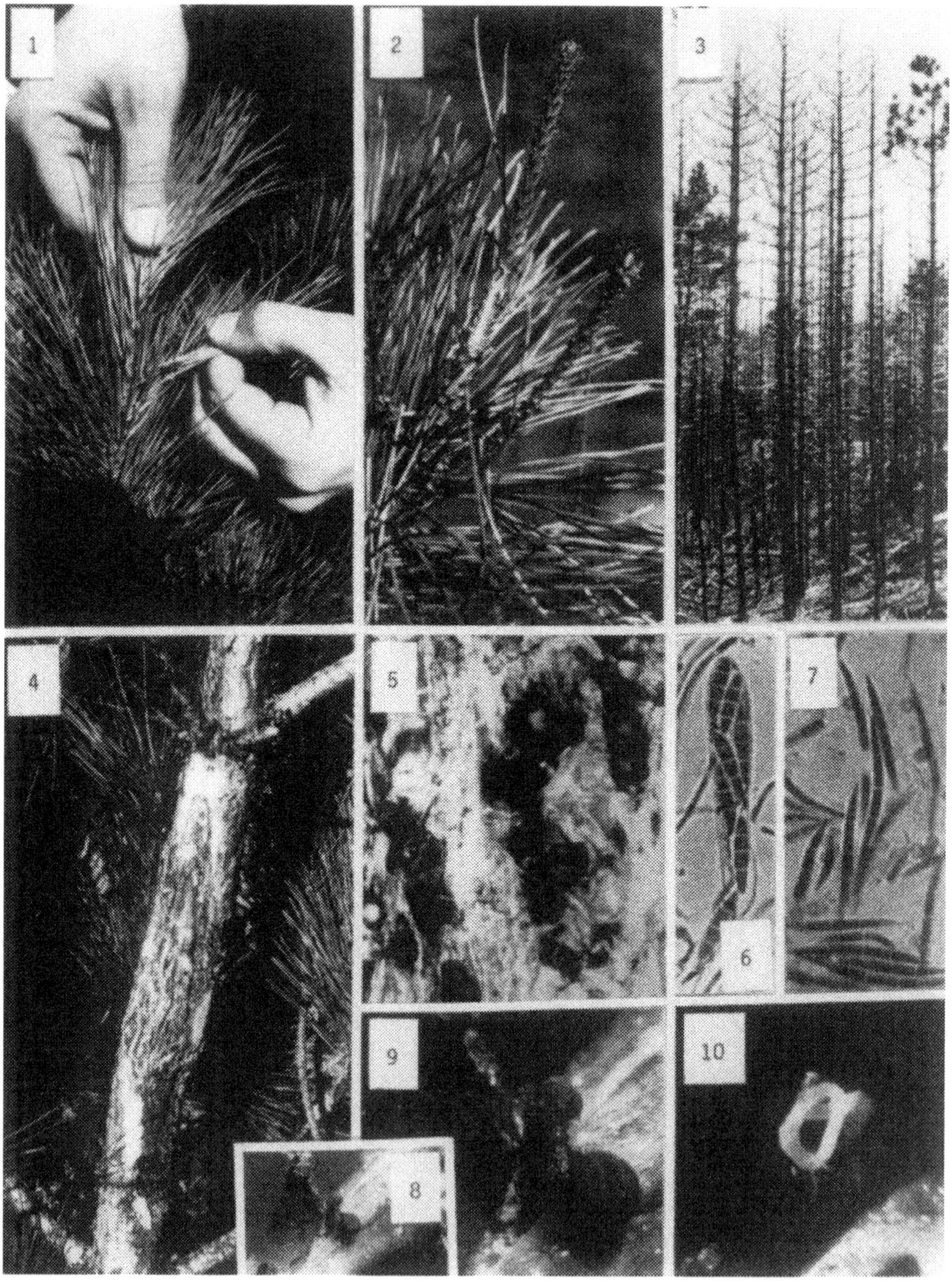
1
2
3
4
5
6
7
8
9
10

Figure 1. "Needle pull test" for diagnosis of Scleroderris canker. In the spring needles on infected branches brown at the base and are easily pulled from the twig.

Figure 2. Gremmeniella abietina infected shoots on lower branches of red pine. The cambium is often yellow-green in color.

Figure 3. "Dieback" phase of Scleroderris canker on large red pines at Boonville, New York.

Figure 4. Stem canker phase of Scleroderris canker. Yellow-green coloration of cambium and wood is often evident around the margin of the canker.

Figure 5. Apothecia of G. abietina.

Figure 6. Ascus and ascospores of G. abietina.

Figure 7. Conidia of Brunchorstia pinea the imperfect stage of G. abietina.

Figure 8. Pycnidia of B. pinea developing under the bract of a needle that has fallen. Pycnidia form on shoots similar to those shown above.

Figure 9. Higher magnification of the pycnidia.

Figure 10. Spore tendril formed from a pycnidium. If suspected shoots are placed overnight in a plastic bag with a wet paper towel, the spore tendrils will develop from mature pycnidia.

WELCOME ADDRESS ON BEHALF OF THE USDA FOREST SERVICE

Arthur L. Schipper, Jr.[1]

Good morning. On behalf of the U.S. Department of Agriculture (USDA), Forest Service, I welcome each of you to this International Symposium on Scleroderris Canker of Conifers.

Dr. Buckman, Deputy Chief for Research in the USDA Forest Service has asked me to express his regrets at being unable to attend, and to stand in for him and welcome you as you embark on what promises to be a most interesting and challenging symposium.

The Forest Service views the Scleroderris canker disease caused by *Gremmeniella abietina* (Lagerb.) Morelet as a real, and a serious, threat to conifer forests and plantations in both the United States and Canada. As you will learn through the papers presented today, tomorrow, and Friday, and on the Thursday field trip, this disease attacks some very important pine species in the northeastern United States and eastern Canada. Its potential for damage to commercial conifer resources in other parts of the North American continent are of great concern.

Research on Scleroderris canker in North America began in 1951 when an unidentified dieback and canker was found on red pine growing in Michigan's Upper Peninsula, and again several years later in Ontario, Canada. Through research in both Canada and the United States, successful disease management strategies were developed for the problem as it occurred in the Great Lakes Region. In 1959, what appeared to be the same disease was found in New York State. However, in New York it was not limited to juvenile trees as in the Great Lakes Region. Instead, mortality of large trees was reported beginning around 1975. Because the course of the disease in New York State was radically different from that in the Midwest, there was great concern about its potential threat to the North American conifer resource. This concern was manifested by the installation of an emergency quarantine of conifers in the affected area and initiation of research by State and Federal scientists.

The Congress of the United States has appropriated over 1 million dollars since 1977 for research on the disease as it occurs in the northeast. These funds, allocated to the North Central Forest Experiment Station, are being used to support cooperative research efforts of university and Forest Service scientists, and state cooperators.

---

1 Staff Research Plant Pathologist, Forest Insect and Disease Research, U.S. Department of Agriculture, Forest Service, Washington, D.C. 20013.

Scleroderris canker research has been an example of effective international cooperative research involving scientists from Environment Canada, Canadian Forestry Service, a number of universities in the northeast and northcentral States, the New York Department of Environmental Conservation, and the USDA Forest Service. We trust that this spirit of cooperation will continue and that future cooperative research on other forest diseases will benefit from this effort as a model.

We welcome this opportunity to participate in a conference that brings together scientists from around the world who are knowledgable about the disease caused by G. abietina. The papers that will be presented here are the result of many years of research.

The internationally acclaimed research scientists included on the program will provide us with state-of-the-art information about the disease, it symptomology, pathology, contributing and mitigating factors, and management tactics. This symposium will help to increase awareness and understanding of the problem, provide information on what can be done now to manage or control the disease, and most important, provide direction to you the scientists for new studies that will lead to an even better understanding of the disease, and new management strategies that will enable resource managers and forest pest management specialists to protect the continent's invaluable coniferous forest resource.

It has been a pleasure to follow reports of progress from the cooperative effort. I look forward to stimulating and informative sessions and the identification of new research thrusts as a result of this symposium.

Again, on behalf of Dr. Robert Buckman, Deputy Chief for Forest Service Research, Dr. Gerald Anderson, Director of Forest Service Forest Insect and Disease Research, and myself, I bid you welcome and convey to you our hopes that you have a fully successful symposium.

WELCOME TO THE SYMPOSIUM FROM THE CANADIAN FORESTRY SERVICE

Lester W. Carlson[1]

The Canadian Forestry Service is extremely pleased to welcome all the participants to this international symposium on Scleroderris Canker of Conifers. It is also our pleasure to cosponsor this meeting with the USDA Forest Service, the IUFRO Scleroderris Canker Work Party, and the State University of New York, College of Environmental Science and Forestry. Some of the participants have traveled far to join us at the Symposium and we hope that they will feel that their effort was justified by the end of the meeting.

Participants at this symposium are here for several reasons, and the concern for the problem of Scleroderris canker certainly has the highest priority. However, some participants have the unique opportunity to obtain continuing Forestry Education credits in conjunction with the Society of American Foresters by attending this symposium. The Canadian Forestry Service wishes to express its thanks to SUNY College of Environmental Science and Forestry for providing this type of educational activity.

The many participants from the Canadian Forestry Service at this symposium demonstrates our concern for the potential devastating effects that Scleroderris canker may have on the Canadian forest industry, especially in eastern Canada.

Forestry has a tremendous impact on the Canadian economy in that one out of ten jobs in our country are related directly or indirectly to the forest industry. Approximately 22 billion dollars per year enter the Canadian economy as a direct result of the forestry sector. This activity leads directly to a positive balance of trade of more than 13 billion dollars per year. Thus, any major threat to this industry has to be considered at all levels of industry and government activity. The portion of wood and pulp supply that depends on the pine species susceptible to Scleroderris canker is equal to 15% of the overall production.

The Canadian Forestry Service research programs on the disease are centered at the Great Lakes Forest Research Centre at Sault Ste. Marie, Ontario and the Laurentian Forest Research Centre, Ste. Foy, Quebec. However, our concerns extend across Canada and each of our six forest research centres and two forest research institutes are concerned and have potential input in solving the problem. We feel that some of the best forest research scientists in Canada have been enlisted to work on the studies of Scleroderris and this

---

[1] Director, Forest Research and Development, Research and Technical Services, Canadian Forestry Service, Ottawa, Canada.

symposium will give them an opportunity to present their research and it will give you the opportunity to contact them and discuss various aspects of the problem.

The Canadian overview and current status of the Scleroderris canker problem will be presented during this meeting. A background brochure on Scleroderris canker in Canada is also available (Gross and Dorworth 1978).

The quarantines that have been imposed by both Canadian and American authorities to impede the spread of the disease will also be addressed. The effectiveness of these quarantines could very well be the subject of extensive discussion throughout the next few days.

This symposium has been placed before you, as plant pathologists, in a format that you should recognize as being classic in how we, as pathologists, approach our research problems. A review of Chapter 1 of "Introduction to Research on Plant Diseases" by Riker and Riker (1936) spells out how a plant disease problem should be reviewed and the outline for this symposium is similar. This familiar approach should allow us to communicate our concerns and research results more efficiently.

The exchange of information and transfer of ideas at a meeting like this is very important. So realizing that lots of information will be passing your way, sharpen your pencils, and your wits.

The CFS again welcomes you and hopes you will find the experience rewarding.

LITERATURE CITED

Gross, H.L. and C.E. Dorworth. 1978. Gremmeniella (Scleroderris) disease of conifers. Can. For. Serv., Great Lakes Forest Research Center, Sault Ste. Marie, Ontario.

Riker, A.J. and R.S. Riker. 1936. Introduction to research on plant diseases. John S. Swift Co. Inc., St. Louis. 117p.

HISTORY OF SCLERODERRIS CANKER IN CANADA

Charles E. Dorworth[1]

ABSTRACT

General opinion supports the view that Scleroderris canker in eastern Canada, in both disease forms, resulted from infections by exotic or imported isolates. Western pathologists believe that the pathogen in the Rocky Mountains is native to the area, where it does not cause economically important damage. Earliest records of occurrence of *Gremmeniella abietina* (Largerb.) Morlet (=*Ascocalyx abietina*) in both eastern and western Canada are from the 1960s. Structure of plantations established prior to 1940 in present infection zones indicates that the pathogen was not present at that time.

INTRODUCTION

In Canada, the history of Scleroderris canker is nearly parallel with that in the United States. Original identification of the pathogen, then known as *Scleroderris lagerbergii* Gremmen, was made by Professor Erik Jorgensen of the University of Toronto who was on secondment with the Canadian Forestry Service at the time. Dr. John Ohman of the United States Forest Service, North Central Forest Experiment Station, performed the first complete analysis of a disease situation involving the organism in North America. Since that time, scientists in the United States and Canada have cooperated closely.

In Central Canada (Ontario) the pathogen was found near Sault Ste. Marie in the mid-1960s. Damage had occurred earlier as well, but apparently not before approximately 1940. This conclusion was indicated by the observation that older plantations of *Pinus resinosa*, in areas suffering maximum damage attributable to Scleroderris canker, do not exhibit stand openings that are signs of early infection. Furthermore, infections are of increasingly more recent origin as one moves further from the site of the original find near Sault Ste. Marie. All evidence indicates that geographic dissemination of the pathogen occurred mostly if not entirely on nursery seedlings followed by natural transmission of the fungus to adjacent plantings and naturally occurring *Pinus banksiana*. Minor infection of *P. strobus* and *P. sylvestris* occurred as well. The source of an extensive infection in the northern Boreal Forest Region of Ontario remains unexplained.

---

[1] Research Pathologist, Department of the Environment, Canadian Forestry Service, Great Lakes Forest Research Centre, P.O. Box 490, Sault Ste. Marie, Ontario, Canada. P6A 5M7

On the basis of the available reports, it would appear that the history of the disease in Quebec does not differ greatly from that in Ontario, with respect to the North American race. In New Brunswick the disease seems to be one of forest plantations, primarily, and dispersion of the fungus occurred largely on infected nursery stock. Much infected nursery stock was dispersed before nursery managers were aware of the presence of this pathogen in North America, let alone of the means of dissemination.

The fungus is rarely reported in Ontario to the south of latitude 44 N, and when it is found, it does not persist. The Forest Insect and Disease Survey Unit of the Canadian Forestry Service in Fredericton, New Brunswick collected the pathogen in the province of Nova Scotia, but the pathogen disappeared within one or two years. The North American race apparently has a southern limit in Canada, at least. The forest area between the east coast and western Ontario, for example, is dominated, in surface structure, largely by the Canadian or Precambrian Shield, and is relatively uniform.

In the west, infection of Pinus contorta and associated species was traced as far back as the 1960s on specimen collections. Whereas damage to pines in the central and eastern portion of the country is often severe, damage to affected species in the Rocky Mountains is not economically important thus far. Our western pathologists are in agreement that G. abietina is a native pathogen in the west. The two west-central provinces of Manitoba and Saskatchewan have not been totally examined. Complete examination of thousands of hectares of forest containing relatively few roads is expensive and difficult. We would suspect, however, that the pathogen, if present, is not causing damage serious enough to be readily observed through aerial survey.

Shortly after the European race became well known in the states of New York and Vermont, it was collected as well in the provinces of Quebec and New Brunswick. Pathologists and foresters with the provincial and federal forestry agencies have worked in close cooperation with personnel of the Canadian Plant Quarantine Division of Agriculture Canada to avoid inadvertent dissemination of the pathogen into non-affected areas of the country. Considerable effort has been put forth in Quebec and New Brunswick to eradicate the pathogen in instances where this seemed practicable. More recently, the pathogen was found as well in the province of Newfoundland, and primarily on planted pines. Thus far, we have not collected the pathogen, as the European race, in the province of Ontario or further west.

Our work in Ontario with provincial forest managers to eradicate or reduce infection by the North American race in critical areas of southern Ontario is still under review. Mr. C.N. Davis of the Great Lakes Forest Research Centre has effectively taken control of this aspect of the program in my absence, and reports that the control work is relatively successful and expanding. In due course, we hope to leave the provincial forestry group with a practical cost analysis and assessment of effectiveness of the methods employed, which are those we presented approximately eight years ago. During the first part of the interval since discovery of the pathogen in Canada, research progressed largely in Ontario, Quebec and New Brunswick. Now we are fortunate to have active participation from Dr. Yasu Hiratsuka in Alberta, who will present new information at this symposium. Despite our efforts, much potential research activity with the European race is constrained by our desire to eradicate the pathogen, where possible, rather than to work with it. In this respect, our situation differs from that in the United States, as we are still finding the pathogen in somewhat discrete pockets, rather than as widespread generalized infection. Members of the Canadian Forestry Service Forest Insect and Disease Survey Unit are in the best position to provide current information on dispersion of the two races in Canada.

SCLERODERRIS CANKER, Gremmeniella abietina (Lagerb.) Morelet, IN CANADA -- AN OVERVIEW

Thomas E. Sterner[1]

ABSTRACT

Scleroderris canker occurs in all provinces of Canada except Prince Edward Island, Manitoba, and Saskatchewan. Three races of the causal fungus, Gremmeniella abietina (Lagerb.) Morelet, have been determined serologically. The North American race is widely distributed. The European race, first detected in Canada in 1978 generally produces symptoms and damage indistinguishable from those of the North American race. To date, steps have been taken to destroy or closely monitor all material confirmed to be infected with the European race.

INTRODUCTION

Scleroderris canker has been detected in all provinces except Prince Edward Island, Manitoba, and Saskatchewan (Figure 1). The North American race, which commonly causes basal stem cankers and lower branch dieback, has been known to be associated with pine planting failures for over 30 years. The European race, first detected in Canada in 1978 occurs in Quebec, New Brunswick, and Newfoundland. In addition, isolates serologically intermediate to the North American and European races have been found at several locations.

DISCUSSION

North American Race

The North American race of G. abietina is widely distributed in Canada and has been a serious problem in nurseries and young plantations for many years. Extensive studies of the disease have provided a number of control recommendations. In eastern Canada, the disease is most damaging to pines, especially red, jack, and Scots pines, and often kills them during their first decade of growth. Once the trees attain a height of about 2 m, they are relatively safe from lethal attack. In western Canada, where lodgepole, ponderosa, and white pines are the principal hosts, the disease has occurred at only a few scattered locations in Alberta and British Columbia. No significant damage has been attributed to this disease in western provinces.

In Ontario, the North American race has been identified as damaging red and jack pines, primarily on poor sites, for over 30 years. Since 1963, the disease has been detected throughout most of the range of pines in the Province

---

[1] Formerly, Director, Forest Insect and Disease Survey, Canadian Forestry Service, Environment Canada, Ottawa, Ontario, KIA 1G5. Currently, Program Manager, Pacific Forest Research Centre, Victoria, British Columbia, Canada.

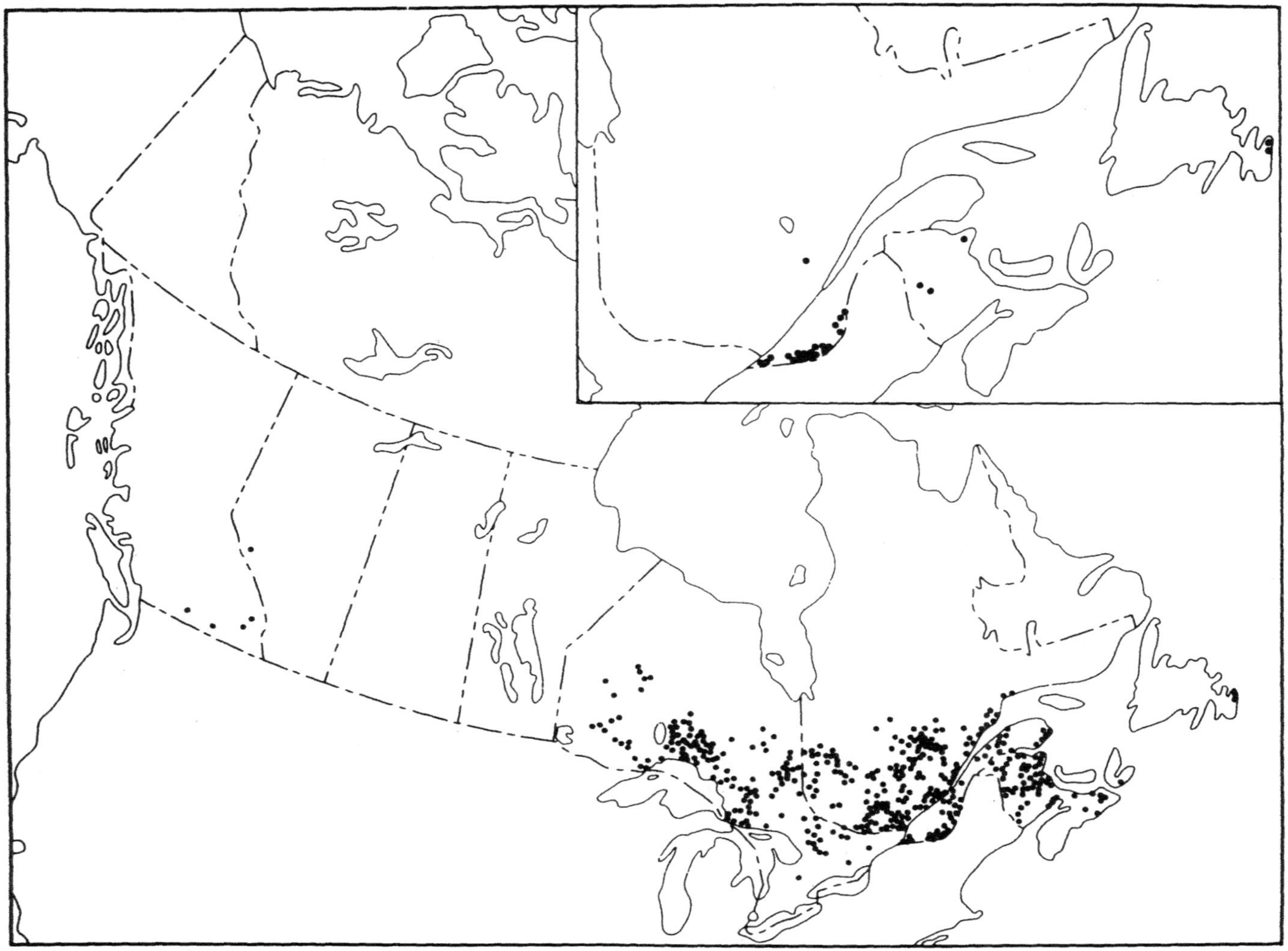

Known distribution of scleroderris canker in Canada. Insert: Locations of collections of European race.

of Quebec. The North American race was first discovered in New Brunswick in 1971 and is now widespread throughout the Province, mostly in jack and red pine plantations. In Nova Scotia, where the disease was first detected in 1972, a few pine plantations suffered limited lower branch mortality during the mid-1970s. No current damage has been found in that Province since 1978.

European Race

In 1975, a disease syndrome different and more serious than that normally associated with Scleroderris canker in North America was reported in New York State. The fungus, which was killing large pine trees, was serologically identified as the European race of *G. abietina*. It has since been detected at scattered locations in Quebec, New Brunswick and Newfoundland (Figure 1). With the exception of one case in Newfoundland, the symptoms and damage caused by the European race in Canada have been indistinguishable from those of the North American race.

In 1978, the European race detected in one plantation in southern Quebec, a few kilometres from the closest infection in New York State. At the end of 1981, this race of *G. abietina* had been identified in several plantations of red, jack, and Scots pines. All infected material found to date has been destroyed. No new infections were found in 1982.

The European race was also confirmed to be present in New Brunswick in 1978. Since then, it and other races serologically identified as intermediate between the North American and European races have been found at other locations. Plantations affected by these races have either been destroyed or are under close surveillance for changes in symptom expression.

Only the European race of *G. abietina* has been found in Newfoundland. Since 1979 it has been detected on red, Scots, jack, and Austrian pines at several locations in and around St. John's Newfoundland. In 1982 it was recorded for the first time outside of the St. John's area, at a location 59 km southwest of the City. Most of these occurrences have been on ornamental trees; however, in 1981 a severe outbreak of the disease was found in a 1.6 ha experimental conifer plantation established in the mid-1950s. Nearly all of the 6,000 red pine trees displayed red needles and shoot dieback throughout the crowns, and more than 30% of the trees were killed. All the infected ornamentals and the entire conifer plantation have been destroyed.

Although the potential of the European race in Canada is not fully known, the situations that occurred in New York State and Newfoundland are cause for concern to those contemplating the future management of pines. For example, in Ontario where this race has not yet been detected, some 6 million cubic meters of pine are harvested on Crown land each year. Most of this harvest is jack pine, but many of the red pines in southern Ontario plantations are approaching merchantable size and the harvesting of this species will likely increase. In addition, there are about 50,000 ha of Scots pine plantations, half of which produce Christmas trees.

Forestry and plant quarantine officials will continue to closely monitor the disease situation in Canada.

SCLERODERRIS CANKER IN THE PROVINCE OF ONTARIO, CANADA

Henry L. Gross[1]

ABSTRACT

The Scleroderris canker disease syndrome is sufficiently distinct, that explicit early reports of dying pines in Ontario in the 1950's can be interpreted as caused by Gremmeniella abietina (Lagerb.) Morelet. The pathogen was isolated in 1962. By 1965, when it was established as the cause of the red pine malady, the North American race of G. abietina was well established in three zones. The distribution pattern implicated two industrial and two provincial tree nurseries. Production of susceptible pine species was shifted to unaffected nurseries, or fungicides were applied to prevent infection. Since 1970 nursery seedlings have had very little influence on spread of the pathogen. Most pine species grown in Ontario are susceptible to infection. Red pine is the most susceptible species and disease control measures are necessary to grow trees of this species throughout the range of G. abietina in Ontario. Initially jack pine was regarded as a highly susceptible species, but by 1977 research data and field evidence indicated that jack pine could be managed on most sites without provision for disease control. Harsh sites and sites with high inoculum densities are exceptions. Annual detection surveys and serological typing of isolates from across the range of G. abietina show that only the North American race is present in Ontario.

INTRODUCTION

Scleroderris canker of pines caused by the fungus Gremmeniella abietina devastated young conifers in Ontario, as well as in other parts of Canada and the northeast and northcentral parts of the United States. In Ontario the disease has been a serious cause for concern because the major species affected, red pine (Pinus resinosa Ait.) and jack pine (P. banksiana Lamb.), with gross standing volumes of 26,000,000 m^3 and 673,000,000 m^3, respectively, represent a valuable forest resource. The impact of the North American race of G. abietina and the potential impact of the European race on this resource are of great interest to pathologists. Moreover, current control practices used in managing the Scleroderris canker disease, as caused by the North American race of the fungus, represent a contribution by pathologists to the Canadian forests.

---

[1] Disease Survey Officer, Department of the Environment, Canadian Forestry Service, P.O. Box 490, Sault Ste. Marie, Ontario, Canada P6A 5M7

HISTORICAL SITUATION

The North American race of G. abietina has probably been present in Canada since the 1950s (Dorworth 1971). Establishing the role of this parasite in the common failure of pine plantings was difficult. G. abietina was facultative in nature and was difficult to isolate. A variety of other fungi were commonly associated with the dead and dying seedlings. Climatic and site factors were also implicated in planting failures, and the mode of action of the North American race was somewhat different from that reported in the European literature (Dorworth 1971). Finally, in 1962 Jorgenson isolated G. abietina and by 1965 (Ohman 1966) had established it as the cause of Scleroderris canker. Previous to this, the pine malady was known by a variety of names, but descriptions of the problem usually noted the presence of cankers, needle base browning, and green stain symptoms (Benzie 1958; Martin 1964). Frost damage was also prevalent at many of the sites when plantings failed, but early studies showed that frost damage was not a necessary factor in the disease process (Dorworth 1970a and 1971). Since then, ecological and epiphytological investigations by Dorworth (1972a, 1973 and 1979) have identified the influence of climatic, edaphic, and topographic factors on infection and colonization of pines by G. abietina.

Surveys by the Forest Insect and Disease Survey (FIDS) Unit of the Canadian Forestry Service (Sippell et al. 1966) soon established that G. abietina was widely distributed in Ontario, with 84 centres of infection known in 1966. The early distribution pattern seemed to indicate that seedlings infected at nurseries were the principal means of spread (Dorworth 1970b and 1971). The initial report of the occurrence of G. abietina in North America (Sippell et al. 1965) noted a serious problem with infected red pine seedlings at the Swastika provincial tree nursery. Punter (1967) noted that 900,000 infected red pines were destroyed at the Kirkwood provincial tree nursery in 1965, and infected pines were detected the following year at industrial nurseries near Kapuskasing and Longlac Ontario. In Ontario, symptoms of the disease on nursery seedlings are difficult to discern at the time seedlings are harvested for shipment. Also, Dorworth (1970b) presented evidence that additional infection often occurred during storage and shipment, and that entire packages of seedlings could be infected by the time seedlings were planted. Thus, infection of planting stock at the nurseries was a major factor in the spread of G. abietina to new areas and the amount of damage that was occurring in new plantations.

Soon after the distribution hazard was recognized in the late 1960s, production of red pine seedlings was shifted to unaffected nurseries in southern Ontario. Nursery inspections by Dorworth in 1970, annual nursery surveys by FIDS staff, and field checks of young plantations indicated that the diseased planting stock problem had been solved by 1970.

Damage caused by the disease continued to increase on young pines near existing centres of infection, however, and the inoculum potential of G. abietina intensified in many areas. Extensive damage occurred to red pine in the 1960s and early 1970s. Entire plantations over 500 ha in size were completely destroyed. Often attempts to replant these areas met a similar fate. Many foresters suspended planting of red pine in the early 1970s, and this step together with the resolution of the nursery problem seems to have reduced the current Scleroderris canker hazard with respect to growing red pine. Now, disease free planting stock, and adequate provision for disease control, as prescribed by Dorworth (1976), have frequently resulted in successful red pine management where there had been failure in the past. Naturally, intrusion by a more virulent race of the pathogen may change this situation.

Early experience with jack pine (Dorworth 1972b and 1973) seemed to indicate that this species was almost as susceptible to Scleroderris canker as was red pine. However, continued observation and research indicated that *G. abietina* was not a serious threat to either planted or natural jack pine (Dorworth 1977, 1982), except on harsh sites and sites with a high inoculum hazard.

Other hosts that have been infected in Ontario are: Scots pine (*Pinus sylvestris* L.), white pine (*P. strobus* L.), lodgepole pine (*P. contorta var. latifolia* Engelm.), mugho pine (*P. mugho* Turro var. *mughus* Zenari), Austrian pine (*P. nigra* Arnold), and white spruce (*Picea glauca* [Moench] Voss) (Myren and Gross 1977). Both white pine and white spruce are commercially important timber species, but Scleroderris canker is not a major problem with these species. Infection of white spruce has been rare and white pine has been affected mostly when high inoculum densities are present on other infested hosts. Scots pine is important as an ornamental tree and for the production of Christmas trees. Most of the Christmas tree plantations are located south of the area affected by *G. abietina*. Consequently, most of this resource is not threatened by the North American race of *G. abietina*. However, even within the range of *G. abietina*, the disease has not been very destructive on Scots pine in Ontario. Other species listed above were planted for research purposes or as ornamentals and only a few trees have been affected.

CURRENT SITUATION

The present distribution of *G. abietina* is reflected in the distribution map for Canada, which for Ontario shows the locations where the fungus has been collected. Curiously, a large part of southern Ontario remains free of the fungus. In fact, several locations where the fungus was introduced on infected planting stock now appear to be free of this pest. In northwestern Ontario, *G. abietina* seems to be spreading. In this area jack pine is the major host species, but the disease does not seem to be a serious threat to management (Dorworth 1982). However, spread through this area probably is important with respect to westward movement of the fungus.

The North American race of *G. abietina* has been in Ontario long enough that we are able to guage its potential influence on the commercially important host species, and to develop management strategies for growing these species. By far the greatest concern now is potential damage by the European race of the fungus. Damage caused by the European race in the northeastern part of the United States and in several plantations of Newfoundland and New Brunswick suggests a considerably more virulent pathogen. Red pine planting programs have not been adjusted because the quarantine policy to date has been effective, research in the United States is expected to provide guidance for control of the European race of *G. abietina*, and experience with the North American race has resulted in successful containment in Ontario.

Annual detection surveys are conducted by the FIDS Unit in southern Ontario so that effective action can be taken to eradicate or suppress the European race of G. abietina as soon as possible. All plantations are checked by aerial reconnaissance and any evidence of damage that could be caused by Scleroderris canker is checked carefully by a ground survey. All tree nurseries and large concentrations of pine are inspected annually, and as a precautionary measure fungicides are applied to seedbeds containing pine. More intensive surveys are conducted in the Cornwall Administrative District, which is adjacent to Quebec and New York. There survey crews, working under contract for the Ontario Ministry of Natural Resources, thoroughly examine all host tree plantations. FIDS assists by training the survey crews and by processing all suspect material.

Neither race of the fungus has been found in parts of Ontario adjacent to New York and southwestern Quebec where the hazard of spread by the European race is greatest. This simplifies the detection aspects of surveys.

In other parts of Ontario, foresters and FIDS staff continue to watch for evidence of the presence of the European race. In 1980 isolates from known disease infection centres were serologically tested. Fortunately, all were of the North American race.

LITERATURE CITED

Benzie, J.W. 1958. A red pine plantation problem in Upper Michigan. USDA Lake States For. Exp. Stn., Tech. Note 524. 2 p.

Dorworth, C.E. 1970a. Scleroderris lagerbergii Gremmen in the pine replant problem in central Ontario. Dep. Environ.,Can. For. Serv., Sault Ste. Marie, Ont. Inf. Rep. O-X-139. 12 p.

Dorworth, C.E. 1970b. Scleroderris canker in Ontario forest nurseries. Dep. Environ., Can. For. Serv., Sault Ste. Marie, Ont., Inf. Rep. O-X-148. 9 p.

Dorworth, C.E. 1971. Diseases of conifers incited by Scleroderris lagerbergii Gremmen: A review and analysis. Dep. Fish. For., Can. For. Serv., Ottawa, Ont., Publ. No.1289. 42 p.

Dorworth, C.E. 1972a. Epidemiology of Scleroderris lagerbergii in central Ontario. Can. J. Bot. 50:751-765.

Dorworth, C.E. 1972b. Scleroderris lagerbergii Gremmen in the boreal forest of Ontario, Dep. Environ., Can. For. Serv., Sault Ste. Marie, Ont., Inf. Rep. O-X-156. 9p.

Dorworth, C.E. 1973. Epiphytology of Scleroderris lagerbergii in a kettle frost pocket. Euro. J. For. Pathol. 3:232:242.

Dorworth, C.E. 1976. Reducing damage to red pine by Gremmeniella abietina in the Great Lakes-St. Lawrence forest region of Ontario. Dep. Environ., Can. For. Serv., Sault Ste. Marie, Ont., Inf. Rep. O-X-252. 22 p.

Dorworth, C.E. 1977. Relative susceptibility of red pine and jack pine to Gremmeniella abietina. Dep. Fish. and Environ., Can. For. Serv., Ottawa, Ont., Bi-Mon. Res. Notes 33:6.

Dorworth, C.E. 1979. Stand reduction of red pine by Gremmeniella abietina. Can. J. For. Res. 9:316-322.

Dorworth, C.E. and C.N. Davis. 1982. Current and predicted future impact of the North American race of Gremmeniella abietina on jack pine in Ontario. Dep. Environ., Can. For. Serv., Sault Ste. Marie, Ont., Inf. Rep. O-X-342. 18 p.

Martin, J.L. 1964. The red pine mortality problem in the Kirkwood Forest Management Unit, Sault Ste. Marie, Ontario. Can. Dep. For., For. Insect Lab., Sault Ste. Marie, Ont., Inf. Rep. 23 p.

Myren, D.T. and H.L. Gross. 1977. Distribution of organisms causing important forest tree diseases in Ontario. Can. Dep. Environ., Can. For. Serv., Inf. Rept. O-X-262. 136 p.

Ohman, J.H. 1966. Scleroderris lagerbergii Gremmen: The cause of dieback and mortality of red and jack pines in Upper Michigan plantations. Plant Dis. Rep. 50:402-405.

Punter, D. 1967. Scleroderris lagerbergii Gremmen, a new threat to nurseries in northern Ontario. For. Chron. 43:161-164.

Sippell, W.L., B.W. Dance, and A.H. Rose. 1965. Ontario Region. In Annual Report of the Forest Insect and Disease Survey, Can. Dep. For., Ottawa, Ont. 44-65.

Sippell, W.L., B.W. Dance, and A.H. Rose. 1966. Ontario Region. In Annual Report of the Forest Insect and Disease Survey. Can. Dep. For. Rur. Dev., For. Br., Ottawa, Ont. 51-74.

SCLERODERRIS CANKER IN THE PROVINCE OF QUEBEC, CANADA

Denis Lachance[1]

ABSTRACT

Gremmeniella abietina was first collected in Quebec in 1964 on white and black spruce growing in natural stands. The fungus was subsequently collected in plantations on jack pine in 1966 and on red, Scots, white, and lodgepole pine the following year. The disease is now found throughout the province in plantations and natural stands. The European race of the fungus was detected in a 10-year-old red pine plantation in southern Quebec in 1978. An intensive survey of the area the following year revealed 19 more infected plantations. To date, the fungus has been found in 18 red, one jack and one Scots pine plantations. Yearly surveillance and sanitation treatments for the European strain have kept the disease at a very low level and may even have eradicated it.

INTRODUCTION

Scleroderris canker has been known for less than 20 years in the province of Quebec, therefore its historical situation is well documented. However, during this short period, the fungus has spread widely and it is now found in almost every area where pine is grown. The discovery of the European race of the fungus in 1978 brought concern of eventual extensive damage in plantations, but this situation has not materialized. However, this disease has had a noticeable effect on reforestation activities in the province.

HISTORICAL SITUATION: NORTH AMERICAN RACE

The first collections of Gremmeniella abietina (Lagerb.) Morelet, (Scleroderris lagerbergii Gremmen) in Quebec were made in 1964, from dead or dying stems, leaders and branches of young black spruce (Picea mariana, Mill.) B.S.P. and white spruce (P. glauca (Moench) Voss) (Smerlis, 1967). These regeneration trees were growing in sparsely stocked, 80-year-old even-aged natural stands of spruce and tamarack in Laurentide Reserve (Laurentide Park), 75 km north of Quebec City. Seventy per cent of the young white spruce and 54% of the black spruce in this limited area were infected, most of them had small cankers on the stem.

G. abietina was subsequently collected on jack pine (P. banksiana Lamb.) in 1966 and on red (P. resinosa Ait.), Scots (P. sylvestris L.), eastern white

---

[1] Research Scientist, Canadian Forestry Service, Laurentian Forest Research Centre, Sainte-Foy, Quebec, Canada. G1V 4C7

(P. strobus L.), and lodgepole pine (P. contorta var. latifolia) in 1967 in a plantation located about 30 km north west of Quebec City. That same year, the disease was reported from a few other plantations of red and jack pine located in central and western Quebec as well as from young jack pines growing in natural stands east and west of the province. In 1968, the disease was recorded from 20 scattered locations; on several pine species in plantations, on jack pine in natural stands, and on white and black spruce in natural stands in one area in central Quebec (Figure 1).

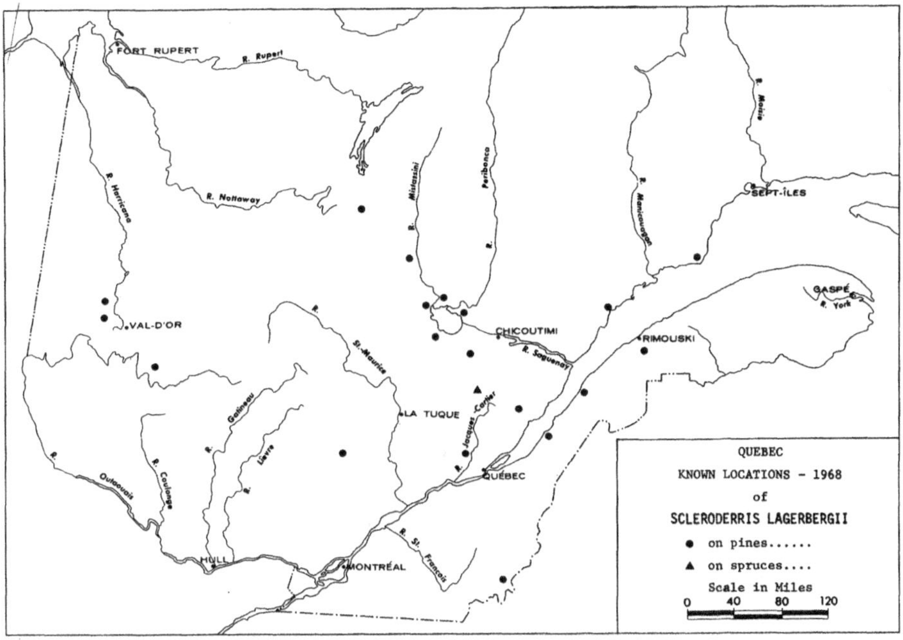

Figure 1. Known distribution of Scleroderris canker in Quebec in 1968.

Ten years after the first collections, Lavallee (1974) published a summary of the situation of the disease. The fungus had then been detected throughout the natural range of jack pine in Quebec. It had been found in several plantations of red, jack and Scots pine. However, the occurrence of the disease was still considered rare along the Ottawa River in southwestern Quebec and south of the St. Lawrence River, considering the large number of pine plantations present in these areas which are located mostly south of the jack pine natural range.

In 1978, our forest insect and disease survey documented an important case of dissemination of the disease through the outplanting of nursery infected red pine seedlings. Seventy-nine per cent of 57 plantations planted with stock originating from a transit nursery, had symptoms of the disease on the seedlings within a year after outplanting. Infection levels varied from 1 to 75%, with an average of 14% infected seedlings. In a large provincial nursery, prophylactic treatments with fungicides were started in 1978 because some diseased seedlings had been detected (and destroyed) there the preceeding year. Beginning in 1979, prophylactic treatments were recommended and have been practiced since in all provincial nurseries for all pine seedling production.

CURRENT SITUATION: NORTH AMERICAN RACE

In less than 20 years, Scleroderris canker has gone from a disease unknown in the province of Quebec, to an almost ever present problem wherever pine trees are pesent. Damage observed varies from year to year possibly because of climatic variation; it varies also with tree age, pine species, and even with the region concerned.

For example, in 1982 a province wide survey of 885 relatively young conifer plantations, mostly red and jack pine, detected Scleroderris canker in 30% of these plantations. Percentage of pine plantations affected varied greatly by region, 67% on the North Shore, 50% in lower St. Lawrence, and 44% in Gaspe peninsula. In southern Quebec, a survey of 1600 pine plantations in 1981 had revealed only 6% of them infected by the fungus. Incidence of the disease in these infected plantations averaged 10.5% of the trees.

Because damage caused by the disease is widespread and sometimes severe, reforestation activities have been modified to either live with or reduce the effect of the problem. Foremost, red pine planting on public lands has ceased since 1978, and has been replaced by white and black spruce, which are now the preferred species. Generally, the production of red and jack pine seedlings in provincial nurseries has remained stable over these years, mainly to satisfy

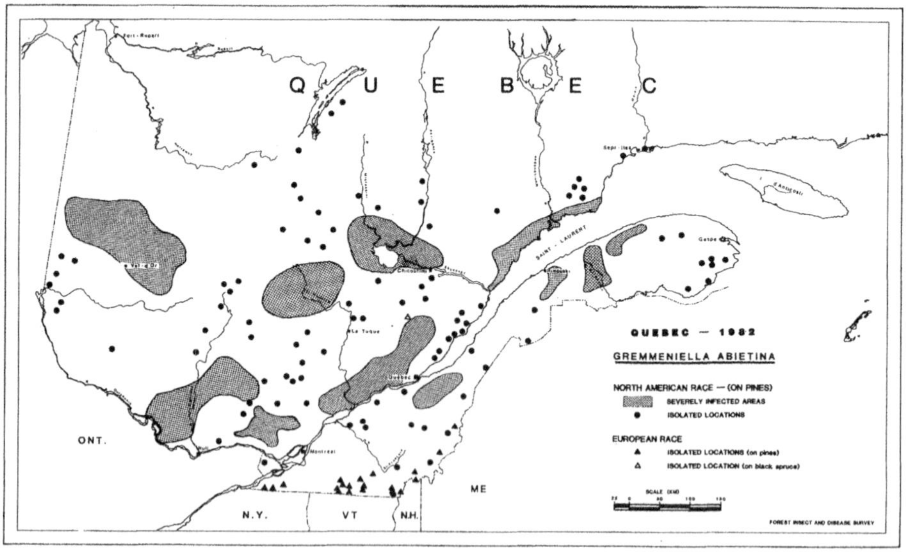

Figure 2. Cumulative occurrences of Scleroderris canker in the province of Quebec up to and including 1982.

the demand of private land owners. Also, pine seedling production in provincial nurseries is now submitted to two yearly insect and disease surveys as well as undergoing a systematic fungicidal spray program, so as to guarantee the outplanting of disease-free seedlings. Parallel to this, all pine seedling lots bought by the Quebec Department of Energy and Resources from private nurseries or from nurseries outside the province, undergo a careful inspection for this disease before being accepted for delivery.

Finally, in 1982, the Outaouais Management Unit of the Quebec Department of Energy and Resources made an intensive survey of 847 red pine plantations and found the disease present in 23% of them. It was also noted during this survey that the disease was not present in pruned plantations. With this information and from our present knowledge of Scleroderris, this management unit has set up a project of pruning treatment in all 5 to 15 year-old red pine plantations located during the survey. Similar actions are becoming common in other areas of the province.

HISTORICAL AND CURRENT SITUATION: EUROPEAN RACE

The European race of G. abietina was first collected in Canada, in 1978 in a 6,000-tree, 10-year-old red pine plantation at Powers Court, Huntingdon County, southern Quebec (Lachance and Benoit, 1978). The location is just a few kilometers north of the U.S.-Canada border and approximately 10 km north of a then known infected plantation in New York state. A tree by tree survey of this newly infected plantation revealed 8 trees with early symptoms of the disease. Following isolation of the fungus, all infected branches were cut and burned, and the plantation was put under quarantine by Agriculture Canada, Plant Quarantine Division.

In the spring of 1979, a detailed survey of all pine plantations located less than 10-15 km away from the U.S.-Canada border, north of the states of New York, Vermont, and New Hampshire, was made by the Forest Insect and Disease Survey Unit of Canadian Forestry Service, in cooperation with that of the Quebec Department of Energy and Resources. Nineteen more plantations were discovered infected with the European race. Seventeen plantations were red pine, one was jack pine, and one was Scots pine. All had very few and very recent infections. Usually less than 10 branches had to be cut in a plantation (of up to 2,000 trees) to eradicate all visible symptoms. Each spring, these 20 plantations where the European race of G. abietina has been found, are examined closely for the presence of symptoms and sanitation treatment is made on the spot if needed. In 1981, symptoms at trace level (2-3 branches/plantation) were found in only three plantations.

In 1979 and 1980, we sent three isolates of G. abietina from black spruce to Great Lakes Forest Research Center for serological race identification. Two of three isolates, growing in different locations in Laurentide Reserve, reacted as European whereas the third isolate had no reaction. On this basis, we report the occurrence of the European race of G. abietina on black spruce in Laurentide Reserve in central Quebec. This situation highlights interest in the delicate and difficult problem of race differentiation. From what we know, this is the same fungus that was reported in black spruce natural regeneration in this region in 1964. The disease situation has not appreciably varied since then.

CONCLUSION

The discovery of Scleroderris canker disease in Quebec in the mid-sixties caused much concern among those planning extensive pine reforestation. The disease caused, and is still causing, noticeable and sometimes extensive damage in several young red and jack pine plantations throughout the province. Red pine, initially envisioned as an exceptional reforestation species has been replaced by white and black spruce. A very positive and stimulating aspect of this disease situation is the relatively prompt action of the forestry community in taking appropriate corrective measures such as prophylactic treatments in nurseries, silvicultural treatments in young plantations, and a gradual shift in the choice of tree species widely planted. Forest pathologists can thus look to the future with confidence since they seem to have influenced forest management decisions as least against this plantation disease.

Finally, the threat posed by the European race of the fungus has not materialized in Quebec. However, we do not know why the situation has remained stable since its discovery in 1978-1979. Therefore the outcome remains uncertain and we must remain vigilant.

REFERENCES

Benoit, P., G. Laflamme, G. Bonneau and R. Picher. 1982. Insectes et maladies des arbres Quebec 1981. Suppl. Foret-Conservation 48(10), 20 pp.

Benoit, P., G. Laflamme, G. Bonneau and R. Picher. 1983. Insectes et maladies des arbres Quebec 1982. Suppl. Foret-Conservation 49(10), 28 pp.

Lachance, D. and P. Benoit. 1978. Quebec Region, In Forest Insect and Disease Survey. Annual Reports 1978 and 1979. Can. For. Serv. pp. 29-37.

Lachance, D., P. Benoit, G. Bonneau, R. Desaulniers and G. Laflamme. 1979. Insectes et maladies des arbres Quebec 1978. Suppl. Foret-Conservation 45(4), 24 pp.

Lachance, D., P. Benoit, G. Bonneau and G. Laflamme. 1980. Insectes et maladies des arbres Quebec 1979. Suppl. Foret-Conservation 46(10), 24 pp.

Lavallee, A. 1974. Le chancre scleroderrien et le reboisement du pin au Quebec. Foret-Conservation 40:23-26.

Martineau, R. and G.B. Ouellette. 1967. Quebec Region, In Annual Report of the Forest Insect and Disease Survey. Can. For. Serv., pp. 34-50.

Martineau, R. and E. Smerlis. 1968. Quebec Region, In Annual Report of the Forest Insect and Disease Survey. Can. Serv., pp. 35-52.

Smerlis, E. 1967. The occurrence and pathogenicity of *Scleroderris lagerbergii* Gremmen in Quebec. Plant Dis. Rep. 51:584-585.

Smerlis, E. 1968. Additional information on the pathogenicity of *Scleroderris lagerbergii*. Plant Dis. Rep. 52:738-739.

SCLERODERRIS CANKER IN THE MARITIME PROVINCES OF CANADA

Laszlo P. Magasi[1]

ABSTRACT

First found in the Maritimes in 1971, the disease is widespread in New Brunswick, was present from 1972 to 1978 in Nova Scotia, and has not yet been found in Prince Edward Island. About one-third of the pine plantations are infected in New Brunswick, the inoculum originating from surrounding areas. Non-North American races have been identified serologically at nine locations since 1978 but European race-like symptoms have not been observed.

INTRODUCTION

Scleroderris canker of pines, caused by the fungus Gremmeniella abietina (Lagerb.) Morelet, has been known in New Brunswick since 1971 (Magasi 1972). It was first found in Nova Scotia in 1972 (Forbes et al. 1973). In spite of frequent and intensive surveys, the disease has not been found in Prince Edward Island, the third province under the jurisdiction for pest surveys by the Maritimes Forest Research Centre. The North American race of the fungus is by far the most common of the various races identified serologically in the Maritimes Region. It is discussed here first by provinces, followed by comments on the occurrence of non-North American races in New Brunswick.

HISTORICAL AND CURRENT SITUATION: NORTH AMERICAN RACE

New Brunswick

The fungus is present in plantations and natural stands of jack pine (Pinus banksiana) and red pine (P. resinosa) and in plantations of Scots pine (P. sylvestris), eastern white pine (P. strobus), Austrian pine (P. nigra austriaca) and pitch pine (P. rigida). The distribution of the disease is general, each circle in Figure 1 representing its occurrence within the UTM grid (10 x 10 km, Universal Transverse Mercator system) without differentiating grids with a single infected location from those where numerous plantations or natural stands are affected. However, 83% of the infected plantations were found in the northern half of the province in 1979, the year of the most recent large-scale survey, indicating a more serious impact there than in the south (Magasi 1980).

---

[1] Head, Forest Insect and Disease Survey, Maritimes Forest Research Centre, Canadian Forestry Service, P.O. Box 4000, Fredericton, N.B., Canada. E3B 5P7

About one-third (34.5%) of the jack pine plantations (the major pine species planted) four years or older are infected in New Brunswick. The following observations suggest that infection originates from the surrounding areas: 1. Planting stock is healthy. In the only case when Scleroderris was identified in a forest nursery the stock was destroyed, 2. No infection has been found in plantations less than 2 years old, and 3. Beyond 3 years of age there is a gradual increase in the frequency of infected plantations (3 yr = 5.3%, 4 yr = 22.2%, 5 yr = 35.3%, 6 yr = 38.5%, 7 yr = 55.5%).

The level of infection (incidence) in affected plantations is extremely variable with up to 60% of the trees infected in jack pine and Scots pine, and up to 100% in red pine plantations. However, incidence in infected plantations less than 10 years old averages 5.5% for jack pine and 9.3% for red pine. The corresponding average incidence in plantations more than 10 years old is 36.8% and 43.1% for jack pine and red pine, respectively.

Nova Scotia

Between 1972 and 1978, the disease was known to occur in a few plantations of red pine, jack pine, and Scots pine, and on a few eastern white pine trees growing wild in some of these plantations (Figure 1). Infected trees exhibited typical disease symptoms, readily yielded *G. abietina* cultures, harbored high numbers of pycnidia, and sustained limited lower branch mortality, but no apothecia have ever been found in Nova Scotia. Since 1978, symptoms have not been seen, all isolation attempts have failed, even from trees known to have been infected previously, and the fungus appears to have died out in the Province. I have no explanation for the disappearance of the disease. Surveys are continuing.

Prince Edward Island

The disease is not known to occur in the Province.

CURRENT SITUATION: NON NORTH AMERICAN RACES

To date, non-North American races of *G. abietina* have been identified serologically at nine locations in New Brunswick (Figure 2). The following summary is based on an annual report of the Forest Insect and Disease Survey (Magasi 1982).

The European race was first found in 1978, at two locations, in samples of red pine in a forest nursery in western New Brunswick and in a plantation in the northeastern part of the Province (Magasi 1979). Eradication and quarantine measure appeared to have controlled the disease. In 1979, it was found and promptly eradicated from a small red pine plantation by the removal of the one infected tree. In 1980, from numerous cultures of *G. abietina*, a single isolate from a red pine plantation and one from jack pine were identified as that of the European race.

G. abietina cultures were identified as "possible European race" at two locations, and as a race "intermediate" between the European and North American races at two other locations.

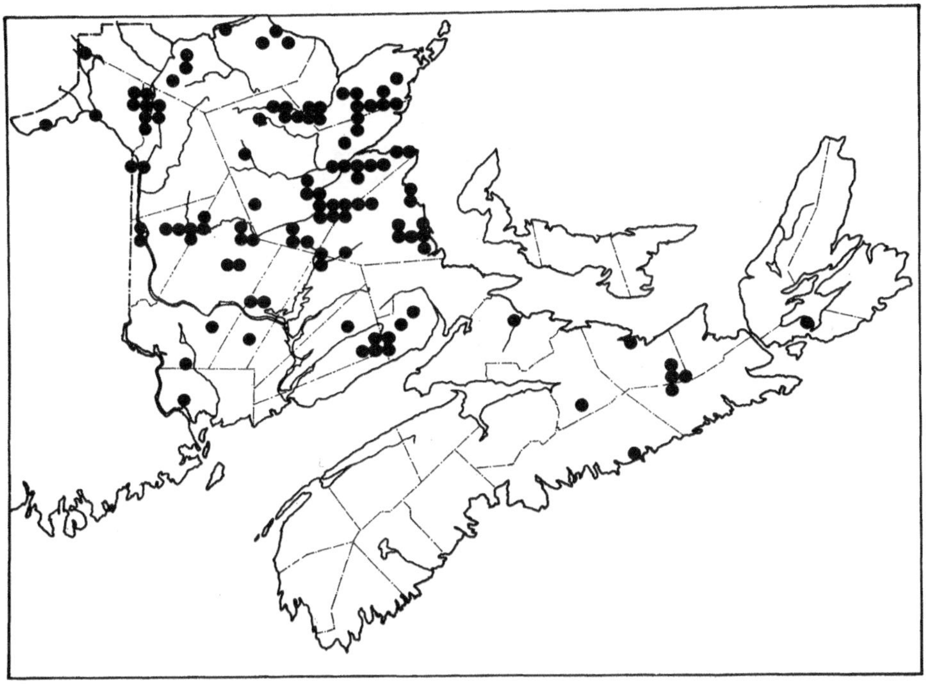

Figure 1. The distribution of Scleroderris canker in the Maritime Provinces of Canada.

Trees are under close surveillance for symptom expression at the six locations where a race of G. abietina other than pure North American has been indicated and where no control measures have been carried out. To date, symptoms on infected trees are indistinguishable from those caused by the North American race.

SUMMARY

Of the three Maritime Provinces New Brunswick is the only one with a "Scleroderris problem," mostly of the North American race of G. abietina. Other races of the fungus have been identified serologically but, to date, no pathological symptoms of these have been observed.

LITERATURE CITED

Forbes, R.S., G.R. Underwood, and G.A. Van Sickle. 1973. Maritimes Region. In Annual Report, Forest Insect and Disease Survey 1972. Can. For. Serv., Ottawa.

Magasi, L.P. 1972. First record of Scleroderris lagerbergii in New Brunswick, Canada. Plant Dis. Rep. 56:245-246.

Magasi, L.P. 1979. Forest pest conditions in the Maritimes in 1978 with an outlook for 1979. Can. For. Serv., Marit. For. Res. Cent. Inf. Rep. M-X-98.

Magasi, L.P. 1980. Forest pest conditions in the Maritimes in 1979 with an outlook for 1980. Can. For. Serv., Marit. For. Res. Cent. Inf. Rep. M-X-106.

Magasi, L.P. 1982. Forest pest conditions in the Maritimes in 1981. Can. For. Serv., Marit. For. Res. Cent. Inf. Rep. M-X-135.

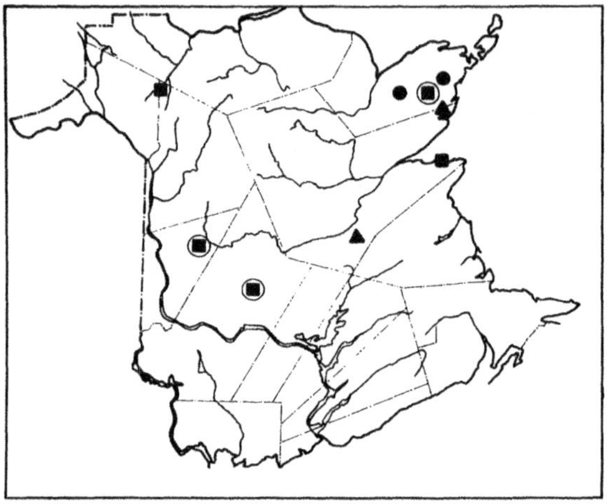

Figure 2. The distribution, type, and disposition of non-North American races of Gremmeniella abietina in New Brunswick. (■ European race; ● possible European race; ▲ intermediate race; ○ eradicated).

SCLERODERRIS CANKER IN NEWFOUNDLAND

J. Hudak and P. Singh[1]

ABSTRACT

Scleroderris canker caused by *Gremmeniella abietina* (Lagerb.) Morelet (= *Scleroderris lagerbergii* Gremmen) was first recorded in Newfoundland in 1979 near St. John's. The disease was detected in 1980 and 1981 at nine other locations in or near the city. Disease development and intensity of infection in a plantation were spectacular. The Province established a quarantine area in 1980 and jointly with Plant Quarantine Division of Agriculture Canada directed control by pruning and eradication. In 1982 the disease was detected at 60 km from St. John's and outside the quarantine area. All cultures of the fungus were serologically identified as the European strain.

---

1 Program Manager and Forest Pathologist respectively. Canadian Forestry Service, Newfoundland Forest Research Centre, P.O. Box 6028, St. John's, Newfoundland, A1C 5X8.

We thank Mr. G.C. Carew for technical assistance. Serological typing of cultures by Dr. C.E. Dorworth of the Canadian Forestry Service and by Dr. D.D. Skilling of the U.S. Forest Service is gratefully acknowledged.

HISTORICAL AND CURRENT SCLERODERRIS SITUATIONS IN THE UNITED STATES

James T. O'Brien[1]

ABSTRACT

Scleroderris canker was first noticed in North America in the Upper Peninsula of Michigan in 1951. When the causal organism, Gremmeniella abietina (North American race), was identified in 1964, the disease was already widespread there and in northern Wisconsin, and in 1969 it was discovered in Minnesota. Most infection was traced to a nursery in Michigan, and spread declined once nursery infection was arrested.

In 1959, the disease, later attributed to the European race, was discovered in New York, where it now occupies 16,300 km^2; it has also been found in Vermont, New Hampshire, and Maine. Quarantines apparently prevent artificial spread, but natural spread will continue.

HISTORY IN THE LAKE STATES

The disease of conifers we now call Scleroderris canker was first noticed in North America in the Upper Peninusla of Michigan in 1951 (Skilling and Cordell 1966). As damage appeared in more and more plantations established in the Upper Peninsula after 1951, concern increased, and in 1957 the USDA Forest Service initiated a research project to find the cause. John Ohman identified the causal fungus, Gremmeniella abietina (Lagerb.) Morelet (then called Scleroderris lagerbergii) in 1964, 2 years after Erik Jorgensen identified the organism in Ontario (Ohman 1966).

A 1965 survey of National Forest lands in Upper Michigan and Wisconsin showed the disease to be widespread in the Upper Peninsula and northern Wisconsin. About 67 percent of the plantations examined -- all established between 1957 and 1963 -- were infected (Skilling and Cordell 1966). Extrapolation from this survey and National Forest records indicated that the infected plantations comprised about 8000 hectares. Skilling and Cordell reported total overall tree losses of 40 percent, one-fourth to one-third of which a later survey indicated could be ascribed to the disease. Surveys of National Forest lands in 1966 did not reveal infection in Lower Michigan or Minnesota.

---

[1] Plant Pathologist, Forest Pest Management, Forest Service, U.S. Department of Agriculture, Durham, NH. For most of the unreferenced material presented, the author is indebted to G. Cooke, P. Manion, D. Fasking and B. Schneider of the State of New York; R. Kelley, A. Gotlieb, and D. Bergdahl of Vermont; and D. Stark of Maine.

In 1966, the disease was found on state land in Lower Michigan and, in 1969, in northeastern Minnesota (Skilling 1970). In 1979, infection was found in recently planted seedlings in Cloquet, Minnesota, 130 km southeast of the nearest previously known infection. The Cloquet infection and the infection in lower Michigan are respectively the western most and southern most known infections in the Lake States.

Most infection could be traced to the Forest Service nursery at Watersmeet, Michigan, though some local natural spread occurred after the trees were planted. Even most of the infected Minnesota plantations had been established with the Watersmeet stock, which had been shipped there in 1959 when the Forest Service nursery in Minnesota could not supply enough seedlings to meet the need.

Thus the evidence suggests that seedlings in the nursery at Watersmeet had become infected by 1950, but the original source of the fungus is not known. The strain of the fungus present in the Lake States is referred to as the North American strain, but there is practically no evidence that G. abietina is native to the Lake States. The opportunities for introduction from elsewhere were many. A large reforestation program was undertaken in the Lake States in the 1930s, and exotic species, such as Scots pine, were included. In Michigan alone, over 430 million conifer seedlings were planted from 1929 to 1944 (Rudolf 1950).

HISTORY IN THE EASTERN STATES

Farther east, another outbreak of Scleroderris canker had begun by 1959, when symptoms were found on trees in a red and Scots pine plantation established in 1956 near Lake Clear in the Adirondack Mountains of New York. The causal organism was identified in 1964 by W.J. French and S.B. Silverborg, who wrote that the disease was present (in 1966) in five plantations within a 16 kilometer radius of Lake Clear (French and Silverborg 1967). The most heavily infected stand reported by French and Silverborg had been established in 1948 to 1950. Subsequently, the disease was found in other areas close to Lake Clear and, in 1971, in the town of Ohio, Herkimer County, about 145 km south of Lake Clear. The plantation was severely damaged, so presumably the disease had been present for some time.

Thus it appears that the fungus had already gained a strong foothold by the late 1950s. As in the Lake States, the original source of infection is unknown. The opportunities for importation were at least as great as in the Lake States. Conifers were planted in the Adirondacks as early as 1870, when T. Dallarme brought seeds and young trees from Germany and planted them on the shores of Round Lake in Oneida county, New York (Paul 1916). The State of New York established a nursery at Lake Clear, probably in the late 1800s, and conifers were planted extensively, particularly after a severe fire in 1903 (1904 report by Forest, Fish and Game Commission). The Lake Clear nursery was closed in 1943 or 1944.

Some investigators suspected a connection between the Lake Clear nursery and spread of the disease but any such link is far from obvious. Nor is there any known connection between northern and southern infections. The infection found in 1971 in Herkimer county was in a plantation established in 1956 with stock from the state nursery at Lowville (established 1928 and closed 1973). Seedlings planted near larger pines north of Boonville (Town of Lyonsdale) in

1961 to 1964 soon died, displaying symptoms of Scleroderris canker. Therefore, the northern infection did not necessarily predate the southern, even though the area of infection in the early years was larger and the present intensity of infection is greater in the north.

In 1972, infection was found in the tops of trees 12 to 15 meters tall in New York, in contrast to our experience in the Lake States, where infection above 2 meters is rare. Setliff and his fellow workers (1975) pointed out that infections high in the larger trees were characteristic of some situations in Europe. Isolates of the fungus were later shown to be serologically identical to isolates from Finland and Norway (Dorworth et al. 1977). The apparently more virulent New York fungus was thereafter referred to as "The European Strain". The North American strain is also present in New York.

An infected area of about 800 km^2 was known in New York in 1966 (French and Silverborg 1967). The true extent was larger, but just how large is uncertain; a reasonably comprehensive survey was not conducted until 1973. The map presented in the survey report indicates an area in the north of about 4000 km^2 and one in the south of about 1800 km^2, for a total infested area of about 5800 km^2. A plant quarantine was enacted in 1977. The quarantine restricted the shipment of coniferous trees (largely Christmas trees) from towns with infected plantations and uninfected "buffer" towns adjacent to them. The original quarantine boundaries encompassed about 10,000 km^2. Within the boundaries, towns with infection comprised about 7300 km^2. By 1980, the quarantine boundaries encompassed some 16,300 km^2,and about 11,000 km^2 was within towns with infection.

In 1971, Scleroderris canker was found in Vermont in a red and Scots pine plantation established in 1960 in the town of Belvidere, Lamoille County (Tattar et al. 1973). Shortly thereafter, infection was found in another plantation in Greensboro, Orleans County. These were the only infections known until 1977, when intensive surveys began. The surveys have shown the disease now present in 92 plantations comprising 309 hectares, and scattered over an area of about 9300 km^2. The quarantine area is roughly 4500 km^2. As in New York and the Lake States, the original source of the Vermont infections is not known. Evidence indicates some spread by infected planting stock, while most was apparently natural spread. The oldest of the four plantations with heavy or severe infection was established in 1960, about the same time the disease was noticed in New York. One would suspect planting stock as the source in this case, because sources of inoculum for natural spread from New York were probably not large in the early 1960s. Of the 92 infections, 46 were classified "trace", which strongly suggests recent infection. Given the massive infection in some New York stands now, New York is probably the source of these recent infections. Serological tests have placed most isolates from Vermont in a category intermediate between the European and North American races (Wendler et al. 1980). However, these tests were more refined than those used for most New York isolates. There is no reason to believe that the isolates of the European strain from New York and the Vermont intermediate isolates are genetically different. Those New York isolates that have been more carefully tested are identical to Vermont isolates.

An infected plantation was found in the town of Jefferson, Coos County, New Hampshire in 1978. The infection was small and of recent origin, so natural spread is indicated. The infection was apparently eradicated. Tests implicated the European strain.

The North American strain was found in Eustis, Franklin County, Maine, in 1978, in a red pine stand of trees from seedling to sawtimber size. The stand is of natural origin (O'Brien and Stark 1979). An infection by the European strain was found in Aurora, Hancock County, in 1980 and another in Washington County in 1982. Both of these appeared to be of recent origin. An infection attributed to an intermediate strain (strong North American characteristics, weak European) was found in 1980 in Coplin plantation, Franklin County, just south of the 1978 find of the North American strain. The disease had apparently been present several years, possibly since the plantation was established (the trees were four to five meters tall in 1980).

When the apparently more virulent European strain of G. abietina was discovered in New York, its potential for spread to uninfected areas created much concern. Quarantines restricting the movement of coniferous trees and parts of trees out of the infected areas were enacted by New York and Vermont in 1977. The Federal government enacted an emergency regulation in 1977 for the same purpose, and later included the town of Jefferson in New Hampshire. The Federal regulation was rescinded in 1983, but the state quarantines remain in effect.

The spread of the European strain into southern Quebec along the United States border and the single infection in New Hampshire almost certainly occurred by natural means. The infection in Hancock County, Maine, 400 km east of the most likely sources of inoculum in New York, is very probably a natural infection. The infection in Washington County also appears to be natural. That spores could be carried 400 km by air currents should come as no surprise. The literature on dispersal of fungi is extensive, and long range dispersal of fungal diseases is well documented. A summary and references are presented by Robert (1962). Blister rust of white pine (caused by Cronartium ribicola), for example, has been spread 300 km and possibly 600 km by air currents (Mielke 1943). And note that a major storm track crosses the United States directly through the Adirondacks and eastern Maine (Johnson and Likens 1972).

As for the scarcity of ascospores in the European strain of G. abietina, ascospores may not be necessary for long range dispersal. Conidia of G. abietina, have been trapped 600 meters from the source, and no doubt under suitable conditions, the conidia could be carried further.

PRESENT STATUS IN THE LAKE STATES

In the Lake States, Scleroderris canker is found in and around plantations, comprising perhaps 10,000 hectares, scattered throughout an area of roughly 20,000 km^2, most of it National Forest land. The area will remain infested indefinitely, and the disease will pose some threat to pine seedlings within it. In fact, a portion of Upper Michigan, some 10,000 hectares in extent, is considered inhospitable to pine seedlings because of the disease (Brown and Orr 1978). In general, however, the disease is in stagnation, possibly even regression, as the trees in infected plantations grow out of susceptible size. Sanitation measures were undertaken in the Michigan nursery in 1966, and beds have been sprayed with fungicides since 1967. Once the production and shipment of diseased planting stock ended, the disease was essentially brought under control.

PRESENT STATUS IN THE EASTERN STATES

The situation in the East is considerably less cheering. The state quarantines are probably adequate, and the likelihood of artificial spread is low. However, natural spread is likely to occur in the future, as it has in the past. When weather conditions similar to those that occurred in the mid 1960s and 1970s occur again, we can expect long range spread, particularly to the north and east of New York. Shorter range spread may occur even when conditions are not quite so favorable for the fungus. More research should be focused on the epidemiological aspects of the Scleroderris problem.

LITERATURE CITED

Brown, H.D. and P.W. Orr. 1978. Forest insect and disease conditions in the United States 1976. Forest Service, U.S. Dep. Agric., Washington, D.C. (U.S. Gov. Print. Off.: 1978-725-103/499).

Dorworth, C.E., J. Krywienczyk and D.D. Skilling. 1977. New York isolates of _Gremmeniella abietina_ (_Scleroderris lagerbergii_) identical in immunogenic reaction to European isolates. Plant Dis. Rep. 61:887-90.

French, W.J. and S.B. Silverborg. 1967. Scleroderris canker of red pine in New York State plantations. Plant Dis. Rep. 51:108-9.

Johnson, N.M. and G.E. Likens. 1972. Atmospheric sulfur: Its effect on the chemical weathering of New England. Science 177:514-516.

Mielke, J.L. 1943. White pine blister rust in western North America. Yale Univ. Sch. For. Bul. 52. 155pp.

O'Brien, J.T. and D.A. Stark. 1979. Scleroderris canker found in Maine. Plant Dis. Rep. 63:194.

Ohman, J.H. 1966. _Scleroderris lagerbergii_ Gremmen: The cause of dieback and mortality of red and jack pines in Upper Michigan plantations. Plant Dis. Rep. 50:402-5.

Paul, B.H. 1916. Reforesting methods and results of forest planting in New York State. Bulletin 374. Agric. Exp. Stn., N.Y. State Coll. Agric., Cornell Univ. 687 pp.

Robert, A.L. 1962. Turbulent transfer of fungus spores. Phytopathology 52: 1095-1100.

Rudolf, P.O. 1950. Forest plantations in the Lake States. USDA For. Serv. Tech. Bull. 1010. 171 pp.

Setliff, E.C., J.A. Sullivan and J.H. Thompson. 1975. _Scleroderris lagerbergii_ in large red and Scots pine trees in New York. Plant Dis. Rep. 59:380-381.

Skilling, D.D. and C.E. Cordell. 1966. Scleroderris canker on National Forests in Upper Michigan and northern Wisconsin. USDA For. Serv. Res. Pap. NC-3. 10 pp.

Skilling, D.D. 1970. Scleroderris canker on red and jack pine in Minnesota. Plant Dis. Rep. 54:132.

Tattar, T.A., H.B. Teillon and E.B. Walker. 1973. Scleroderris canker on red and Scots pine in Vermont. Plant Dis. Rep. 57:338.

Wendler, P.B., A.R. Gotlieb and D.R. Bergdahl. 1980. A newly discovered serotype of *Gremmeniella abietina*. Can. J. Bot. 58:1923-1928.

BRUNCHORSTIA DIEBACK IN EUROPE

John N. Gibbs[1]

ABSTRACT

Gremmeniella abietina, or its imperfect state Brunchorstia pinea is well distributed in Europe. Damage is most severe in young plantations of Pinus nigra in the north and west of the Continent. Within these plantations a high proportion of the crop may show crown dieback. In recent years serious disease has also appeared in some stands of Pinus sylvestris.

There is some evidence that a distinct race of G. abietina occurs in the Alps but more work is required if the full extent of variation in the fungus is to be determined. This work may shed light on the differences in fungus behavior which occur in different parts of its range. Of special interest is the fact that the frequency of apothecia production is very much lower in some areas than others.

Infection of current years shoots of P. nigra occurs in early to mid-summer, although symptoms are not usually expressed until the following winter. The influence of environmental factors on the host pathogen relationship during this long period merits investigation.

Studies in various countries on disease distribution within plantations have commonly shown that damage is more severe on trees that are shaded or sheltered.

INTRODUCTION

Scleroderris canker is just one name given to the disease of conifers caused by the fungus which is now known as Gremmeniella abietina (Lagerb.) Morelet. In Britain it is called Brunchorstia dieback and this name illustrates two important features of the disease as it occurs in Europe. Firstly the form of the fungus that is associated with the disease is often the imperfect stage Brunchorstia pinea (Karst.) Hohn., and secondly, the disease is usually characterized by the death of shoots or parts of shoots. In severe attacks most, or all, of the tree may die.

THE CAUSAL ORGANISM AND ITS DISTRIBUTION

There are fortunately few tree pathogens which have been the subject of such confusion as has G. abietina (for a full account see Dorworth 1971,

---

[1] Principal Pathologist, Forestry Commission Research Station Alice Holt Lodge, Wrecclesham, Farnham, Surrey U.K. - GU10 4LH.

Morelet 1980a). The first description of the fungus and disease was provided by Karsten (1884) in Finland and in Scandinavia there has long been a clear recognition that *B. pinea* is the imperfect state of *G. abietina*. However for most of Europe the waters were seriously muddied by Schwarz (1895) who reported that *B. pinea* was the conidial stage of the common ascomycete *Cenangium ferruginosum* Fr. This blunder led to the publication of many erroneous reports on the disease; and it was only with the pains-taking studies of Gremmen (1965) that the record was set straight. From the data available to him, Gremmen concluded that the disease showed two distinct areas of distribution - one being in the Alps and the other embracing Scandinavia, Great Britain and part of Northern Germany and the Netherlands. Subsequent work has expanded the known distribution of the disease, and in the process united the two areas identified by Gremmen. Thus Siepmann, Lang and Schonhar (1975) have published a map of West Germany showing that *G. abietina* is present throughout the country from the Baltic coast to southern Bavaria. Morelet (1980b) has reported the disease from many departments in Eastern France, and it is also known from Czechoslovakia (Jancarik and Urosevic 1973) Poland (Donaubauer 1974) and parts of the eastern USSR (Dorozhkin and Fedorov 1977; Krutov 1979). As is often the case, it is difficult to decide whether the new records reflect a spread of the disease or the activity of mycologists. In France at least, the fungus seems to have appeared fairly recently (Morelet 1980b) and this may be true of other areas as well.

In North America, Schwarz's error had unfortunate consequences also (Dorworth 1971) and the disease has been known with certainty only since 1965. In Japan, *G. abietina* has recently been described from *Abies* (Yokota, Uozumi and Matsuzaki 1974).

HOST RANGE

G. abietina is now known to be capable of infecting species of *Abies*, *Larix*, *Picea*, *Pinus*, *Tsuga* and *Pseudotsuga*, but pine and spruce are the genera most seriously affected (Skilling 1981).

In Europe the Austrian pine (*Pinus nigra* var *nigra*) Arnold and the Corsican pine (*P. nigra* var *maritima*) Melv. have been widely planted outside their natural range, for example in Scandinavia, Great Britain, the Netherlands and north Germany; and it is in such plantations that *G. abietina* has had its most severe effect. Many records attest to the fact that the damage is particularly serious in young crops - between 8 and 25 years old (Roll-Hansen 1972; Siepmann, Lang and Schonhar 1975; Morelet 1980b). The fungus has been found within the natural range of *P. nigra* var *nigra* in Austria but disease losses have not been great (Donaubauer 1968, 1974).

There are many records of *G. abietina* on *Pinus sylvestris* L. but damage has often been insignificant. There are however exceptions in the losses which occurred in many nurseries and plantations of Scots pine in Scandinavia during the period after 1960. In these situations *G. abietina* may not have been acting alone; and Roll-Hansen (1972) has drawn attention to the possible compounding effects of other fungi, climatic damage and plantation management systems. Climatic factors have also been implicated in the severe damage to 15-30 year old Scots pine stands in Finland which occurred in the late 1970s (Kurkela 1981). During the last few years *G. abietina* has been associated with serious dieback in 25 year old Scots pine plantations in southeast Scotland and northeast England.

Among the other European pines, P. cembra L. may be heavily attacked, as may P. uncinata Mirb., the tree-like form of P. mugo Turra. There are also records for a range of other native and exotic species (Stephan 1979; Morelet 1980b). Of special importance is P. contorta Loud., which is a major plantation tree in several northern countries. Recent reviews by von Weissenberg (1975) and Roll-Hansen (1978) have indicated that early fears about the susceptibility of this species have not been confirmed.

Norway spruce (Picea abies L.) has generally been considered a much less important host than pine, with disease often occurring only where the spruce had been growing under or adjacent to severely affected pines (Gremmen 1966a). Recently however G. abietina has been implicated in widespread damage to 15-20 year old plantations of Norway spruce in southwestern Sweden. A survey in 1977 indicated that 3-4% of the trees were affected (Barklund and Rowe 1981).

VARIATION IN THE PATHOGEN

The literature on variation in G. abietina has recently been reviewed by Dorworth (1981). For many years it has been suggested that a distinct race of the fungus exists on P. cembra in the Alps (Ettlinger 1945). This race, which has been named Brunchorstia pinea var cembrae by Morelet (1980a), has conidia which are typically 5-7 septate and up to 73 u long. This compares with isolates from elsewhere ('var pini') which are usually 3-septate and up to 55 u in length. Donaubauer (1974) reported that isolates from P. nigra and P. cembra retained their spore characteristics following inoculation into both hosts. Cultures of var cembrae are reported to be paler and to grow faster than those of var pini (Morelet 1980a). Dorworth and Krywienczyk (1974) conducted cultural and immunological studies on the fungus and reported that a sample of European isolates showed consistent differences from a sample of North American isolates. It was suggested that they be considered as separate races. The difference between European and North American isolates have been substantiated by later work (Dorworth 1981) and its importance has been highlighted by the discovery that a severe outbreak of the disease in New York State was cause by a form of the fungus with the serological characteristics of the European strain (Skilling 1977). Subsequent work in North America has indicated that the two races may differ in other respects. Thus the 'European strain' has generally not been found to produce apothecia in North America, while the 'North American strain' produces them abundantly. This is consistent with our knowledge of the behavior of the fungus in some parts of Europe where apothecia are often either absent (as in Britain) or rare as in France (Morelet 1980b) or West Germany (Siepmann 1972). Roll-Hansen and Roll-Hansen (1973) have drawn attention to the contrasting situations in southwest Norway where apothecia are rare, and the rest of the country where they are much more common.

There is also some evidence for differences in pathogenic behavior. The North American strain, is typically restricted to the lowermost 2 m of the crown, with the death of individual branches leading to the formation of cankers on the stem. By contrast, the European strain, both in Europe and North America, frequently causes dieback to the upper crown. (Skilling 1977; Dorworth 1981). In addition Skilling (1981) has indicated that the European strain may have a larger host range than the North American strain and may also be able to cause infection under a wider range of environmental conditions.

In his review Dorworth (1981) indicated a hope that substantial progress towards an understanding of the variation in the fungus would shortly be made. As many consider that all forms of G. abietina currently found in North America may have originated in Europe, it seems likely that much of the research will need to be done on this Continent.

AVAILABILITY OF INOCULUM

In Europe pycnidia are commonly found during the spring and summer on pine shoots showing death of buds and needles, and they may be observed for at least 12 months thereafter (Siepmann 1972). In those parts of Europe where apothecia are formed, they occur in the summer of the second year. Gremmen reported in the Netherlands that in 1966 and 1967 conidia were produced between February and July and ascospores between November and July.

The extent to which a pine may carry its own source of inoculum is highlighted by some recent work. Siepmann (1975) has shown that G. abietina can readily be isolated from dead suppressed branches in isolated stands of P. nigra showing no dieback. The fungus was most readily obtained from recently dead tissue near live-dead junctions and from recently formed necrotic lesions. Subsequent inoculation experiments showed that isolates from these 'naturally pruned branches' were as pathogenic as those taken from twigs with obvious dieback (Siepmann 1978).

INFECTION

For many years it was thought that some kind of damage to the tree was necessary before infection could occur. However, it is now known that while G. abietina may sometimes invade wounds - for example on Scots pine in Norway (Roll-Hansen 1964) and Finland (Kurkela 1981) - such prior damage is not essential. This was first established by Gremmen (1968) working with young Corsican pine that had been brought into the greenhouse. He showed that if young shoots were inoculated with conidia or ascospores in April and May, disease symptoms appeared from December onwards. Subsequently in Britain Young and Strouts (1972) obtained disease on 9 out of 10 potted Corsican pines inoculated with conidia in May and 5 out of 10 inoculated in July. Siepmann (1978) also achieved successful infection of field-grown Black pines in each of the years 1974 to 1976 by inoculating with conidia at the beginning of July. In 1974 and 1975 the same trees were inoculated again in August.

Seasonal susceptibility was further examined by Gibbs (unpublished data). During 1980 2-year-old Corsican pine seedlings were inoculated with conidia in the middle of May, June, July or August. The following spring, disease developed on 17 out of 20 trees inoculated in May and on all 20 trees inoculated in June. There were no symptoms on trees inoculated in July or August.

In Europe little critical work on the environmental conditions necessary for infection has been carried out. Gremmen (personal communication) found that high humidity was required for at least three days after inoculation. The trees inoculated by Young and Strouts (1972) and by Gibbs (unpublished data) were standing in shade and were enclosed in a polythene bag for a week after inoculation. Siepmann (1978) sprayed his trees with water continuously for 12-14 days and then for 8 hours a day for a further 10-19 days.

With a knowledge of the susceptibility of young shoots, several workers have suggested that disease severity may depend, in part, upon summer rainfall. Thus Donaubauer (1972) tentatively linked the serious disease in Austria in 1969 to heavy rain and low temperatures in mid-June 1968. In the same way the marked decline in disease observed during 1971 in several stands of young Corsican pine in Thetford Chase, East Anglia, could well have been due to the very dry May-June of 1970, when rainfall was 60% below average (Young and Strouts 1972).

HOST INVASION AND SYMPTOM DEVELOPMENT

In *P. nigra* there is a long latent period between infection of the young shoot in May or June and the killing of host tissue. The first symptoms are usually observed in November or December when some buds show fresh resin bleeding (Read 1966). Examination of these buds reveals small patches of necrotic tissue, usually near the base. These often originate in the living part of the bud-scales and develop progressively to kill the bud (Siepmann 1972, 1976). Discrete areas of necrotic tissue can also be found in positions remote from buds. Some of these are initiated at the base of the short shoots while others occur within the pulvini. Siepmann (1972) noted that these lesions were rare in January but that they were more common from March onwards. Some lesions remained restricted but others expanded and coalesced.

The status of the fungus during the latent period has been studied in artificially inoculated *P. nigra*. It appears that during the first few months the fungus may exist as mycelium on protected surfaces of the shoot (Lang and Schutt 1974). However by August it has established itself in the dead cells of the epidermis and hypodermis of the long shoots, and in the dead part of the bud scales (Siepmann 1976). During late autumn or winter invasion of the living tissue occurs.

Brunchorstia dieback only becomes evident to the casual observer in spring when needles on necrotic shoots turn brown and buds do not flush. By this time the fungus is well established throughout the cortical and xylem tissue and its presence may be revealed by a yellow-green color in the vicinity of the vascular cambium.

In mild attacks, new shoots are often produced during the course of the summer from buds on surviving portions of the long shoot. These may become diseased in their turn, but occasionally they form the basis for sustained recovery growth. In more severe attacks, older internodes die back. It is sometimes stated that *G. abietina* grows down into these older internodes, but no clear data on this point have been published. With the North American strain of the fungus, where damage often involves lateral branches on young trees, it has been established that *G. abietina* can grow back down the branch during the winter, at approximately one internode per year, until it reaches the main stem. It is under these conditions that the characteristic 'Scleroderris canker' is formed (Dorworth 1971 and personal communication).

Virtually no research has been conducted on the possible effect of environmental factors on the host/pathogen relationship during the period between the initial infection in May and full symptom expression nearly a year later. From field studies of young *P. resinosa* in Canada, Dorworth (1971) suggested that frost might have more influence on post-infection colonization of host tissue than on infection itself and this is the kind of observation that should be followed up by detailed investigation.

PATTERNS OF DISEASE AND DAMAGE

As might be expected from a knowledge of disease epidemiology, many studies have indicated that the disease is most serious where the trees are shaded or sheltered. In northern Holland, Gremmen (1966b) described a 15 year old stand of Corsican pine in which the disease was confined to the western side. He attributed this to the effect of shelter from an adjacent stand of 50 year old Scots pine on the micro-climate at the shoot surface. In northeast England Read (1966) studied disease distribution in young Corsican pine plantations in valleys running east to west. The disease was consistently severe on north-facing slopes and in valley bottoms, while trees on south-facing aspects remained vigorous and healthy. He considered that the influence of shading on host vigor was more important than its effect upon spore germination and infection (Read 1968). It should be noted, however, that Read was not aware of the importance of early summer infection in the etiology of the disease. Further south, in Thetford Chase, East Anglia, the late C.W.T. Young conducted a detailed survey of the disease in a 4.6 ha stand of 17 year old Corsican pine. In the autumn of 1970 plots, each of six trees, were established at 12 m intervals throughout the plantation. Each tree was then allocated a disease rating on the scale 0 (healthy) to 3 (dead) and a mean figure for each plot calculated.

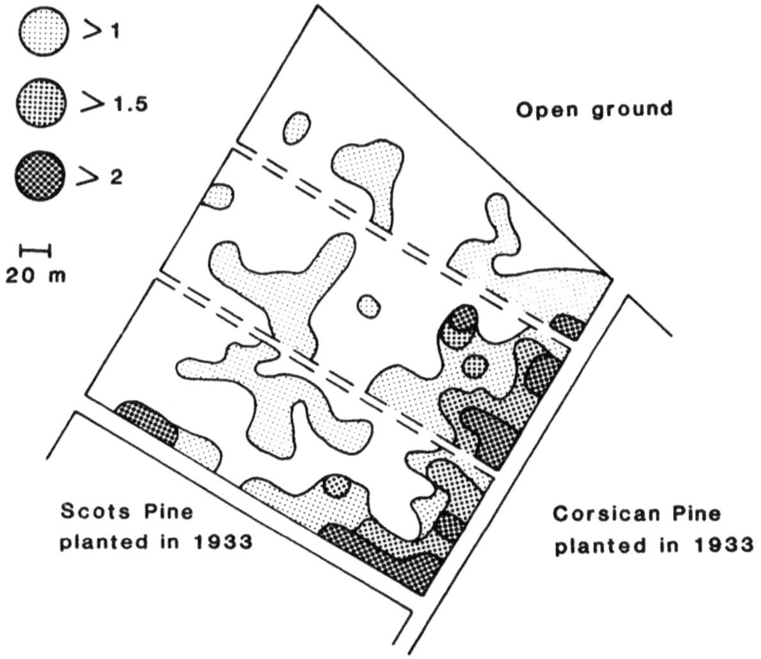

Figure 1. Distribution of Brunchorstia dieback in a 17 year old plantation of Corsican pine, Thetford Forest, 1970. The stippling indicates different degrees of damage - for further details see text.

From Figure 1 it can be seen that the damage was much worse in the southern corner of the plantation where the crop was shaded by older pine. In 1975 Butin and Hackelberg (1978) undertook a retrospective analysis of the course of a *G. abietina* epidemic in a 3.3 ha *P. nigra* plantation in the Harz. By determining the age of the trees at the time of death, they concluded that the first trees to die were concentrated along the southern edge of the plantation where the trees were shaded by an old beech forest.

Some very intriguing unpublished data were obtained by C.W.T. Young in a study of a 16 year old Corsican pine plantation at Kings Forest, East Anglia. Both healthy trees and those which suffered from Brunchorstia dieback in 1970 showed a progressive decline in radial growth during the years 1966-9 (Figure 2). In 1970 the diseased trees deteriorated sharply and these trees barely grew at all in 1971. The healthy trees also produced very little radial increment in 1970 but showed a sharp improvement in 1971. This effect was also shown by healthy trees well away from the affected part of the plantation and was thus not due to compensatory growth. A very similar trend with a marked improvement in 1971 is shown in data from girth band measurements on 400 Corsican pine in a series of mensuration sample plots in Sherwood Forest, Nottinghamshire (Figure 3); and, while the reasons for this pattern of growth are unclear at present, they indicate that care must be taken in interpreting growth ring data. In their Harz study Butin and Hackelberg (1978) noted that on many trees the growth rings decreased in width in the late 1960s and then increased again. In view of the information presented here, their assumption that these trees had suffered from Brunchorstia dieback and then recovered can be called into question.

CONCLUSION

Brunchorstia dieback merits an important place among European conifer disease. There are several reasons for this. Firstly, it is of considerable economic significance. It has greatly restricted the use of Black pine in the cooler and wetter parts of Europe, and even in areas where good crops can be grown it remains a threat, as the widespread damage in the period 1969-70 has

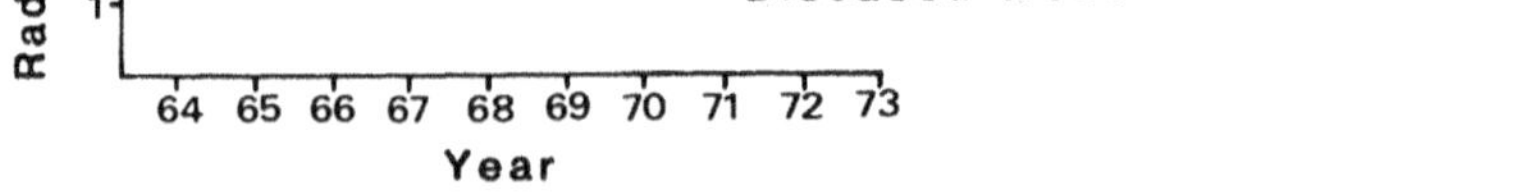

Figure 2. Radial increment in Corsican pine planted in 1955 and felled in late 1971, Kings Forest, East Anglia.

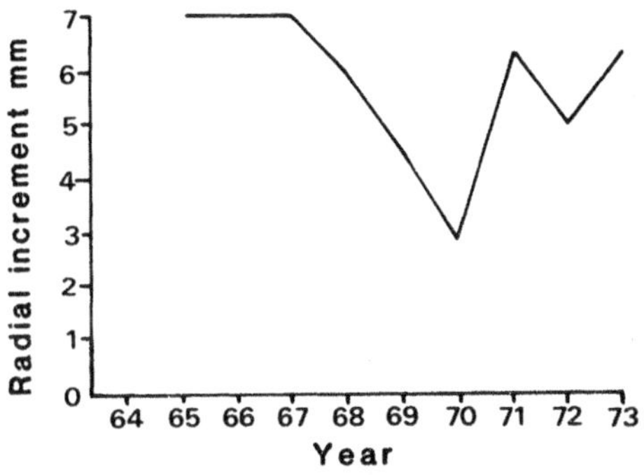

Figure 3. Radial increment in the 400 best trees in sample plots of Corsican pine, planted in 1945, Sherwood Forest.

shown. Although less destructive on Scots pine, there are many records of heavy losses in nurseries, and some in plantations.

Secondly, Brunchorstia dieback is of great biological interest. There are many shoot diseases of pine, but few in which a long latent period during the growing season of the host is followed by tissue colonization in the dormant season. As indicated above, research is needed on the influence of environmental factors both on infection and on host invasion. Answers are also urgently required to questions about the extent of variation in G. abietina and its geographic origin.

LITERATURE CITED

Barklund, P. and J. Rowe. 1981. Gremmeniella abietina, a primary parasite in a Norway spruce die-back. Eur. J. For. Pathol. 11: 97-108.

Butin, H. and L. Hackelberg. 1978. Über den Verlauf einer Scleroderris lagerbergii Epidemie in einem Schwarzkiefernbestand. Eur. J. For. Pathol. 8: 369-379.

Donaubauer, E. 1968. Zum Vorkommen von Brunchorstia pinea (Karst.) v. Hohn. in Österreich. Anz. Schäedlingskd. 41: 24-26.

Donaubauer, E. 1972. Environmental factors influencing outbreak of Scleroderris lagerbergii Gremmen. Eur. J. For. Pathol. 2: 21-25.

Donaubauer, E. 1974. Über das Kieferntriebsterben in Österreich. In '100 Jahre Forstliche Bundesversuchsanstalt'. Forstl. Bundesversuchsanst. Wien 67-98.

Dorozhkin, N.A. and V.N Fedorov. 1977. [Scleroderris canker of pine in Byelorussia] Dokl. Akad. Nauk B.SSR 21: 1040-1042.

Dorworth, C.E. 1971. Diseases of conifers incited by Scleroderris lagerbergii Gremmen - A review and analysis. Can. For. Serv. Publ. 1289: 42 pp.

Dorworth, C.E. 1981. Status of pathogenic and physiologic races of Gremmeniella abietina. Plant Dis. 65: 927-931.

Dorworth, C.E. and J. Krywienezyk. 1975. Comparisons among isolates of Gremmeniella abietina by means of growth rate, conidia measurement and immunogenic reaction. Can. J. Bot. 53: 2506-2525.

Ettlinger, L. 1945. Über die Gattung Crumenula sensu Rehm, mit besonderer Berücksichtigung der Crumenula-Triebsterbens der Pinus - Arten. Beitr. Kryptogamenflor Schweiz 10: 1-75.

Gremmen, J. 1965. Brunchorstia pinea (Karst.) Höhn., en ernstige ziekte van de Oostenrijkse en Corsicaanse den. Ned. Bosbouw Tijdschr. 37: 87-98.

Gremmen, J. 1966a. Aantasting van fijnspar door Brunchorstia pinea. Ned. Bosbouw Tijdschr. 38: 454-457.

Gremmen, J. 1966b. Wat is de Werkelijke oorzaak van de Brunchorstia ziekte van de Oostenrijkse en Corsicaanse den? Ned. Bosbouw Tijdschr. 38:304-309.

Gremmen, J. 1968. Bijdrage tot de biologie van Brunchorstia pinea (Karst.) Höhn., de oorzaak van het taksterven bij Oostenrijkse en Corsicaanse den. Ned.ouw Bosb.ouw Tijdschr. 40:221-231.

Jancarik, V. and B. Urosevic. 1973. First record of Brunchorstia pinea (Karst.) v. Höhn. in a forest nursery in Czechoslovakia. Eur. J. For. Pathol. 3: 121-123.

Karsten, P.A. 1884. Fragmenta mycologic XIV Hedwigia 23:57-63.

Krutov, V.I. 1979. [On the parasitic mycoflora of artificial pine phytocoenoses in the felling areas in the Karelian ASSR and Murmansk region.] Mikologiya I Fitopatologiya 13:342-348.

Kurkela, T. 1981. [Canker and dieback of Scots pine at pre-commercial stage caused by Gremmeniella abietina] Folia For. 485: 12 pp.

Lang, K.J. and P. Schutt. 1974. Anatomishce Untersuchungen zur Infectionsbiologie von Scleroderris lagerbergii Gr. Eur. J. For. Pathol. 4:166-174.

Morelet, M. 1980a. La maladie a Brunchorstia 1. Position systematique et nomenclautre due pathogene. Eur. J. For. Pathol. 10:268-277.

Morelet, M. 1980b. La maladie a Brunchorstia 2. Repartition en France Eur. J. For. Pathol. 10:354-359.

Read, D.J. 1966. Dieback disease of pines with special reference to Corsican pine. I. The nature of the disease symptoms and their development in relation to the crown and to aspect. Forestry 39:151-161.

Read, D.J. 1968. Dieback disease of pines with special reference to Corsican pine. III. Mycological factors. Forestry 41:72-82.

Roll-Hansen, F. 1964. Scleroderris lagerbergii Gremmen and girdling of Pinus sylvestris L. Medd. Nor. Skogforsoeksves. 19:156-175.

Roll-Hansen. F. 1972. Scleroderris lagerbergii: resistance and differences in a attack between pine species and provenances. Eur. J. For. Pathol. Z: 26-39.

Roll-Hansen, F. 1978. Fungi dangerous to Pinus contorta with special reference to pathogens from North Europe. Eur. J. For. Pathol. 8:1-14.

Roll-Hansen, F. and H. Roll-Hansen. 1973. Scleroderris lagerbergii in Norway: Hosts, distribution, perfect and imperfect state and mode of attack. Medd. Nor. Skogforsoeksves. 30:443-459.

Schwarz, F. 1895. Die Erkrankung der Kiefern durch Cenangium abietis. Beitrag zur Geschichte einer Pilzepidemie Verl. Gustav Fischer. Jena: 127 pp.

Siepmann, R. 1972. Frucht körperbildung und zum Infektionsverlauf bei Scleroderris lagerbergii Befall an Schwarzkiefer (Pinus nigra Arnold). Forstwiss. Centralbl. 91:153-160.

Siepmann, R. 1975. Scleroderris lagerbergii Gr. Als Schwäche parasit in Gesunden Schwarzkiefernbeständen (Pinus nigra Arnold) Eur. J. For. Pathol. 5:137-142.

Siepmann, R. 1976. Ein Beitrag zur Infectionsbiologie des durch Scleroderris lagerbergii Gr. Verursachten Schwarskieferntriebsterbens. Eur. J. For. Pathol. 6:103-109.

Siepmann, R. 1978. Anfälligkeit verschiedener Schwarzkiefern-Herkunfte bei Scleroderris lagerbergii - Befall. Eur. J. For. Pathol. 8:280-284.

Siepmann, R. K.J. Lang and S. Schonhar. 1975. Verbreitung der durch Scleroderris lagerbergii Gr. Verursachten Schwarzkieferntriebsterben in der Bundes Deutschland. Eur. J. For. Pathol. 5:185-189.

Skilling, D.D. 1977. The development of a more virulent strain of Scleroderris lagerbergii in New York State. Eur. J. For. Pathol. 7:297-302.

Skilling, D.D. 1981. Scleroderris canker - The situation in 1980. J. For. 79: 95-97.

Stephan, B.R. 1979. Beobachtungen uber das vorkommen und die Variation von Scleroderris lagerbergii Gremmen In Norddeutschland. Z. Mykol. 45:45-53.

von Weissenberg, K. 1975. Pathogens observed on Lodgepole pine grown in Finland. Eur. J. For. Pathol. 5:309-317.

Yokota, S., T. Uozumi and S. Matsuzaki. 1974. Scleroderris canker of Todo fir in Hokkaido, northern Japan. Eur. J. For. Pathol. 4:65-74.

Young, C.W.T. and R.G. Strouts. 1972. Dieback of Corsican pine. Rep. For. Res. Lond. 1972: 94-95.

DIFFERENCES IN DISEASE SYNDROME WITH PATHOGEN RACE AND WITH HOST AND ENVIRONMENT

Charles E. Dorworth[1]

ABSTRACT

Of the three components involved in the interaction that results in Scleroderris canker, the host is the most stable. Particular provenances can be chosen and used as the basis upon which the effects of diverse forms of the pathogen are manifested. Of the environmental elements germane to the problem, climate is the most difficult to predict even over long intervals. Short-term extreme variation may predispose plants to extraordinary degrees of infection and/or colonization by certain facultative saprophytes such as Brunchorstia pinea. The arena for elucidating Scleroderris canker in terms of interacting disease components should be international.

INTRODUCTION

The title of this segment, interpreted broadly, includes many of the questions we might consider today. Darroll Skilling and his colleagues chose the title, I believe, and it is an appropriate selection. Wherever pathology is taught we have, in effect, the same design as is represented in this title: pathogen + host + environment = DISEASE, with each of the contributing components at a particular level or within a prescribed range. We need only place quantitative and qualitative values before the elements comprising the disease interaction, whereafter impact or disease syndrome display becomes predictable. The reverse applies as well, more or less, so that we are permitted to work inductively or deductively.

It was my privilege this year to share the considerations of a number of colleagues, and many of you are here today. If I seem to avoid or to minimize certain points in discussion, you will appreciate that I do not wish to anticipate the presentations of these colleagues. I benefited greatly from the candid manner in which many of you discussed your work and I hope to establish a balance between my obligation to the Symposium and my desire to have certain excellent considerations presented by the scientists who developed them.

To begin, I will erect certain premises and utilize these on an a priori basis for further discussion even though I do not presume, certainly, that you necessarily agree with these premises. Premise 1. Gremmeniella abietina (=Ascocalyx abietina) is a facultative saprophyte. As such, it responds directly and positively (as enhanced symptom display) to environmental circum-

[1] Research Pathologist, Canadian Forestry Service, Great Lakes Forest Research Centre, P.O. Box 490, Sault Ste. Marie, Ontario, Canada. P6A 5M7

stances that are suboptimum for development of the host. Premise 2. The pathogenic races, morphologic races or varieties, and the physiologic races described to date are equally valid although they are not interchangeable on the basis of our present information. Premise 3. The serologic technique used to differentiate races, while possibly imperfect in detail and certainly manifesting the problems of a technique in the process of development, does describe serovars which are physiologic races by definition (Dorworth 1981) and pathogenic races by weight of evidence accumulated in the course of 12 years of research.

PROBLEMS OF MEASURING EFFECTS OF DISEASE

Our branch of pathology suffers from an inherent constraint: that of isolating a rather large host plant which matures over a number of years and may encounter considerable variation in environmental conditions during maturation. We have two alternatives: we can control the environment strictly around a limited number of smaller plants or distribute field experiments widely and evaluate results of experimental efforts statistically. The first method has an inherent disadvantage because the instruments are expensive. It is not practical to maintain a very large plant over an extended period, much less a number of large plants. The second alternative requires attention to our perennial problem: that of locating areas in which to conduct inoculations while ensuring that no damage occurs to the surrounding forest. This point becomes of particular importance when we require simultaneous inoculations between native and imported pathogens. It explains why we developed a chemical technique to evaluate differences among isolates of B. pinea. We require an island on which to conduct inoculations. Occasionally research people and others lose control of experimental forest fires, with dramatic and sometimes tragic results. Whereas fires are dramatic, our work with micro-organisms appears insidious to the public. What is more important, we are in the business of enhancing the forest, and introduction of microorganisms which we could probably not evict thereafter would be ethically unacceptable.

VALUE OF THE DISEASE COMPONENTS IN DISEASE MEASUREMENT

Again, I speak of disease evaluation from the standpoint of an experimental biologist. It is also possible to approach this matter from the viewpoint of natural observation, but I leave that to people with more experience in the technique. A good example here: Skilling noted the similarity between the new Scleroderris canker problem in New York and that in Europe (Skilling 1977) before this was verified through serology. Additional examples of the observational approach and of other experimental work will be presented today. These are all useful.

Perhaps you can agree that the host represents the most stable of the disease triangle components. Species responses have been noted in the past (Dorworth 1971, Donaubauer 1972) and Skilling will discuss a portion of his recent work here. We are not dealing with pure host lines, as agricultural pathologists might. Provenances of trees are rather stable, nonetheless, as environmental segregates of the species. We can transfer provenances from place to place and reproduce the parent forest within the limits imposed by a changed environment. This does not detract from the value to be gained in host genetic work.

The second most stable element is probably the pathogen. One can recognize different races of the pathogen as well as the fact that mutation is manifested most readily in organisms with brief life cycles. Still, one suspects that pathogen races would be segregated in appropriate ecologic niches much as is the case with the hosts, in the natural forest. All we lack are sufficient data to make clear predictions about the occurrence and activity of the pathogen. In the broad sense, we do this now. Our field personnel in Canada, at least, are told to look for upper crown dieback in particular when they survey for the presence of the European race of G. abietina. We require first a means of stabilizing isolates in the laboratory for comparative experiments and simply as references. We began this work in Sault Ste. Marie again this year, and are prepared to cooperate with anyone who is interested. It is necessary to characterize the pathogen isolates by serology or morphology or other means. This provides a common language for description of the segregates of the pathogen on a worldwide basis. If it becomes possible to perform proper comparative inoculations on differential hosts, it will be possible to note quantitatively the effects of various isolates. Thereafter, detection of new isolates or of mutations will evolve directly from standard procedures. To some extent at least, personal decisions are minimized at this level of research. The mystery and fascination are reduced but the work becomes more useful.

If it is possible to elaborate the effects upon the pathogen of changing environment, including the effects of changed patterns of forest management, then it becomes possible to predict alterations in pathogen populations. Again, we have considerable data from laboratory experiments with respect to growth rate and culture formation. In laboratories of medical microbiology it is possible to predict changes in populations of, for example, influenza virus. Those interested in apple scab can predict the time of spray requirement to within an hour if necessary. Finally, we have a forest disease on which a great deal of work has been done at the initial stage of concept formulation. This would seem to be the appropriate stage at which to apply principles of agricultural pathology to our work, where these are useful. It is my contention that this is the forest disease that most closely approximates diseases of agricultural crops (Dorworth and Davis 1982). Again, this removes some of the mystique with which we surround our work in the forest.

Finally, it is necessary to evaluate the effects of climate upon the pathogen-host interaction. Again, such work is possible if we have the equipment, the time and the money to ensure vast numbers of replicates. Eventually we would have sufficient data to construct superb three-dimensional projections that will permit disease predictions to be made with some accuracy. Then we can relax and forget the matter because, with present technology, we are unable to combat the disease over large forest areas. The active participation of chemical synthesis workers is needed to give us compounds that are environmentally safe, inexpensive to apply and, what is most important, that permit translocation by the plant over several seasons. Climate is still a partly unpredictable element over the lifespan of these large plants. A single climatic event may undo years of careful forest management. Approximately 10 years ago widespread mortality of Pinus sylvestris was registered from Minnesota eastward to New Brunswick, with millions of dollars in attendant loss. Most of us suspect that an inclement period in the previous winter and spring predisposed these trees to infection and colonization by Cenangium ferruginosum. The economic loss was great, but the event may not be repeated for another 20 years. It was difficult to reconstruct completely the circumstances that promoted this mortality and to conduct experiments which might accurately demonstrate climatic predisposition would require financing beyond our capabilities.

VARIATION OF DISEASE SYNDROME ON DIFFERENT HOSTS

At present it is possible to say directly that Pinus resinosa is the host most seriously affected by Brunchorstia in both its forms. In Europe, Pinus nigra, of one type or another, is affected to an equal extent on occasion, but usually where it exists in an environment suboptimum for development. Pinus contorta, P. banksiana and P. sylvestris display variable resistance depending upon both inherent genetic factors and climatic factors. These species may be killed by infection or may merely serve as substrata on which the pathogen is maintained as a source of later infection. Picea spp. are seldom damaged severely, although this is the genus on which the first collection of Crumenula abietina -- the original name for the perfect state -- was made, and there is one record of the fungus on Abies. The situation is different in Japan where Abies sachalinensis is heavily damaged. Yokota (1975) noted this as a consequence of top infection of the host in situations where it was later depressed beneath snow level during the winter. Beyond that, it seems that the old observation that pines are the principal host is valid.

VARIATION IN DISEASE SYNDROME WITH DIFFERENT PATHOGEN RACES

The original observation that the North American race kills tissues beneath a point 2 metres from the ground, whereas the European race kills tissues at any height above ground, is still valid. The main stem may or may not become involved in the event of infection by either race. Capretti and Dorworth are currently conducting congruent upper crown inoculations with the North American race in Canada and with the European race in Italy to reconfirm that point. In addition, we have some well replicated experiments utilizing pure lines of the pathogen from four hosts for inoculation upon Pinus nigra, P. sylvestris and Picea abies. These are the same isolates that Capretti and Dorworth used to develop antisera for race determination in Italy. At the present state of our knowledge, the major differences between the two races seem to be that a) the European race has a far broader host range than does the North American race, and b) the European race is capable of infecting or colonizing tissues at heights well above 2 metres.

An argument prevailed in the past to the effect that these races were not validly described because primary correlation had not been achieved between serovar and pathogenic race through inoculation. In fact, much of the weight of this counter-argument came from those of us who used the technique actively. We were quite candid in disclosing our problems and in criticizing our procedures, whereas those who generated the sometimes intense negative criticism were usually not researchers. Generally they used one portion of our commentary, the autocritique, to demonstrate their point. However, we have tested enough isolates of this pathogen over the past 12 years to demonstrate that the serovar is adequate to the pathogenic race. Jonas Salk did not reinoculate his original host species either, as far as I know. I declare the argument terminated. It has wasted enough of our time. If that does not satisfy one and all, interested parties may obtain some of our isolates and disprove this contention. I shall be the first to acknowledge a piece of good work disproving my thesis.

SPECIFICATION OF DISEASE SYNDROME WITH HOST AND WITH PATHOGEN RACE

Financial support for research on G. abietina has continued for a number of years. I suggest that we operate as though this expression of official interest is permanent by conducting some extremely well coordinated tests at the international level. We have in fact begun to do so. There are young pathologists at work in Italy, Britain, Sweden and North America -- principally here in the United States. Europeans have North American species, particularly Pinus contorta, under cultivation in areas where Brunchorstia prevails; and Capretti now has P. resinosa and P. banksiana at the early stages of growth in Italy, near Florence. The inoculation series I mentioned earlier seem especially promising. Capretti has also initiated experiments to determine LID_{50} for the two described European varieties of Brunchorstia pinea on Pinus nigra in a test that will complement the work done at Cornell University. There are other examples, and there may be work in progress that has escaped my attention. Skilling's recent efforts to define host ranges for the pathogen are quite complete. Capretti now has sera for four discrete isolates of Brunchorstia. Thus, we have the immediate potential for comparing serologic and certain etiologic work on the two continents. These diverse elements might be brought together as efficiently as possible, in order that good work can be utilized. Thereafter, it would be possible to pursue further cooperative work, possibly under the auspices of IUFRO, in order to obtain the maximum degree of information about this disease while we have the opportunity.

LITERATURE CITED

Donaubauer, E. 1972. Distribution and hosts of Scleroderris lagerbergii Gremmen in Europe and North America. Eur. J. For. Pathol. 2:6-10.

Dorworth, C.E. 1981. Status of pathogenic and physiologic races of Gremmeniella abietina. Plant Dis. 65:927-931.

Dorworth, C.E. and C.N. Davis. 1982. Current and predicted future impact of the North American race of Gremmeniella abietina on jack pine in Ontario. Dep. Environ., Can. For. Serv., Sault Ste. Marie, Ont. Inf. Rep. O-X-342.

Skilling, D.D. 1977. The development of a more virulent strain of Gremmeniella abietina in New York State. Eur. J. For. Pathol. 7:297-302.

Yokota, S. 1975. Scleroderris canker of Todo-fir in Hokkaido, Northern Japan. III. Dormant infection of the causal spruce fungus. Eur. J. For. Pathol. 5:7-12.

PATHOGENICITY AND HOST RANGE OF RACES OF *Gremmeniella abietina* IN HOKKAIDO

Shun-ichi Yokota[1]

ABSTRACT

Isolates of *Gremmeniella abietina* from *Abies sachalinensis* infected 13 tree species when they were artificially inoculated to 27 tree species in 5 genera. A heavy infection selectively occurred only to *Abies* species. Further, in cross inoculation tests using isolates from different hosts (*Abies sachalinensis*, *Pinus strobus* and *P. resinosa*), the isolates from *A. sachalinensis* indicated selective pathogenicity only to *A. sachalinensis*, the original host. While the isolates from both pines indicated strong pathogenicity non-selectively to all three species tested regardless of the original host. Results indicated that there are at least two pathological races of *G. abietina* in Hokkaido, Japan; the Asian race represented by the isolates from *A. sachalinensis* and the race from *P. strobus* isolates which is closely related to the North American race represented by the isolate from *P. resinosa*.

INTRODUCTION

Under natural conditions, *Gremmeniella abietina* (Lagerb.) Morelet attacks mainly *Abies sachalinensis* Mast. in Hokkaido, Japan. However, in the very limited plantations of Tokyo University Forest in Hokkaido, *A. alba* Mill and *Pinus strobus* L. were found severely damaged by this pathogen although these cases were exceptional. So far, no tree species other than these hosts have been found attacked by the pathogen. Disease symptoms and damage on hosts infected with this pathogen have been previously reported (Yokota 1975-a, -b; Yokota et al. 1974-a, -b, 1975). However, comparisons between isolates from *A. sachalinensis* and from *P. strobus* have never been made.

---

1 Director, Kyushu Branch, Forestry and Forest Products Research Institute, Kurokami, Kumamoto 860, Japan.

The author wishes to express his special appreciation to Dr. S. Takai, Great Lakes Forest Research Centre, Sault Ste. Marie, Ontario, Canada, for valuable suggestions and review of the manuscript. Thanks are due to Dr. C.E. Dorworth, Great Lakes Forest Research Centre of Canada, and Mr. I. Takahashi, Tokyo University Forest, Furano, Hokkaido, for supply of the isolates used for the inoculation experiments.

There are three physiological races of this pathogen (North American, European and Asian races). Based mainly on immunogenic comparison, the isolates from Abies were characterized as the Asian race (Dorworth and Krywienczyk 1975).

This paper deals with determination of the host range for G. abietina and comparison of pathogenicity of isolates from A. sachalinensis classified as the Asian race, P. strobus unclassified for the racial type and P. resinosa classified as the North American race.

METHODS

An agar block (1-2 mm square) of the mycelial colonies of G. abietina, isolated from different host sources, was placed over a fresh bark wound (similar size to that of inoculum) which was made on a 1-year-old shoot. Several shoots on the same tree were inoculated in order to eliminate individual difference.

The inoculation was conducted in the fall. Results were checked in the following spring for the numbers of shoots cankered and/or killed, and the size of the cankered lesion. Details of the cultures used for the inoculation are shown in Table 1.

Table 1. Details of cultures used for inoculation experiments.

Cultures	Origin
D70-1	Mono-ascosporous, June 25, 1969, on 10-yr-old Abies sachalinensis Mast. at Kuttchian
D70-3	Multi-pycnosporous, produced on the culture of D70-1 on Jan. 25, 1971
SC-8	Multi-pycnosporous, June 29, 1973, on 14-yr-old Pinus strobus L. at Yamabe, Furano (I. Takahashi)
SC74-1	Multi-pycnosporous, July 1, 1974, on 7-yr-old A. sachalinensis at Kimobetsu
SD74-1	Multi-ascosporous, July 6, 1974, on 15-yr-old P. strobus at Yamabe, Furano
SA-2	Multi-ascosporous, July 22, 1977, on 10-yr-old A. sachalinensis at Takadomari
SC-1	Mono-ascosporous, P. resinosa Ait., in Ontario, Canada (C.E. Dorworth)

To determine the host range of the isolates, 27 tree species as shown in Table 2, were artificially inoculated. To examine selective pathogenicity of isolates from different host sources, cross inoculation were conducted with combinations of G. abietina isolates from A. sachalinensis, P. strobus, and P. resinosa.

RESULTS

Host Range

Table 2 shows the host range of the causal fungus through artificial inoculations. Of 27 tree species tested, 13 species were determined as hosts of

Table 2. Host range of *Gremmeniella abietina* isolates from *Abies sachalinensis*, *Pinus strobus*, and *P. resinosa* by artificial inoculation[1].

Trees tested	Isolates from: *A. sachalinensis* D70-1	*A. sachalinensis* SA-2	*P. strobus* SD74-1	*P. resinosa* SC-1
Abies sachalinensis Mast.	4/5;8	4/6;(1);14[2]	2/6;19	0/6
A. alba Mill	4/4,(2);31			
A. balsamea Mill	0/5			
A. holophylla Maxim.	4/5,(3);27			
A. mariesii Mast.		5/5,(4);16		
A. fraseri (Pursh) Poir		5/5,(4);15		
Picea glehnii Mast.	0/5			
P. abies Karst.	0/5	0/5	1/5,(1)	3/5,(3)
P. glauca Voss	0/5			
P. mariana B.S.P.	0/5			
P. jezoensis Carr.		0/5	5/5,(4);28	5/5,(5)
P. rubens Sarg.		2/5;17		
Larix leptolepis Gord.	0/5	0/3	2/3,(1);26	1/3,(1)
L. gmelini Gord.	0/5			
L. olgensis A. Henry var. *koreana* Nakai	0/5			
Pinus monticola Dougl.	3/5;8			
P. resinosa Ait.	0/5	5/6,(1);25	6/6,(6)	6/6,(6)
P. sylvestris L.	0/5	0/5		
P. thunbergii Parl.		3/3;18	3/3,(3)	3/3,(3)
P. banksiana Lamb.		5/6;16	6/6,(6)	6/6,(6)
P. strobus L.		2/6;12	6/6,(6)	6/6,(4);19
P. pumila Regel		2/6;18	6/6,(6)	6/6,(1);22
P. pentaphylla Mayr		3/5;11		
P. koraiensis Sieb. & Zucc.		0/5		
P. peuce Griseb.		0/5		
Psudotsuga menziesii Franco	0/5			

[1] D70-1 was inoculated on Oct. 26, 1972; SA-2, SD74-1, and SC-1 were inoculated on Nov. 9, 1979.
[2] This means "6 branches were inoculated and 4 were infected, 1 of them was killed; mean length of cankered lesion of 3 shoots was 14 mm".

the isolates from A. sachalinensis. In particular, the isolates from A. sachalinensis appeared highly virulent to Abies species, however, not pathogenic to Larix species and Pseudotsuga menziesii. Among Picea species tested, only P. rubens was infected by SA-2 and among Pinus species more than a half were infected. The isolates from P. strobus and P. resinosa, which were used for comparison with those isolated from A. sachalinensis, infected almost all tree species inoculated, particularly Pinus species resulting in high incidence of shoot mortality.

Cross Inoculation

Table 3 shows the results of cross inoculation experiments. D70-3 and SC74-1 from A. sachalinensis showed strong pathogenicity to only A. sachalinensis, very weak to P. strobus, and none to P. resinosa. Isolates SC-8 and SD74-1, from P. strobus showed strong pathogenicity to all three species tested, and many inoculated shoots were killed. The isolate SC-1 from P. resinosa also was strongly pathogenic to all three species tested, although it was somewhat weaker than the P. strobus isolate, judging from the number of dead shoots and mean length of infected lesions.

Table 3. Results of cross inoculation experiments.

Isolates \ Trees tested	10-yr.-old A. sachalin.	5-yr.-old P. strobus	20-yr.-old P. resinosa
D70-3 (from A. sachalin.)	6/12;13	5/20;8	0/19
SC-8 (from P. strobus)	12/12,(1);21	18/20,(9);40	17/19,(14);40
SC-1 (from P. resinosa)	7/12;14	17/19,(5);18	18/19,(15);40

Inoculation: Nov. 9, 1973

Ditto (2)

Isolates \ Trees tested	5-yr.-old A. sachalin.	5-yr.-old P. strobus	5-yr.-old P. resinosa
SC74-1 (from A. sachalin.)	21/22,(7);12	1/25;5	0/13
SD74-1 (from P. strobus)	25/25,(13);10	36/36,(35);50	13/15,(13)
SC-1 (from P. resinosa)	15/21,(6);11	11/28,(7);10	17/21,(17)

Inoculation: Nov. 8, 1974

Table 4 shows the over-all characteristics of G. abietina from A. sachalinensis and P. strobus in Hokkaido. No difference in morphological characteristics were found in the perfect stage of the isolates. The size of conidia and the number of septa, however, differed; mean length of conidia for the isolates from A. sachalinensis was 40-56 um with 7 septa, whereas mean length for isolates from P. strobus was 31-33 um, mostly with 3 septa. Other conspicuous differences between the isolates include pigment formation under the cankered lesion, infection site, canker extension, and host range.

Table 4. Comparison of Gremmeniella abietina from A. sachalinensis and P. strobus.

Items \ Combination	G. abietina- A. sachalinensis	G. abietina- P. strobus
Conidia	40-56 um (mean), mostly 7-septate	31-33 um (mean), mostly 3-septate
Immunogenic reaction	Differs from European and N. American race (Asian race)	Not examined yet (N. American race?)
Infection	June - August	June - August
Pigment	None	Green - yellow
Needle Infection	Does not occur	Occurs
Disease development	Canker occurs on branches less than 2 years of age	Canker extends every year
Host range	Strong pathogenicity only to Abies by artificial inoculation	Strong pathogenicity to many tree species by artificial inoculation

DISCUSSION

Dorworth (1971) listed 20 species of the genera Picea, Pinus, and Pseudotsuga as hosts of G. abietina. Donaubauer (1972) also listed 24 species of the genera Pinus, Picea, Pseudotsuga, and Larix as the host. However, no species of the genus Abies was listed by these authors. Under natural conditions, however, the main host of this pathogen is A. sachalinensis in Hokkaido, Japan. The isolates from A. sachalinensis were characterized as the Asian race through immunogenical, physiological and morphological examinations by Dorworth and Krywienczyk (1975). Therefore, the presence of the Asian race is assumed. To date, only one species of Abies, A. fraseri, is found to be infected in the United States by the European race of the pathogen (Skilling, personal communication April 20, 1979). The author, however, considers this may be an exceptional case.

From the present inoculation experiments, at least 13 tree species, Abies sachalinensis, A. alba, A. holophylla, A. mariesee, A. fraseri, Picea rubens, Pinus monticola, P. thunbergii, P. banksiana, P. resinosa, P. strobus, P. pumila, and P. pentaphylla have been shown to be hosts of G. abietina isolated from A. sachalinensis.

No infection occurred on three species of Larix inoculated with isolates from A. sachalinensis in the present study. However, Larix sp. and L. sibirica

Ledeb. were infected by Brunchorstia pinea, the imperfect stage of the pathogen, in Estonia (Hanso 1973), L. sib irica was infected in Norway (Roll-Hansen 1972), and L. leptolepis was infected by the European race of the pathogen in North America (Dorworth et al. 1977). It is noteworthy that isolates from A. sachalinensis, particularly the isolate D70-1, showed a narrow host range and caused severe damage only to Abies, whereas isolates from P. strobus and P. resinosa showed a fairly wide host range and severely affected not only all Pinus species tested but to a lesser degree also infected Picea abies, P. jezoensis, and Larix leptolepis. The contrast clearly indicates a difference in pathogenicity between the isolates from A. sachalinensis and those from the two pine hosts.

This is likewise supported by the results obtained from cross inoculation experiments. The isolates from A. sachalinensis were clearly limited to A. sachalinensis. On the other hand, the isolates from pines showed a strong pathogenicity on both Pinus, and Abies.

There were also morphological differences between the isolates. Mean conidial length and number of septa were 48 um, 7 septa for the isolates from A. sachalinensis and 32 um, 3 septa for the isolates from P. strobus. In Europe, it is reported the G. abietina conidia from P. cembra are larger, with around 7 septa, than conidia on other Pinus species (Ettlinger 1945; Donaubauer 1974; Barvacovi et al. 1979). Donaubauer (1974) suggests a morphologically differentiated cembra-form. The conidia from A. sachalinensis coincided in size with those from P. cembra.

Dorworth and Krywienczyk (1975) classified G. abietina into 3 physiological races, European, North American (Lake States') and Asian races, based on immunogenic reaction, morphology and cultural characteristics. Five isolates from A. sachalinensis in Hokkaido were clàssified as Asian races (1975). The isolates from P. strobus have not been determined for the racial type. However, as summarized in Table 4 several characteristics including morphology, infection sites, symptoms and host range, strongly suggest that G. abietina from P. strobus is closely related to the North American race. Therefore it is suggested that at least two physiological races of G. abietina, Asian and possibly North American, occur in Hokkaido, Japan.

CONCLUSIONS

Artificial inoculation experiments with isolates from Abies sachalinensis on 27 tree species, demonstrate at least 13 tree species as host. In particular, trees of the genus Abies were severely affected. No infection was found on 3 species of Larix and Pseudotsuga menziesii. Isolates from both Pinus strobus and P. resinosa showed strong pathogenicity and nonselectively to the nine tree species inoculated.

Based on the results of cross inoculation experiments, together with those on the host range and comparison of overall characteristics between the isolates from A. sachalinensis and P. strobus, it may be concluded that there are at least two physiological races of G. abietina in Hokkaido, one on A. sachalinensis, the Asian race, and the other on P. strobus, closely related to the North American (Lake States') race.

LITERATURE CITED

Barvacovi, A., P. Capretti, and F. Moriondo. 1979. Diffusione e danni di Brunchorstia pinea (Karst.) Höhn. su popolamenti naturali e artificiali di conifere in Italia. Ann. Accad. Ital. Sci. For. 28:3-41.

Donaubauer, E. 1972. Distribution and hosts of Scleroderris lagerbergii in Europe and North America. Eur. J. For. Pathol. 2:6-11.

Donaubauer, E. 1974. Über das Kieferntriebsterben in Österreich. 100 Jahre Forstliche Bundesversuchsanstalt:67-98. Dorworth, C.E. 1971. Diseases of conifers incited by Scleroderris lagerbergii Gremmen: A review and analysis. Dept. Fish. For., Can. For. Serv. Publ. No. 1289: 42 pp.

Dorworth, C.E. and J. Krywienczyk. 1975. Comparison among isolates of Gremmeniella abietina by means of growth rate, conidia measurement, and immunogenic reaction. Can. J. Bot. 53:2506-2525.

Ettlinger, L. 1945. Über die Gattung Crumenula sensu Rehm, mit besonderer Berucksichtigung des Crumenula-Trieb-sterbens der Pinus Arten. Beitr. zur Kryptogemenflora der Schweiz 10:1-73.

Hanso, M. 1973. A brief outline of the research work done in Estonia on the fungus Scleroderris lagerbergii Gremmen. Proc. First Session Working Party Canker Diseases (Scleroderris), Subject Group S2-06, Div. II, IUFRO, Minneapolis, Minnesota, U.S.A. Sept., 1973:8-9.

Roll-Hansen, F. 1972. Scleroderris lagerbergii: Resistance and differences in attack between pine species and provenances. A literature review. Eur. J. For. Pathol. 2:26-39.

Yokota, S. 1975-a. Scleroderris canker of Todo-fir in Hokkaido, northern Japan III. Dormant infection of the causal fungus. Eur. J. For. Pathol. 5:7-12.

Yokota, S. 1975-b. Scleroderris canker of Todo-fir in Hokkaido, northern Japan IV. An analysis of climatic data associated with the outbreak. 5:13-21.

Yokota, S., T. Uozumi, and S. Matsuzaki. 1974-a. Scleroderris canker of Todo-fir in Hokkaido, northern Japan I. Present status of damage, and features of infected plantations. Eur. J. For. Pathol. 4:65-74.

Yokota, S., T. Uozumi, and S. Matsuzaki. 1974-b. Scleroderris canker of Todo-fir in Hokkaido, northern Japan II. Physiological and pathological characteristics of the causal fungus. Eur. J. For. Pathol. 4:155-166.

Yokota, S., T. Uozumi, and S. Matsuzaki. 1975. Scleroderris canker of Todo-fir in Hokkaido, northern Japan V. Relationship between disease development and forest type. Eur. J. For. Pathol. 5:356-366.

METABOLITES OF *Gremmeniella abietina*

William A. Ayer and Yasuyuki Hiratsuka[1]

ABSTRACT

Gremmeniella abietina was grown on a medium of 10% V-8 juice containing added glucose, and the metabolites produced were isolated by extraction with ethyl acetate or methylene dichloride. Chromatographic separation of the crude metabolites gave a blue compound, a strongly fluorescent compound, a red compound, a dark purple compound, and a yellow compound. The structures of these compounds will be reported, along with some of their biological activities. A comparison of the metabolite spectra of various North American and European strains of *G. abietina* will be presented.

INTRODUCTION

The metabolites produced when the fungus *Gremmeniella abietina* is grown in liquid culture have not previously been identified. In order to determine what compounds are responsible for symptom expression and to see whether there is a correlation between metabolites produced and the virulence of particular strains, we have studied the metabolites produced by liquid cultures of the fungus.

RESULTS AND DISCUSSION

The first strain of *G. abietina* studied (Canadian Forestry Service (CFS) strain C-699) was isolated from New Brunswick pine trees. It was grown at 16-17C on a medium of 10% V-8 juice containing 1% added glucose (Dorworth 1972), in still culture and in a stirred an aerated fermentation apparatus. Best yields of metabolites were obtained from the still cultures. Extraction of the filtered culture broth yielded *ca.* 0.04 g/l of crude metabolites, while Soxhlet extraction of the mycelium provided substantially more material (*ca.* 0.45 g from the mycelium growing in 1 liter of medium. Separation of the crude metabolites by flash chromatography gave, in order of elution (solvent system

[1] Professor of Chemistry, University of Alberta, Edmonton, Alberta, Canada T6G 2G2 and Research Scientist, Northern Forest Research Centre, Canadian Forestry Service, Edmonton, Alberta, Canada T6H 3S5.

We thank the Natural Sciences and Engineering Research Council of Canada for financial support, Professor Jon Clardy for the X-ray structure of compound 9, and Professor Gordon Wood for the FDMS of compound 7.

$CHCl_3$, Skellysolve B, HOAc (10:10:1)), a blue compound, a colorless strongly fluorescent compound, a red compound, a dark purple compound, a yellow compound, and a mixture of brownish materials which have not yet been completely separated and identified.

The colorless, strongly fluorescent compound, mp 250-253C, $C_{18}H_{16}O_2$, shows carbonyl absorption in the IR at 1705 and 1665 cm^{-1} and readily forms a diacetyl derivative absorbing at 1778, 1760, and 1724 cm^{-1}. The ^1H NMR spectrum showed clearly defined signals for all sixteen protons: methyl singlets at δ1.30 and 1.54; a secondary methyl doublet (δ1.50) coupled to a methine hydrogen (δ 4.80) geminal to oxygen; an aromatic methyl at δ 2.80; an aromatic hydrogen (δ 6.80); and phenolic hydrogens at δ 11.37 and 11.60. A search of the literature revealed that this compound is the anhydride 1, first isolated from oxidation of atrovenetin (2) (Barton et al. 1959), but later isolated from cultures of *Penicillium herquei* (Narasimhachari and Vining 1963) and subsequently from other fungi. The anhydride 1 isolated from *G. abietina* is enantiomeric with that previously isolated.

The dark burgundy compound, which causes a green discoloration on contact with the skin and a blue-green discoloration of paper, was shown by high resolution electron impact mass spectrometry (EIMS) and chemical ionization mass spectrometry (CIMS) to possess the molecular formula $C_{19}H_{16}O_6$. The ^1H NMR spectrum was virtually identical with that of the anhydride 1. The ^{13}C NMR spectrum, determined in DMSO-d_6, shows carbonyl peaks at δ 198 and 196 (those of the anhydride 1 appear at ~ δ 166) and a signal at δ 88 which is not present in the spectrum of the anhydride. When the spectrum is determined in CD_2Cl_2, an additional carbonyl signal appears at δ 176 and the intensity of the signal at δ88 is diminished. The data suggested that the compound is the triketone 3, present mainly as the hydrate 4. This compound, atrovenitinone, has previously been prepared (Narasimhachari and Vining 1963), by oxidation of atrovenitin (2) with benzoquinone. It was characterized as the crystalline ethanolate 5. Crystallization of the dark purple compound from ethanol provided pale yellow needles of 5, identical in all respects with an authentic sample provided by Prof. Vining. Compound 4 has not previously been isolated from natural sources. When the solvent system used for the separation of 3 contained acetone, the acetone adduct 6 (as a mixture of epimers) was obtained instead of compound 3. Treatment of 3 with acetone in the presence of a catalytic amount of acetic acid also provided 6. When a solution of compound 3 was placed on freshly peeled pine, the characteristic blue-green color of Scleroderris infected wood developed.

The blue compound proved rather unstable and difficult to purify, but after repeated silica gel chromatography and crystallization from ethanol, deep blue crystals were obtained. Efforts to determine the molecular formula of this compound, which we believe to be responsible, at least in part, for the bluish-green coloration of the wood of Scleroderris infected pine, were initially unsuccessful. The highest mass peak in the EIMS appeared at m/z 337 and corresponded to the ion $C_{19}H_{18}O_5$, Another prominent peak occurred at m/z 326 ($C_{19}H_{18}O_5$). The CIMS did not afford a recognizable parent ion. However field desorption mass spectrometry (performed by Prof. G. Wood, University of Windsor) suggested a molecular weight of 663 amu. Addition of the two high mass EIMS fragments then suggested the molecular formula is $C_{38}H_{33}NO_{10}$ (MW 663). The ^1H NMR spectrum is similar to that of the anhydride 1 and the triketone hydrate 4 with most signals doubled. The ultraviolet-visible spectrum

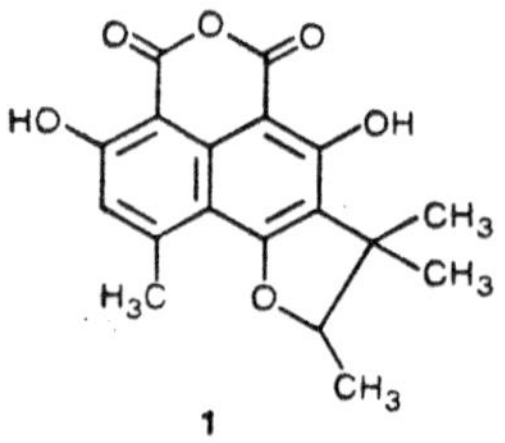

1

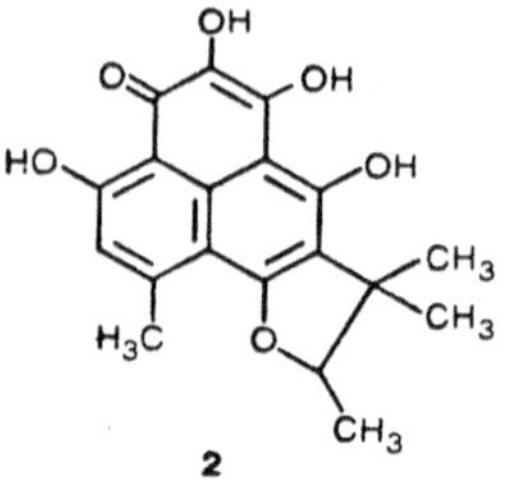

2

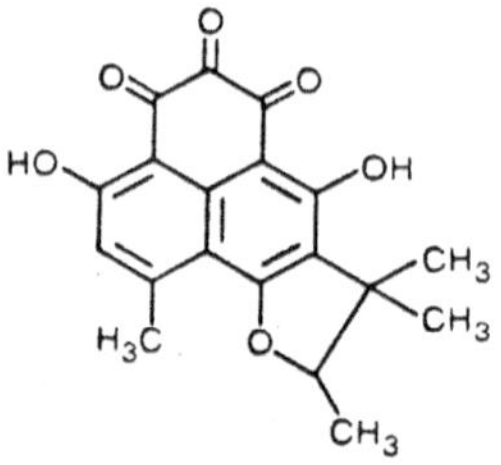

3

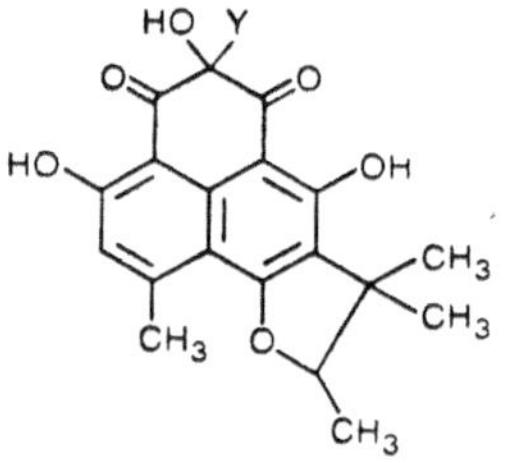

4 Y = OH
5 Y = OCH_2CH_3
6 Y = CH_2COCH_3

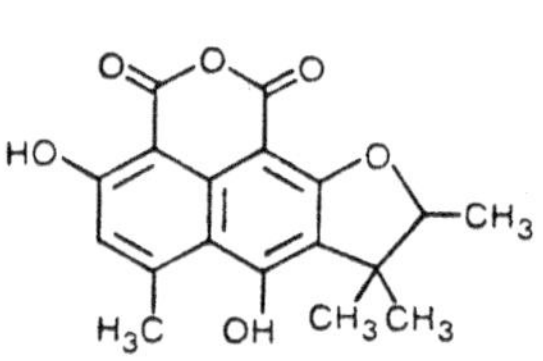

8

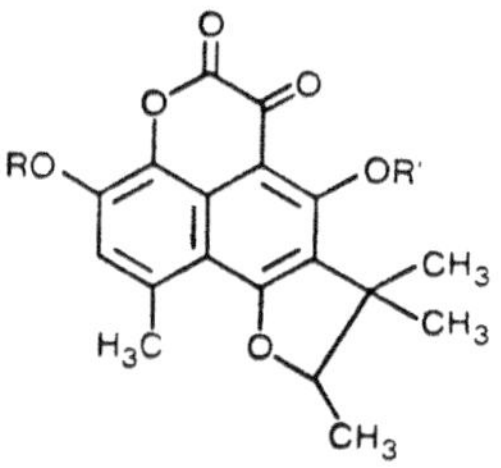

9 R = R' = H
9a R = Ac, R' = H
9b R = R' = Ac

($CHCl_3$) shows $_{max}$ 268, 368, 405 (sh) and 608 nm. Narasimhachari and Vining (1963) had previously suggested that the characteristic green pigment of Penicillium herquei might be formed by a ninhydrin-like reaction between atrovenitenone (4) and amino acids, but they were unable to completely characterize the pigment. In our hands, treatment of a buffered solution of glycine (see Scheme 1).

SCHEME 1

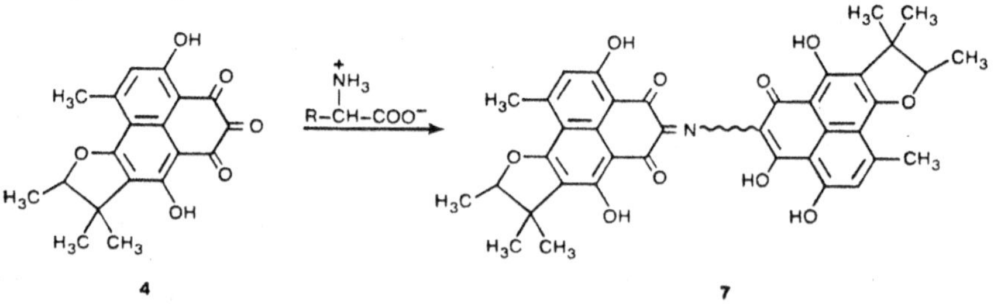

4 7

With atrovenitinone (4) in aq. dioxane and purification of the product by flash chromatography gave blue crystalline material identical in all respects with the compound isolated from G. abietina. The pigment, for which we suggest the trivial name Scleroderris blue, thus possesses structure 7 (Scheme 1).

The yellow compound, mp 239-240C, is isomeric with the anhydride 1 ($C_{18}H_{16}O_6$) and initially we suspected it to be the isomeric ether 8. The ^{1}H NMR spectrum was very similar to that of 1, and treatment with acetic anhydride-pyridine afforded a monoacetyl derivative. The ^{13}C NMR spectrum, however, shows carbonyl absorption at δ 186.2 and 170.0, which is not in good agreement with the anhydride structure. Also, the IR spectrum shows carbonyl absorption at 1750 cm^{-1}, unlike that of the anhydride. An X-ray crystallographic study of the monoacetate, performed by J. Clardy and E. Arnold of Cornell University, revealed that the yellow compound possesses the biogenetically novel internal phenylglyozylate lactone structure 9 (the monoacetate is 9a), the ketonic carbonyl accounting for the δ 186.2 signal in the ^{13}C NMR spectrum. Treatment of 9 with acetic anhydride-sulfuric acid at 60C furnished the diacetyl derivative 9b.

The red compound, mp 210-212·, $C_{18}H_{16}O_5$, has one oxygen less than the anhydride 1 and the yellow compound 9. The ^{1}H NMR spectrum is very similar to that of 1, but the ^{13}C NMR spectrum shows ketonic carbonyls at δ 189.6 and 186.2. The IR spectrum shows carbonyl absorption at 1714 and 1690 cm^{-1}. This compound readily forms a diacetyl derivative, the IR of which shows carbonyl absorption at 1780 and 1740 cm^{-1}, simlar to that of acenaphthenquinone 10. This evidence suggested that the red compound has structure 11 (the diacetate is 11a) and this was confirmed by alkaline peroxide oxidation to give, after

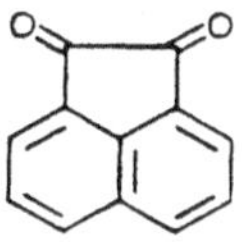

10

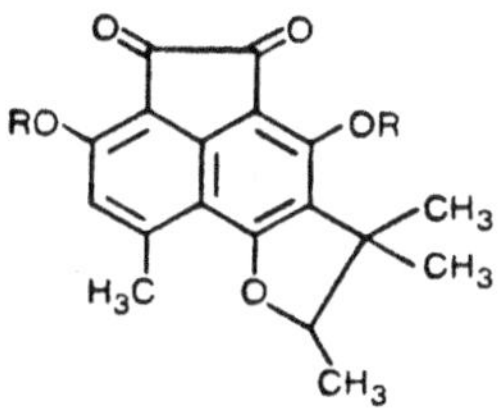

11 R = H
11a R = Ac

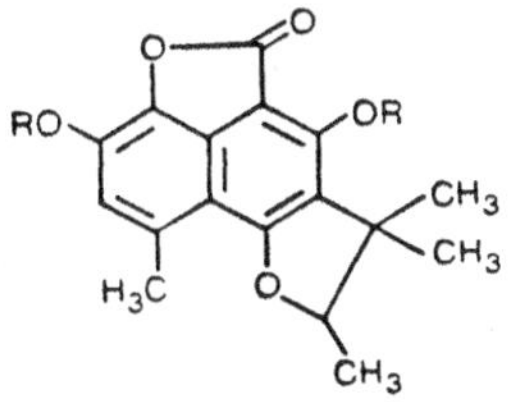

12 R = H
12a R = CH_3

acidification, the anhydride 1, along with the lactone 12, the product of Dakin oxidation of 11. Treatment of 12 with CH_2N_2 provides the ether 12a.

The red compound represents the first acenaphthenquinone isolated from natural sources. We believe that red compound 11 may be the biogenetic precursor of the anhydride 1 and the yellow compound 9.

We have compared, using high pressure liquid chromatography (HPLC), the metabolites of three other strains of G. abietina with those of strain C-699. The metabolites of strain C-659 ("North American strain", from Ontario, SC-1 of Dorworth and Krywienczyk 1975) are very similar to C-699. However, the metabolite spectra produced by strain C-706 ("European strain" from Sweden, SS-1 of Dorworth and Krywienczyk, 1975) and strain C-707 ("New York strain", from New York, SUS-8 of Dorworth et al. 1977) are quite different from that of strain C-699. It is interesting to note that strain C-699 (SC-73 of Dorworth) was isolated from New Brunswick and serologically reacted positively to "European strain" similar to "New York strain". The anhydride 1 is produced by all four strains. Identification of the other metabolites produced by the different strains is in progress.

LITERATURE CITED

Barton, D.H.R., P. deMayo, G.A. Morrison, and H. Raistrick. 1959. The constitutions of atrovenetin and of some related herqueinone derivatives. Tetrahedron 6:48-62.

Dorworth, C.E. 1972. Epidemiology of Scleroderris lagerbergii in central Ontario. Can. J. Bot. 50:751-765.

Dorworth, C.E. 1981. Status of pathogenic and physiologic races of Gremmeniella abietina. Plant Dis. 65:927-931.

Dorworth, C.E. and J. Krywienczyk. 1975. Comparison among isolates of Gremmeniella abietina by means of growth rate, conidia measurement, and immunogenic reaction. Can J. Bot. 53:2506-2525.

Dorworth, C.E., J. Krywienczyk and D.D. Skilling. 1977. New York isolates of Gremmeniella abietina (Scleroderris lagerbergii) identical in immunogenic reactions to European isolates. Plant Dis. Rep. 61:887-890.

Narasimhachari, N., and L.C. Vining. 1963. Studies on the pigments of Penicillium herquei. Can. J. Chem. 41:641-647.

GENETIC VARIATION IN *Gremmeniella abietina*

Glenn R. Furnier, Neil A. Anderson and Darroll Skilling[1]

ABSTRACT

Heterogenic incompatibility occurs in *Gremmeniella abietina*. Hyphal anastomosis occurred between European and North American isolates. Immunogenic tests of single ascospore cultures from a single apothecium collected in the field suggests hybridization between European and North American strains. Conidia produced from single ascospore cultures typing as either European or North American from the above apothecium, were used to inoculate seedlings of several species of pine, spruce, fir, larch, and Douglas-fir. Both strains infected all hosts, except that white fir was not infected by the European strain. The North American strain generally caused greater percent infection than the European strain.

INTRODUCTION

Gremmeniella abietina (Lagerb.) Morelet (*Scleroderris lagerbergii* Gremmen) causes a canker and dieback disease of conifers. The main disease symptoms are dead branch tips and branch and stem cankers (Skilling 1981). Dorworth and Krywienczyk (1975) have identified three strains of this fungus based on cultural characteristics and immunogenic reaction. The Asian strain, found primarily in northern Japan, attacks branches occurring only below the snow line (Yokotia et al. 1974). The North American strain attacks only young trees. Trees over 2 m in height recover and survive such attacks. The European strain can attack and kill trees of all ages (Skilling 1981).

---

[1] Research Assistant and Professor, University of Minnesota, Department of Plant Pathology, St. Paul, MN 55108 and Plant Pathologist, North Central Forest Experiment Station, St. Paul, MN 55108.

The authors wish to thank Drs. Charles E. Dorworth, Yasu Hiratsuka, and Finn Roll-Hansen for advice and for isolates of *Gremmeniella abietina*. We also wish to thank Bruce Berry for assistance in the laboratory, and Denise DuPont and Marianne Kienzler for performing immunogenic strain typing of isolates of *G. abietina*. This study was supported by the USDA Forest Service, North Central Forest Experiment Station, RWU-NC-2205.

Setliff et al. (1975) found G. *abietina* killing red pine (*Pinus resinosa*) and Scotch pine (*Pinus sylvestris*) trees of all ages in northern New York in 1975. Dorworth et al. (1977) found that isolates from this area gave immunogenic reactions identical to European isolates. This European strain has since been found in Vermont, New Hampshire, Maine, Quebec, New Brunswick, and Newfoundland, and poses a major threat to the coniferous forests of North America (Skilling 1981).

Very little is known about the genetics of G. *abietina*. An important question is whether there are any barriers to hybridization between the European and North American strains. Since the North American strain produces wind-disseminated ascospores much more frequently than the European strain, hybridization between them could increase the rate of spread of isolates possessing the European strain's greater virulence.

Many fungi possess incompatibility systems that restrict mating between different populations of a species. These barriers may act early, preventing heterokaryon formation, or may occur at some later stage. Zones of inhibition, barrage reactions, and other reactions have been displayed by many fungi possessing incompatibility systems when grown on agar plates (Anagnostakis 1977; Esser and Raper 1965). The first objective of this study was to determine the incompatibility system in G. *abietina*.

During 1980 an unusual infection by *Gremmeniella abietina* was found on pole-size red and jack pine (*Pinus banksiana*) on the Tug Hill plateau in northern New York State. Apothecia of G. *abietina* were present on these infected trees. This is normally a characteristic of the North American strain of G. *abietina*; however branches were infected up to 7 m in the crown. Infection at this level is usually found only with the European strain of G. *abietina*. The second objective of this study was to attempt to determine if isolates from ascospores of their apothecia were a hybrid between the European and North American strains.

METHODS

Information on the location and host from which each isolate of G. *abietina* was obtained is presented in Table 1. Single ascospore cultures were paired on a V8 juice-charcoal-malt extract agar medium (V8CM), containing 200 ml/l V8 juice, 2g/l charcoal, 30g/l malt extract, and 15g/l agar. Twenty single ascospore cultures were obtained from three apothecia (SC-100, SC1001, 11-81). To determine the segregation pattern of an incompatibility system, diallel pairings were made with the set of 20 single ascospore cultures from each apothecium. Pairings were also made between single ascospore cultures of SC-1000 and 11-81 to determine if they shared common incompatibility alleles. In addition, pairings were made between isolates representing Asian, European, and North American strains. All pairings were scored independently by two observers.

The contact zones of all pairings were observed under a compound microscope to detect hyphal anastomoses or aversion. An autoclaved dialysis membrane, four cm square, was placed on the surface of a V8CM agar plate. Mycelium of the paired cultures was placed approximately two cm apart on the membrane. The membrane permitted the fungus to obtain nutrients from the agar, but restricted fungal growth to the surface of the membrane. When hyphae from the paired isolates met, that area of the membrane was cut out and placed on a

Table 1. Origin of isolates of *Gremmeniella abietina*. Letters under strain designate strain as determined by immunologic methods (A = Asian, E = European, N = North American, H = putative European X North American hybrid).

Isolate	Strain	Place of Origin	Host
SC-1	N	Ontario, Canada	*Pinus resinosa*
SC-5	N	Ontario, Canada	*Pinus resinosa*
SC-656	N	Alberta, Canada	*Pinus contorta*
SC-657	N	Ontario, Canada	Unknown
SC-1000	N	Alberta, Canada	*Pinus contorta*
SC-1001	N	Alberta, Canada	*Pinus contorta*
SH-1	E	Loenen, The Netherlands	*Pinus nigra* var. *corsicana*
SJ-1	A	Hokkaido, Japan	*Abies sachalinensis*
SJ-5	A	Makkari, Japan	*Abies sachalinensis*
SUS-2	N	Michigan, U.S.A.	*Pinus banksiana*
SUS-10	E	New York, U.S.A.	*Pinus resinosa*
11-81	H	New York, U.S.A.	*Pinus banksiana*
19-28	E	Quebec, Canada	*Picea mariana*
75-31/12	E	More and Romsdal,Norway	*Pinus mugo*

glass slide with the hyphae intact. Water and a coverslip were placed over the membrane and the hyphae observed with a compound microscope.

In 1979, the immunology laboratory of the North Central Forest Experiment Station (NCFES), St. Paul, MN, received several isolates of *G. abietina* from New York and Vermont that gave positive immunogenic reactions with both European and North American strains (D.D. Skilling unpublished data). Previously, isolates had given a positive immunogenic reaction with one strain or the other, but never both. Among these new isolates was field isolate 11-81. To further investigate this phenomenon, six single ascospore cultures derived from a single apothecium in the 11-81 collection on jack pine, were sent to the NCFES immunology laboratory for strain identification.

During the spring of 1981 two single ascospore cultures were grown on a medium consisting of red pine twigs, wheat, and V-8 juice. These cultures were incubated under fluorescent light at 15 C for 4 weeks or until large numbers of conidia were present.

In May, 1981 two inoculation areas were prepared in the area where the putative hybrid infection was originally found. In each area small plots 0.7 x 0.7 m were planted with 100 2-0 seedlings of red, white, Scots, and ponderosa pine (*Pinus ponderosa*); Englemann (*Picea engelmanni*), Colorado blue (*Picea pungens*), Norway (*Picea abies*), and white spruce (*Picea glauca*); shasta red (*Abies magnifica*), white (*Abies concolor*), grand (*Abies grandis*), and balsam fir (*Abies balsamea*), Japanese larch (*Larix leptolepis*); and Douglas-fir (*Pseudotsugo menziesii*). Fifty 2-0 seedlings of sitka spruce, *Picea sitchensis*, were also planted in each area.

On July 17, 1981 at 16:15 one series of host seedlings was inoculated with conidia of the North American hybrid culture No 11-81-3. The other area was inoculated with European culture No 11-81-9. Temperature at time of inoculation was 15C, relative humidity was 100 percent, wind was 3-6, gusting to 11 km per hour. The sky was cloudy with a light mist. The seedlings were brushed prior to inoculation to remove larger water droplets. Each 100 seedlings received approximately 280 thousand conidia in 100 ml of distilled water. Application was with a one-liter hand spray bottle. The Sitka spruce received about 140 thousand conidia. Germination of conidia used in the inoculation test was found to be approximately 50 percent. A second inoculation was applied on October 16, 1981 at 14:30. Temperature was 8C, relative humidity 100%, wind 0-1, gusting to 8 km per hour. Sky was cloudy with light mist. Inoculation rates and strains were the same as in the July application. Germination of conidia was again approximately 50 percent. No covering or additional moisture was applied to the seedlings after either inoculation.

RESULTS

A cleared zone indicating cell death was found only when the black spruce isolate 19-28 and two Asian isolates Sj-1 and Sj-5 were paired with other isolates listed in Table 1. The interaction between the North American and European isolates indicated some incompatibility but no cleared zone was noted. Thus it appears that there is no major barrier to nuclear and cytoplasmic exchange. All possible crosses of isolates in Table 1 were not made but three tentative incompatibility groups of isolates were observed: 19-28; Sj-1 and Sj-5; and the North American, European and putative Hybrid isolate.

The reactions between single ascospore cultures in the diallel pairings were not very strong, but within SC-1000 and 11-81 a weak reaction occurred as shown in Tables 2-5. No reaction was observed in the pairings between single ascospore cultures of SC-1001. Results of the SC-1000 and 11-81 diallels and allelic assignments based on a single gene heterogenic incompatibility system are presented in Tables 2-5. Eighty-two percent of the SC-1000 pairings and 96 percent of the 11-81 pairings agree with the allelic asignments. The observed number of cultures possessing each allele in SC-1000 differs significantly ($p < .005$) from the expected 1:1 ratio.

Single gene, two gene, and heterogenic and homogenic hypotheses were considered. In both segregating apothecia, the single gene heterogenic incompatibility hypothesis best fits the data. The inconsistencies in some of the pairings may be the result of misscoring due to the weakness of the reaction. The single ascospore pairings between cultures of SC-1000 and 11-81 were all the incompatible type reaction, indicating that these two apothecia did not share common alleles.

Microscopic observations of hundreds of pairings failed to reveal any aversion between hyphae of different isolates. This was the situation both in pairings between and within strains. In all pairs, the hyphae of different isolates made contact with each other. In some pairings it appeared that anastomoses had occurred between hyphae of different strains.

Single ascospore cultures 6, 8, and 9 of isolate 11-81 gave immunogenic reactions similar to the European strain, while the immunogenic reaction of single ascospore cultures 3, 7, and 15 were similar to the North American strain.

Table 2. Results of diallel pairings of 20 single ascospore cultures of <u>Gremmeniella abietina</u> isolate SC-1000 (+ = "compatible type" reaction, - = "incompatible type" reaction).

1	+																			
2	-	+																		
3	-	+	+																	
4	-	+	+	+																
5	+	-	-	-	+															
6	-	+	+	+	-	+														
7	-	-	+	+	-	+	+													
8	-	+	+	-	-	-	-	+												
9	-	+	+	+	-	+	+	-	+											
10	-	+	+	+	-	-	+	+	-	+										
11	-	+	+	+	-	+	-	+	+	-	+									
12	-	+	+	+	-	+	-	+	+	-	+	+								
13	-	+	+	+	-	+	-	-	-	-	+	+	+							
14	-	+	+	+	-	+	-	-	-	-	+	+	+	+						
15	-	+	+	+	-	+	-	+	-	-	+	+	+	+	+					
16	-	+	-	+	-	-	-	+	-	-	-	-	-	-	-	+				
17	-	+	+	+	-	+	-	+	+	+	+	+	+	-	+	-	+			
18	-	+	+	+	-	+	+	-	+	+	-	-	+	-	+	-	-	+		
19	-	+	+	+	-	+	-	+	+	-	+	+	+	+	+	-	+	-	+	
20	-	+	+	-	-	+	+	+	+	+	+	+	+	+	+	-	+	-	+	+
C	1	2	3	4	5	6	7	8	9	10	11	12	13	14	15	16	17	18	19	20

Table 3. Designation of alleles controlling "incompatible type" reactions based on diallel pairings of <u>Gremmeniella abietina</u> isolate SC-1000. Pairing of cultures with like alleles yields compatible type reaction. Pairing of cultures with unlike alleles yields incompatible type reaction.

Allele	Cultures Possessing Allelle
A1	1, 5, 16
A2	2, 3, 4, 6, 7, 8, 9, 10, 11, 12, 13, 14, 15, 17, 18, 19, 20

Table 4. Results of diallel pairings of 21 single ascospore cultures of Gremmeniella abietina isolate 11-81 (+ = "compatible type" reaction, - = "incompatible type" reaction).

1	+																			
2	+	+																		
3	-	+	+																	
4	-	-	-	+																
5	+	+	+	-	+															
6	+	+	+	-	+	+														
7	-	-	-	+	-	-	+													
8	-	-	-	+	-	-	+	+												
9	+	+	+	-	+	+	-	-	+											
10	-	-	-	+	-	-	+	+	-	+										
11	-	-	-	+	-	-	+	+	-	+	+									
12	-	+	-	+	-	-	+	+	-	+	+	+								
13	-	+	+	-	+	+	-	-	+	-	-	-	+							
15	-	-	-	+	-	-	+	+	-	+	+	+	-	+						
16	+	+	+	-	+	+	-	-	+	-	+	-	-	-	+					
17	-	-	-	+	-	-	+	+	-	+	+	+	+	+	-	+				
18	+	+	+	-	+	+	-	-	+	-	+	-	+	-	+	-	+			
19	-	-	-	+	-	-	+	+	-	+	+	+	-	+	-	+	-	+		
20	-	-	-	+	-	-	+	+	-	+	+	+	-	+	-	+	-	+	+	
21	-	+	+	-	+	+	-	-	+	+	-	-	+	-	+	-	+	-	-	+
C	1	2	3	4	5	6	7	8	9	10	11	12	13	15	16	17	18	19	20	21

Table 5. Designation of alleles controlling "incompatible type" reactions based on diallel pairings of Gremmeniella abietina isolate 11-81. Pairing of cultures with like alleles yields compatible type reaction. Pairing of cultures with unlike alleles yields incompatible type reaction.

Allele	Cultures Possessing Allele
A3	1, 2, 3, 5, 6, 9, 13, 16, 18, 21
A4	4, 7, 8, 10, 11, 12, 15, 17, 19, 20

During July of 1982 the test seedlings were examined for infection by G. abietina. The results of this examination are shown in Table 6. Both strains were pathogenic on all host seedlings except that no infection was found on white fir inoculated with the European strain. For most species the North American strain caused a higher percent infection than the European strain. The trees were left in the beds for an additional year to determine if apothecia will develop on the infected trees. If apothecia develop, the ascospores will be typed by serology to determine what strains develop in the second generation.

Table 6. Hybrid strain host range study of seedlings planted and inoculated 1981. Readings made July 1982.

Species	North American Strain			European Strain		
	Total trees	Scleroderris infected		Total trees	Scleroderris infected	
	No.	No.	%	No.	No.	%
Engelmann spruce	90	69	77	90	64	71
Japanese larch	53	20	38	59	36	61
shasta red fir	95	10	11	92	19	21
sitka spruce	42	26	62	48	17	35
ponderosa pine	96	94	98	88	86	98
Colorado blue spruce	98	11	11	99	11	11
white fir	90	5	6	87	0	0
grand fir	99	7	7	100	1	1
Norway spruce	97	19	20	98	11	11
white spruce	92	10	11	98	4	4
Douglas fir	98	54	55	99	26	26
balsam fir	99	7	7	97	4	4
Scots pine	86	47	55	92	41	45
white pine	99	41	41	99	10	10
red pine	85	71	84	70	59	84

DISCUSSION

Fungal incompatibility systems can be homogenic or heterogenic. In a homogenic system, isolates with the same alleles are incompatible. This promotes outbreeding. In a heterogenic system, isolates with different alleles are incompatible. This promotes inbreeding. Heterogenic incompatibility affects the ability of two isolates to form a stable heterokaryon and in all cases, has been shown to be controlled by several genes. Sexual incompatibility controls the ability of two isolates to hybridize. Most sexual interactions are controlled by homogenic incompatibility. In heterothallic basidiomycetes, where a stable heterokaryon precedes production of the sexual stage, heterogenic incompatibility can act as a barrier to sexual reproduction by incompatible isolates (Esser and Raper 1965). In contrast, heterogenic incompatibility is not a barrier to sexual crosses in the ascomycete <u>Endothia parasitica</u>. Crosses between vegetatively incompatible isolates can occur by fertilization of trichogynes of one isolate by spermatia of another (Anagnostaskis 1977).

Homogenic incompatibility systems often operate in addition to a heterogenic incompatibility system. Such is the case with <u>Podospora anserina</u> (Esser and Raper 1965) and <u>Thanatephorus cucumeris</u> (Anderson et al. 1972; Puhalla and Carter 1976). The heterogenic incompatibility system acts as an isolating mechanism, breaking the species into a number of races. The homogenic incompatibility system promotes outbreeding within each race.

Results from intra diallel pairings of single ascospore cultures of isolate SC-1000 and 11-81 suggest that a heterogenic incompatibility system may operate in *G. abietina*. The incompatibility reaction is not manifest in aversion of hyphae between isolates. Rather, incompatibility occurred after anastomosis. Because selfed-pairings show a compatible type reaction, a heterogenic system is proposed. In addition, the observation of an incompatible type reaction in all single ascospore culture pairings between SC-1000 and 11-81 supports a heterogenic system. If this were a homogenic system, then these results would have to be explained by postulating that all cultures of SC-1000 from Alberta and 11-81 from New York possess the same allele at the incompatibility locus. This seems less likely than the heterogenic case in which these widely separated isolates would possess different alleles. Therefore, a heterogenic incompatibility system is proposed. The segregation of alleles in SC-1000 and 11-81 apothecia can be explained by apothecium formation occurring without a stable heterokaryon intermediate.

Interstrain anastomoses appear to have taken place, but this does not rule out an incompatibility reaction taking place at a later stage. Proof of the ability to form a stable interstrain heterokaryon awaits a successful complementation test. Inability to generate auxotrophs has prevented such a test to date (Furnier 1980).

The immunology results on single ascospore cultures of isolate 11-81 suggest that hybridization between the European and North American strains has taken place in nature. The antigenic determinants used in strain typing are segregating in this apothecium, indicating its hybrid nature.

Infection studies were made using two single ascospore cultures from a single apothecium. One gave an immunogenic reaction with the European strain, the other with the North American strain. All host species were infected except that the European strain did not infect white fir. In general, the single ascospore strain that typed positive with the North American strain caused more disease than the "European" strain.

The implications of interstrain hybridization in *G. abietina* in North America are not entirely clear. The major differences between the European and North American strains are the former's ability to infect large trees and the latter's greater production of the wind-disseminated ascospores. Hybridization could yield progeny with a reduced ability to infect large trees and reduced production of the sexual stage. The opposite situation could, however, be just as likely, yielding types infecting large trees and with the ability to produce large numbers of ascospores. This could increase the rate of spread of types infecting large trees, posing an earlier than expected threat to forests currently outside the areas affected by the European strain. The likely result will be hybridization and introgression that yield types with many different combinations of characteristics.

LITERATURE CITED

Anagnostakis, S.L. 1977. Vegetative incompatibility in *Endothia parasitica*. Exp. Mycol. 1:306-316.

Anderson, N.A., H.M. Stretton, J.V. Groth and N.T. Flentje. 1972. Genetics of heterokaryosis in *Thanatephorus cucumeris*. Phytopathology 62:1057-1065.

Dorworth, C.E. and J. Krywienczyk. 1975. Comparisons among isolates of Gremmeniella abietina by means of growth rate, conidia measurement, and immunogenic reaction. Can. J. Bot. 53:2506-2525.

Dorworth, C.E., J. Krywienczyk and D.D. Skilling. 1977. New York isolates of Gremmeniella abietina (Scleroderris lagerbergii) identical in immunogenic reaction to European isolates. Plant Dis. Rep. 61:887-890.

Esser, K. and J.R. Raper. 1965. Incompatibility in fungi. Springer-Verlag, New York. 124 p.

Furnier, G.R. 1980. Investigation of incompatibility in Gremmeniella abietina. M.S. Thesis, Univ. Minnesota, St. Paul. 38 p.

Puhalla, J.E. and W.W. Carter. 1976. The role of the H locus in heterokaryosis in Rhizoctonia solani. Phytopathology 66:1348-1353.

Setliff, E.C., J.A. Sullivan and J.H. Thompson. 1975. Scleroderris lagerbergii in large red and Scots pine trees in New York. Plant Dis. Rep. 59:380-381.

Skilling, D.D. 1981. Scleroderris canker - the situation in 1980. J. For. 79: 95-97.

Yokota, S., T. Uozumi and S. Matsuzaki. 1974. Scleroderris canker of Todo-fir in Hokkaido, northern Japan. I. Present status of damage, and features of infected plantations. Eur. J. For. Pathol. 4:65-74.

THE USE OF POLYACRYLAMIDE GEL ELECTROPHORESIS FOR RAPID DIFFERENTIATION OF Gremmeniella abietina ISOLATES

Nicole Benhamou, G.B. Ouellette, Alain Asselin, and E. Maicas[1]

ABSTRACT

Soluble proteins from crude mycelial extracts of Gremmeniella abietina isolates from Europe, North America and Japan were compared using non gradient and gradient polyacrylamide gel-slab electrophoresis. With Coomassie blue, a total of 18 bands were recognized when results from all isolates were grouped (standard type). With reference to this standard type, isolates could be classified into six groups: 1) Isolates from Europe, New York State (US7-US11), and some from red pine from Quebec, are characterized by the presence of band 12 and the absence of bands 4 and 10. 2) Isolates from Ontario (C1) and others from pine from Quebec, show the presence of band 4 and the absence of bands 10 and 12. 3) Isolates from Japan lack band 12. 4) Isolates from New Brunswick and some from Quebec (MER) show the greatest number of bands and combining both European and American characteristics. 5) Isolates from spruce from Quebec are relatively similar to the isolates from the second group but distinguishable by at least three additional bands. 6) Isolates from balsam fir from Quebec are characterized by the lack of several bands and especially band 13. The importance and the interest of these two tests as complementary or even as a possible substitute to serology is discussed.

INTRODUCTION

For several years, the criteria used for differentiating races or strains of Gremmeniella abietina (Lagerb) Morelet were based on symptomatology of infected hosts, conidium morphology and geographic and site differences. (Ettlinger 1945; Gremmen 1968; Dorworth 1971). In 1975, by means of immunological tests, Dorworth and Krywienczyk described the occurrence of three serotypes of G. abietina, heterologous among, but homologous within, the geographic

---

[1] Research assistant and research scientist, Canadian Forestry Service, Laurentian Forestry Service, Laurentian Forest Research Center, P.O. Box 3800, Sainte-Foy, Canada. G1V 4C7 and Professor and graduate student, Faculte des Sciences de l'Agriculture et de l'alimentation, Pavillon Comtois, Laval University, Sainte-Foym, Quebec, Canada. G1K 7P4

Appreciation is expressed to Miss Rachel Pelletier and Mr. Jean Grenier for their excellent technical assistance and to Mr. Claude Moffet and Mrs. Joan Murphy for their appreciable help. The constant encouragement of Dr. Andre Lavallee, Program Manager of Laurentian Forest Research Center, was greatly appreciated.

range of this fungus in Europe, North America, and Asia. Later, by comparing symptomatology, growth and spore dispersal patterns, Skilling (1977) suggested that most of the isolates found associated with mortality of large trees in New York State, were similar to the European isolates. Dorworth et al. (1977) confirmed Skillings' report, and their serological study provided strong evidence that the New York isolates were identical to the European ones and different from the standard North American type. Serological methods were, therefore, considered to be of great interest for differentiating among these various isolates. Considering the threat in the New York state plantations, Canada and the United States applied quarantine regulations to stop import of trees from areas where the European race occurred. Because of the strong governmental measures and the complexity of the serological techniques (Dorworth et al. 1982), it became necessary to develop other techniques for differentiating G. abietina isolates. This need was also stressed by the variability of the results obtained by serology with G. abietina isolates from spruce from Quebec (D. Lachance, personnal communication), and the difficulties encountered assigning some Canadian isolates, e.g. those from New Brunswick, with a particular race. For these reasons and others, electrophoretic techniques were considered as a possible substitute to serology. For the last fifteen years, gel electrophoresis of soluble fungal proteins has been used as an aid to resolve taxonomy problems (Glynn and Reid 1969), or to evaluate relationships between species (Dorworth 1974) or among several strains of one species (Webb et al. 1972; Torp and Andersen 1982). The objective of the present study was to evaluate the use of polyacrylamide gel-slab electrophoresis, for a rapid and reliable differentiation of races of G. abietina. We also tried to gain more information about some Canadian isolates which could not be identified through serology.

MATERIAL AND METHODS

Origin of Isolates and Cultural Methods

Isolates of G. abietina from different geographical areas, except those from Quebec, were obtained from the culture collection of Dr. C.E. Dorworth, Great Lakes Forest Research Centre, Sault Ste. Marie, Ontario (Dorworth 1982) (Table 1). Isolates were grown at 4C on agar medium made of 25 ml of Campbell's V8 juice; 15 g Bactoagar; 7.5 g malt extract and 475 ml of distilled water, dispensed at 25 ml per Petri dish. Pieces $1cm^2$, cut from actively growing mycelium on agar were transferred in 125 ml Erlenmeyer flasks containing 45 ml of an autoclaved liquid growth medium [containing for one liter]: 100 ml of each of the following: 0.5% KH_2PO_4; 0.5% NaH_2PO_4; 0.5% $MgSO_4$ 0.5% KCl; 2% asparagine; 1% glucose; 2% $FeCl_2$; 4.5% HCl plus distilled water. This 45 ml liquid medium was sterilized and complemented with 5 ml of a filter sterilized vitamin solution (100 ml of 12.10^{-5}% biotine; 4 ml of 25.10^{-4}% thiamine; 4 ml of 25.10^{-4}% pyridoxine). The mycelium was allowed to grow at 14C for 4 to 5 weeks and then ground in a sterilized homogenizer. Freshly prepared flasks of autoclaved liquid medium (same composition as above), were inoculated with 1 ml of the ground mycelial suspension and maintained at 14C for at least 4 but not more than 8 weeks.

Preparation and Gel Electrophoresis of Crude Mycelial Extracts

The mycelium was harvested by filtration through a Buchner funnel, washed with distilled water and ground to a powder with autoclaved acid-washed silica gel (Silicon dioxide S-150) using a prechilled mortar and pestle (Bucher 1974).

Table 1. Origin of isolates of Gremmeniella abietina tested by polyacrylamide gel-slab electrophoresis

Isolate No.	Place of Origin	Host	Serological Identification
US 2	Michigan - USA	*Pinus banksiana*	Am
US 6	New York - USA	*Pinus strobus*	Am
US 7	New York - USA	*Pinus resinosa*	E
US 8	New York - USA	*Pinus resinosa*	E
US 9	New York - USA	*Pinus resinosa*	E
US 10	New York - USA	*Pinus resinosa*	E
US 11	New York - USA	*Pinus sylvestris*	E
F 2	Virolahti-Finland	*Pinus sylvestris*	E
F 3	Ylikiiminki-Finland	*Pinus sylvestris*	E
F 4	Virolahti-Finland	*Larix sibirica*	E
N 2	Ostfold - Co- Norway	*Picea abies*	E
N 3	Reykjavik - Norway	*Pinus aristata*	E
G 1	Germany	*Pinus strobus*	E
J 1	Hokkaido - Japan	*Abies sachalinensis*	As
J 2	Hokkaido - Japan	*Abies sachalinensis*	As
C 1	Ontario - Canada	*Pinus resinosa*	Am
79 - 253	Armtown-Quebec-Canada	*Pinus resinosa*	E
79 - 353	St.Lambert-Quebec-Canada	*Pinus resinosa*	-
79 - 347	East Pimade-Quebec-Canada	*Pinus resinosa*	E
79 - 384	Sutton-Quebec-Canada	*Pinus resinosa*	E
79 - 3456	MTF-Quebec-Canada	*Pinus resinosa*	Am
79 - 257	Power court-Quebec-Canada	*Pinus resinosa*	E
79 - 376	Sutton-Quebec-Canada	*Pinus resinosa*	E
79 - 1039	Mont-Oxford-Quebec-Canada	*Pinus resinosa*	E
79 - 333	Power court-Quebec-Canada	*Pinus resinosa*	E
79 - 622	Pont rouge-Quebec-Canada	*Pinus resinosa*	Am
79 - 623	Chute Panet-Quebec-Canada	*Pinus sylvestris*	-
79 - 419	Lac Gauthier-Quebec-Canada	*Pinus resinosa*	-
79 - 1041	Memphremagoz-Quebec-Canada	*Pinus resinosa*	E
MER 3677	MTF-Quebec-Canada	*Pinus resinosa*	Am
MER 961	MTF-Quebec-Canada	*Pinus resinosa*	-
MER 1973	Hereford-Quebec-Canada	*Pinus resinosa*	E
MER 1887	Hereford-Quebec-Canada	*Pinus resinosa*	E
NB 138	New Brunswick-Canada	*Pinus resinosa*	Problem
NB 192	New Brunswick-Canada	*Pinus resinosa*	Problem
NB 202	New Brunswick-Canada	*Pinus resinosa*	Problem
NB 047	New Brunswick-Canada	*Pinus resinosa*	Problem
81 - 867	Montmorency-Quebec-Canada	*Picea mariana*	Problem
79 - 606	Launiere Parc-Quebec-Canada	*Picea mariana*	Problem
79 - 211	Lac Gregory-Quebec-Canada	*Abies balsamea*	Problem
79 - 218	Lac Jumeau-Quebec-Canada	*Abies balsamea*	Problem
79 - 502	Zacharie-Quebec-Canada	*Pinus banksiana*	E
79 - 077	St.Magloire-Quebec-Canada	*Pinus sylvestris*	Intermediate
79 - 063	Launiere Parc-Quebec-Canada	*Picea mariana*	E
C 36	Valcartier-Quebec-Canada	*Pinus contorta*	-
C 94	Newfoundland-Canada	*Pinus sylvestris*	-
C 77	New Brunswick-Canada	*Pinus resinosa*	-

E: European race - Am: North American race - As: Asian race

The soluble proteins were extracted by adding to the mortar 0.3 ml cold 0.1 M Tris-glycine buffer pH = 7.5, containing 10% sucrose and 1 mM of a protease inhibitor (Phenyl methylsulfonylfluoride: PMSF;). The material was centrifuged at 10,000 x G for 15 min at OC and the supernatant was collected with a Pasteur pipette avoiding the presence of a lipid layer floating on the surface. Soluble proteins were separated by polyacrylamide gel-slab electrophoresis. Best results were obtained using a 14% acrylamide separating gel: 9.4 ml of acrylamide (30%) - bisacrylamide (0.8%) solution; 5 ml of Tris-HCl buffer, pH 8 ; 5.5 ml of distilled water; 0.2 ml of ammonium persulfate 10%; 8 ul of Temed. The stacking acrylamide was made with: 1.7 ml of an acrylamide (30%) - bisacrylamide (1.037%) solution; 1.25 ml of Tris-H_3PO_4 buffer: pH = 6.7; 7 ml of distilled water; 0.3 ml of ammonium persulfate; 40 ul of Temed. For each experiment, 120 ul of crude extracts of G. abietina isolates was placed directly into the preformed slots.

A voltage of 50v was applied for 30 min, then increased to 150v for about 4 hours. After the electrophoretic run, the polyacrylamide gel was carefully removed and stained, for 30 min, with 2% Coomassie brilliant blue R-250 in 50% methanol, 40% distilled water and 10% acetic acid. Excess coloration was eliminated by incubation of the gel in a 50% methanol, 40% distilled water, 10% acetic acid solution for 2 hours. The gel was then stored in a 7% acetic acid and dried later under vacuum. Silver staining was used in some cases according to Morrissey (1981).

Gradient Gel Electrophoresis

Gradient polyacrylamide gel-slab consisted of a 7.5-15% separation polyacrylamide gel. The 15% gel solution (15.5 ml of an acrylamide (30%) bisacrylamide (0.8%) solution; 3.85 ml of Tris HCl buffer pH 8.9; 8.95 ml of sucrose 50%; 2.7 ml of distilled water; 110 ul of ammonium persulfate; 8 ul of Temed) was progressively diluted with the 7.5% solution (7.9 ml of acrylamide (30%) - bisacrylamide (0.8%); 3.85 ml of Tris - HCl buffer pH 8.9; 2.6 ml of sucrose 60%; 16.5 ml of distilled water; 110 ul of ammonium persulfate; 8 ul of Temed) and poured at one drop per second with a peristaltic pump.

After polymerization of the saccharose gradient gel, a 5% concentration gel consisting of 3.34 ml of acrylamide (30%) bisacrylamide (1.037%); 2.5 ml of Tris H_3PO_4 buffer pH 6.7; 13.95 ml of distilled water, 0.8 ml of ammonium persulfate and 80 ul of Temed, was added. A constant current of 3 mA was applied for about 16 hours. After the electrophoretic run, the gel was removed and treated with Coomassie brilliant blue R-250 as described above.

RESULTS

Electrophoretic tests carried out with many different G. abietina isolates showed that the patterns of protein bands obtained from mycelial extracts could vary with culture age. The best patterns were obtained with 4 to 8 week-old (5 week-old cultures were mostly used) cultures and 14% polyacrylmaide gels.

From the analysis of the results obtained with Coomassie blue staining, a standard type pattern (ST), grouping all identifiable bands contained in all the isolates tested (representatives of which are shown in Figure 1), was established and constituted our point of reference. Band 7 was one of the most pronounced bands found which was constantly associated with all these isolates;

therefore whatever sight variations in band positioning might have occurred between runs this band constituted a reliable marker for identifying other bands. Frequently, a variability in band thickness and color intensity was observed, indicating a variation in the amounts of extracted proteins. These **differences were not investigated and our study was based on the comparative**

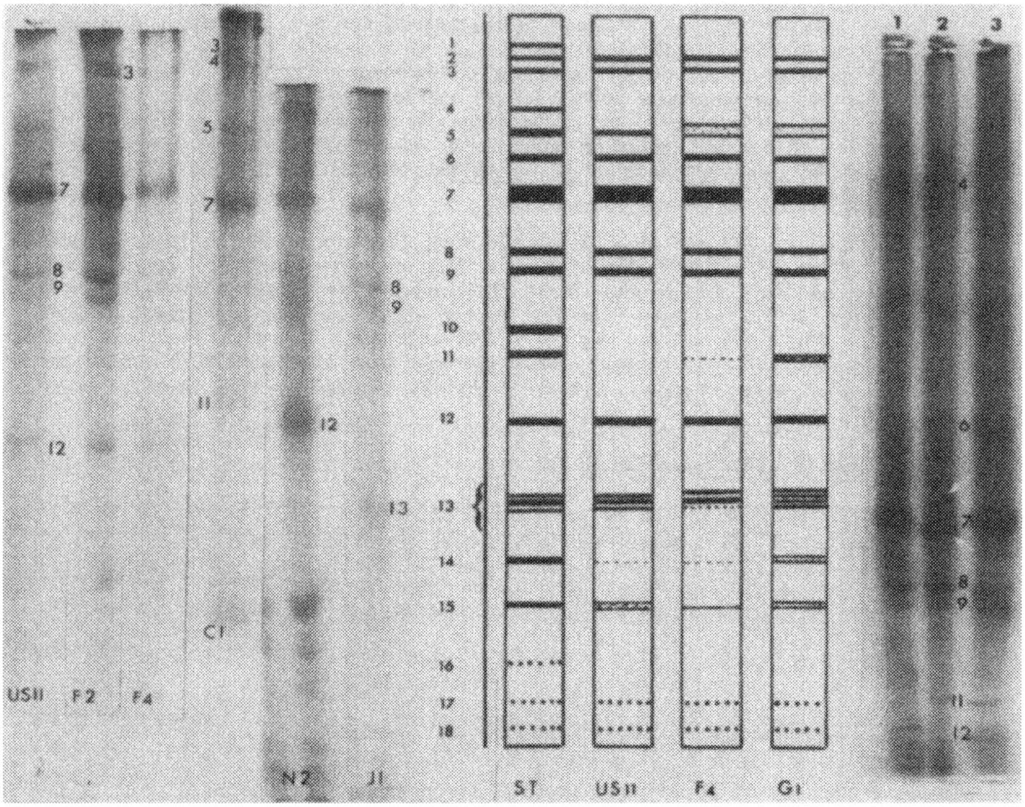

Figure 1. Illustrations of typical protein patterns otained with Commassie blue and 14% acrylamide gel (left part) and gradient gel (right part). Left part: the numbers from top to bottom, in the direction of the separation front, correspond to the main bands identified and considered to be of significance in the present study. Note: band 12 missing in C1 and J1; band 11 missing in F4, N2, and J1; presence of band 4 in C1 and and not in others. Although band 13 is visible only in J1, it was also present in the others (except N2) but did not reproduce well here. Right part: numbers 1, 2 and 3 correspond respectively to isolates US11, 81-867, and C77. The numbered bands are those believed to correspond to the same bands shown in the left part. Note: absence of band 12 in 81-867; and absence of band 11 in US11. The band identified as 4 is present in 81-867 but not in others: the faint line visible at this level in US11 is probably a background effect as it really could not be detected in the gel itself. Middle part: a histogram representation of some of the isolates constituting Group I.

evaluation of the number and the positioning of identifiable electrophoretic bands. Bands 16-18 were considered as non specific and their presence or absence was considered non reliable.

Analysis of representative electrophoretic patterns for soluble proteins indicated that the *G. abietina* isolates tested could be divided into six well defined groups (Table 2). Pine isolates (79-253, -353, -347, -384, -3456, -257, -374, -376, -1039, 1071 and 81-133 from Quebec exhibiting affinities with one another, with the other pine isolates from New York (US7, 8, 9 ,10, 11) or Europe (F2, F3, N3, G1) and with a larch isolate from Europe (F4), were placed in the first group. Qualitative comparisons with the standard type, showed that bands 4 and 10 were constantly missing. Band 11 was prominent in some isolates (G1, 79-1039), but was, most often, very weak, and sometimes not recognizable. Bands 12 and 13 were always present.

Other pine isolates from Quebec (79-622, -623, -419, -1064) gave patterns similar to that of isolate C1 (a pine isolate from Ontario, identified as American by serology) and formed the group II. This group was characterized by the lack of bands 10 and 12 while band 13 was always present.

In group III, fir isolates from Japan (J1, 2) gave similar patterns showing the total absence of band 12, the weakness of bands 6 and 11 and the constant occurrence of bands 10 and 13.

Four pine isolates from Quebec (MER3677, 961, 1973, 1887) and four from New Brunswick (NB 138, 192, 202, 047) formed the group IV. They were demarcated from the other groups by patterns exhibiting the greatest number of bands. Only band 14 was missing.

Table 2. Classification of isolates of *Gremmeniella abietina* according to patterns obtained after polyacrylamide gel electrophoresis

	Bands														
	1	2	3	(4)	5	6	7	8	9	(10)	(11)	(12)	(13)	14	15
Group I - Isolates from Europe, New York, Quebec	-	+	+	(-)	+	(+)	+	+	+	(-)	(±)	(+)	(+)	±	+
Group II - Isolates from Ontario, Quebec	-	+	+	(+)	+	(+)	+	+	+	(-)	(+)	(-)	(+)	+	+
Group III - Isolates from Japan	-	+	+	(+)	+	(-)	+	+	+	(+)	(±)	(-)	(+)	+	+
Group IV - Isolates from Quebec (MER), New Br.	±	+	+	(+)	+	(+)	+	+	+	(+)	(+)	(+)	(+)	-	+
Group V - Isolates spruce from Quebec	-	+	+	(+)	+	(+)	+	+	+	(-)	(+)	(-)	(+)	+	+
Group VI - Isolates balsam fir and one from pine from Quebec	-	±	+	(-)	+	(+)	+	+	+	(-)	(+)	(-)	(-)	+	+

Data in parenthesis are considered as the most reliable results in grouping isolates.

Examination of electrophoretic patterns from spruce isolates from Quebec (81-867; 79-606) revealed differences with groups I and III but had a close relationship with group II. They were, however, placed in a fifth group because bands 10 and 12 were missing. Two light bands, not observed in other isolates, appeared between bands 12 and 13.

Group VI is composed of G. abietina isolates from balsam fir (79-211,-218) and pine (79-502) from Quebec. The electrophoretic patterns of these three isolates were easily distinguishable by the lack of bands 4, 10, 12 and especially band 13, which was present in the other groups.

From all the European isolates tested, the spruce isolate N2 from Norway, revealed an atypical pattern. Bands 7 and 12 were always recognizable, but not bands 8, 9, 11 and 13.

Patterns of soluble proteins obtained from US2 and US6 were totally dissimilar from those resulting from all the other US isolates . By the number and relative positions of the bands, they could be more or less considered as related with group II. The pine isolate 79-077 from Quebec lacked bands 1-4, 8-10 and 14 and had no affinity with any of the other isolates.

Silver staining was found to be efficient and yielded the same patterns as with Coomassie blue. However results were not always consistent. With improvements, this technique could be very useful since its sensitivity is much greater than the Coomassie blue staining method.

Isolates: 81-867 from spruce from Quebec, US11, C36, C94, C77 from pine respectively from New York, Quebec, Newfoundland and New Brunswick produced various patterns with saccharose gradient polyacrylamide gel-slab electrophoresis. Electrophoretic patterns confirm the constant presence of one large band, corresponding to band 7 as obtained with polyacrylamide gel-slab electrophoresis. However the Rf of this band was displaced towards the anode, bands beyond 13 becoming somewhat crowded at the run end. The main differences described above for the same isolates were well confirmed: the band corresponding to band 12 was present in US11, C94, and C77, but not in 81-867 (from spruce) and C36; three extra bands, one as band 4 and two as satellites to band 7 in 81-867, separated it from C36 (these two satellite bands were also absent in the others); C94 differed from the others by lacking a band corresponding to band 6 (which however was weak in US11 and 81-867) and by having a band corresponding to band 4 (in 81-667) and another one corresponding to band 3 (in US11).

DISCUSSION AND CONCLUSION

We have shown that non gradient and gradient polyacrylamide gel-slab electrophoresis can be used to discriminate among isolates of G. abietina. Electrophoretic comparisons of the different isolates (Table 1) tested generally confirmed the inital results obtained using serology (Dorworth 1974; Dorworth and Krywienczyk 1975; Dorworth 1982) concerning the existence of three distinct races of this fungus and the affiliation of the New York isolates to the European race (Skilling 1977; Dorworth et al. 1977). The protein patterns obtained from European isolates, except for isolate N2, were dissimilar to those from North American strains represented by C1. Thus, among other possible differences, band 4 only present in the American and band 12 only present in the European isolates, were considered as good criteria of differentiation between these two groups. The absence of band 12 (and generally band 6) and the presence of band 10 in Japanese isolates made them easily distinguishable from both European and American types. They were also different from all other isolates (Table 1) and therefore could be identified without any problem. In the light of these results, three groups of isolates with distinctive protein patterns are proposed and correspond to previous groupings obtained by serology. The New York isolates, except US2, US6, and a large number of isolates from red pine from southern Quebec, exhibited very close affinities with the European type. A few isolates from this same part of Quebec gave patterns similar to the standard American type C1. These results do not indicate, however, that the European strain is more abundant in Quebec, because only a limited number of isolates from the northern part of this Province were tested.

The evidence for the establishment of three more groups came from the examination of particular patterns obtained with some red pine isolates from New Brunswick and spruce and balsam fir isolates from Quebec which did not show homology with any of the groups already described. Patterns of some isolates from New Brunswick and Quebec (MER) revealed the interesting feature to combine both bands 12 and 4 considered to be specific for European and American isolates respectively. This may explain why, using serology, the identification of these isolates was problematic and not consistant (D. Lachance, personal communication). Since characteristics of both European and American isolates are present with these isolates, it is not surprising that they might be antigenically indistinguishable. By non gradient and gradient electrophoretic tests, isolates from spruce from Quebec were found to be dissimilar to European ones. Although they showed a relative homology with the American race, they had to belong to a separate group because of the presence of a few additional bands in their patterns. It was not possible to explain why these isolates were sometimes identified as European by serology, since, band 12, characteristic for all the European isolates was never present. Isolates from balsam fir gave patterns different from all others, even with fir isolates from Japan. The lack of a great number of bands and especially band 13 which was found associated to other isolates, constituted a good criterium of identification and classification of these isolates in a particular group.

The present data indicates that a good correlation could be established between results obtained with serology and electrophoresis tests with *G. abietina* isolates. There are, however, a number of discrepancies that deserve further consideration. Thus, isolates C94 from Newfoundland, C77 from New Brunswick, 79-077 from Quebec which was serologically classified as intermediate, failed to fit in any one of the well defined groups. More studies are necessary to identify these isolates. To determine the extent of Scleroderris canker in Quebec, *G. abietina* isolates from different sites should be electrophoretically tested.

Present results have shown precise and specific differences among the various isolates of *G. abietina*. However, before being able to regard this method as an ideal substitute to current methods of identification, the proposed criteria of differentiation should be confirmed by other more refined tests such as immunoelectrophoretic techniques.

LITERATURE CITED

Bucher, J.B. 1976. Anwendung der diskontinuierlichen Polyacrylamid gelelektrophorese in der Taxonomie der Gattung *Nodulosphaeria Rbh* (Ascomycetes). Jahrang 119(2):125-164.

Dorworth, C.E. 1971. Diseases of conifers incited by Scleroderris lagerbergii Gremmen: A review and analysis. Can. For. Serv. Publ. 1-42.

Dorworth, C.E. 1974. Comparison of soluble proteins of Ascocalyx abietis and Gremmeniella abietina by serology and electrophoresis. Can. J. Bot. 52: 919-922.

Dorworth, C.E. and J. Krywienczyk. 1975. Comparisons among isolates of Gremmeniella abietina by means of growth rate, conidia measurement and immunogenic reaction. Can. J. Bot. 53:2506-2525.

Dorworth, C.E., J. Krywienczyk and D.D. Skilling. 1977. New York isolates of Gremmeniella abietina (Scleroderris lagerbergii) identical in immunogenic reaction to European isolates. Plant. Dis. Rep. 61:887-890.

Dorworth, C.E., D.P. Webb, J. Krywienczyk and D. McNamara. 1982. Comparisons between serologic and gas chromatographic technique for characterization of Gremmeniella abietina and related species. Eur. J. For. Pathol. 12:-209-217.

Ettlinger, L. 1945. Uber die Gattung Crumenula sensu Rehm., mit besonderer Berucksichtigung des Crumenula - Triebsterbens der Pinus -Arten. Beitr. Kryptogamenflora Schweiz. 10:1-73.

Glynn, A.N. and J. Reid. 1969. Electrophoretic patterns of soluble fungal proteins and their possible use as taxonomic criteria in the genus Fusarium. Can. J. Bot. 47:1823-1831.

Gremmen, J. 1968. Le chancre Scleroderris du pin noir en Europe et en Amerique du Nord. Bull. Mycol. France. 84:165-150.

Morrissey, J.H. 1981. Silver stain for proteins in polyacrylamide gels: a modified procedure with enhanced uniform sensitivity. Anal. Biochem. 117:307-310.

Skilling, D.D. 1977. The development of a more virulent strain of Scleroderris lagerbergii in New York State. Eur. J. For. Pathol. 7:297-302.

Torp, J. and B. Andersen. 1982. Two dimensional electrophoresis of proteins from cultures of Erysiphe graminis sp hordei. Physiological Plant Pathol. 21:151-160.

Webb, H.M., A. Gafoor, and J.B. Heale. 1972. Protein and enzyme patterns in strains of Verticillium. Trans. Br. Mycol. Soc. 59:393-402.

DISPERSAL OF CONIDIA OF *Gremmeniella abietina* RELATED TO WEATHER

D.R. Bergdahl[1]

ABSTRACT

Condia of *Gremmeniella abietina* (Lagerb.) Morelet were trapped near infected red pine (*Pinus resinosa* Ait.) using rotorods and silicone coated glass microscope slides. Continuous trapping was done during the primary period of condia dispersal in Greensboro, Vermont. Numbers of conidia collected were related to weather conditions. Most conidia were trapped during periods of precipitation, but some were trapped during dry conditions. The dispersal of condia during dry conditions was associated with higher wind speeds and low relative humidities.

INTRODUCTION

Gremmeniella abietina (Lagerb.) Morelet is an ascomycete that produces both sexual and asexual spores on infected tissues of the host. Sexual spores (ascospores) are produced in an apothecium and are actively liberated and then dispersed by wind during and immediately following periods of precipitation as described in epidemiological studies by Dorworth (1972) and Skilling (1972).

The asexual stage of *G. abietina* is *Brunchorstia pinea* (Karst.) Hoehn. Conidia (macroconidia) of *B. pinea* are usually produced in pycnidia that extend above the epidermis of the host, but cryptopycnidia have been found on several species of *Pinus* (Cauchon and Lachance 1980, Azjchowski and Bergdahl, unpublished data). Very little is known about the cryptopycnidial state, but it is known to produce conidia that appear similar to those produced in pycnidia. These spores germinate and produce normal cultures of *G. abietina* (Cauchon and Lachance 1980). The conidia are passively liberated in a slime flux from both pycnidia and cryptopycnidia and are believed to be primarily rainsplash disseminated during periods of precipitation, but little is known about the patterns of conidia dissemination related to environmental conditions.

---

1 Associate Professor, Department of Forestry, Aiken Center, University of Vermont, Burlington 05405.

This research was supported in part by funds provided by the USDA Forest Service North Central Forest Experiment Station, RMU FS-NC-2205 and by the McIntire-Stennis Cooperative Forestry Research Program (USDA/CSRS and The University of Vermont).

Microconidia have also been found in cultures representing North America, European and intermediate serotypes (races) of G. abietina. These microconidia have also occasionally been found in pycnidia on several species of Pinus in North America (Zajchowski and Bergdahl, unpublished data) and have been reported in Europe by Roll-Hansen and Roll-Hansen (1973). These microconidia rarely germinate and their biological function is unknown (Zajchowski and Bergdahl 1982). The microconidial stage was not sampled in this study.

The asexual form of G. abietina predominates in New York and Vermont as it does in Great Britain (Read 1968). The European race and intermediate race as described by Wendler et al. (1978) are found in Vermont and New York (Manion, personal communication). In these areas the ascigerious stage of G. abietina is reported to be rare and therefore must not be important in the epidemiology of this fungus. Since little was known about the dispersal of the asexual spores of the European race of G. abietina in North America, the objective of this study was to determine what environmental conditions were associated with the dispersal of conidia.

METHODS

A twenty-year-old plantation of red pine in northern Vermont (Greensboro) was selected for this study on conidia dispersal. This plantation was infested with the European race of G. abietina and previous studies indicated that high numbers of conidia were being dispersed. All of the trees were infected.

Silcone-coated microscope slides and rotorods were used to trap conidia in the field. The method of slide trapping was similar to the method described by Ostry and Nicholls (1982). The rotorod sampling system was primarily used as a backup to the slides, but this system also allowed the sampling of greater volumes of air per unit time. Twenty sampling points were established for each system of trapping. Conidia of G. abietina were continuously trapped June 5-29, 1982 as previous studies indicated this was the peak time for conidia dispersal at Greensboro. This period of trapping was then subsequently divided into 93 individual trapping events which were related to specific weather conditions. Specific weather data were continuously recorded during the period of conidia trapping using the following instrumentation: Precipitation was monitored with a remote recording tipping bucket rain gauge. Temperature and relative humidity was recorded with an electronic remote recording temperature relative humidity sensor. Wind was monitored with a skyvane wind indicator and anemometer. (Weather Measure Inc., Sacramento, CA).

To determine spore numbers the conidia collected on the glass slides and rotorods were stained with lactophenol-cotton blue and counted using a compound microscope. Numbers of conidia were recorded as spores/mm^2 and then related to the specific weather data collected during the period of conidia trapping.

RESULTS

Conidia of G. abietina were trapped during 53 of 93 trapping periods from June 5-29, 1982. Most conidia were trapped during periods of precipitation, but some were trapped when conditions were dry (Figure 1). Some precipitation occurred during 29 of the 53 periods when spores were trapped, but during the remaining 24 periods no precipitation occurred (Table 1). Peak numbers of conidia trapped did not correspond with peak amounts of precipitation, but high numbers of spores were always associated with precipitation.

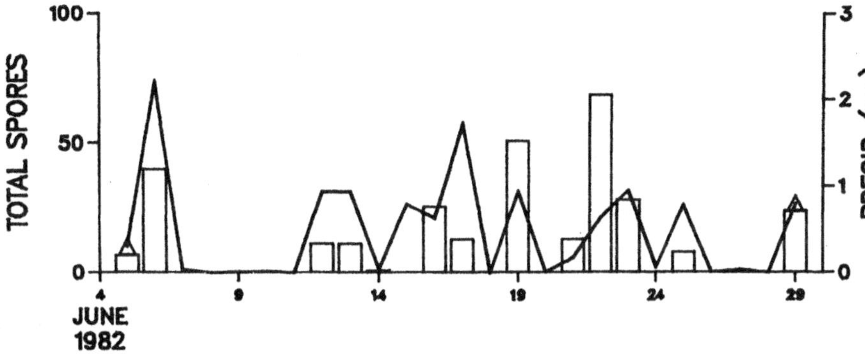

Figure 1. Number of conidia of Gremmeniella abietina trapped and amount of precipitation recorded in a plantation of Pinus resinosa near Greensboro Vermont June 5-29, 1983.

Table 1. Periods of conidia trapping of Gremmeniella abietina in a plantation of Pinus resinosa in Greensboro, Vermont during June 1982. Spore numbers are related to amount of precipitation, wind speed, temperature and relative humidity.

	Number of Trapping Periods		
	40	24	29
Number of Spores			
Range (sp/mm^2/hr)	0	.02-2.46	.12-25.20
Average (sp/mm^2/hr)	0	.21	4.34
Precipitation (cm)			
Range	0	0	.02-2.06
Average	0	0	.31
Wind Speed (km/hr)			
Range (high)[a]	0-32	0-29	0-27
Mean (high)[a]	16	15	11
Temperature (°C)			
Range (high)[a]	10-32	8-30	12-22
Mean of medians	18	16	15
Relative Humidity (%)			
Range (high)[a]	45-100	48-100	90-100
Mean of medians	71	78	95

[a] = high wind speed, temperature or relative humidity.

During the 29 periods when there was precipitation and conidia were trapped, the numbers of conidia trapped averaged 4.34 spores/mm^2/hr (range 0.12--25.20), precipitation averaged .31 cm (range 0.02-2.06), high wind speed averaged 11 km/hr (range 0-27), temperature averaged 15C (range 12-22) and relative humidity averaged 95% (range 90-100) (Table 1).

During the 24 periods without precipitation when conidia were trapped, the numbers of conidia averaged 0.21 spores/mm^2/hr (range 0.02-2.46), high wind speed averaged 15 km/hr (range 0-29), temperature averaged 16C (range 8-30) and relative humidity averaged 78% (range 48-100).

During the 40 periods when there was no precipitation and conidia were not trapped, high wind speeds averaged 16 km/hr (range 0-32), temperature averaged 18C (range 10-32) and relative humidity averaged 71% (range 45-100).

Conida were not trapped when high and average temperatures were $\geq$ 32 and 30C, respectively. Conidia were trapped only 3 of 11 times when the high temperatures exceeded 25C. For these 3 periods high wind speeds averaged 29 km/hr and relative humidity averaged 60% (range 48-98). Also, 67% of the conidia trapped at temperatures above 25C were trapped when the relative humidity was less than 50%.

DISCUSSION

Previous studies on the epidemiology and dispersal of G. abietina in North America have mainly concerned the ascigerous state and weather conditions required for ascospore discharge and movement. The conidial state of the European race of G. abietina predominates in the areas of infestation of this race in North America. Since the ascigerous state is rare, the dispersal of the conidial state becomes an important consideration. This study has shown that most conidia are dispersed during periods of precipitation as was mentioned by Dorworth (1972) and Skilling (1972). This study has also shown that conidia can be dispersed during dry weather. The mode of dispersal during dry weather remains unknown, but probably involves some form of redistribution phenomenon involving previously dispersed conidia. The actual viability of the conidia that were trapped during dry weather is unknown, but the report by Zajchowski and Bergdahl (1983) indicates that conidia remain viable on the foliage of red pine in the field for long periods of time. The phenomenon of dry dispersal and longterm viability of conidia in the field therefore may be important in advancing our understanding of the epidemiology of the European race of G. abietina in North America.

LITERATURE CITED

Cauchon, R. and D. Lachance. 1980. Research on cryptopycnidia for rapid diagnosis of Gremmeniella abietina. Can. J. Plant Pathol. 2:232-234.

Dorworth, C.E. 1972. Epidemiology of Scleroderris lagerbergii in Central Ontario. Can. J. Bot. 50:751-765.

Ostry, M.E. and T.H. Nicholls. 1982. A technique for trapping fungal spores. U.S.D.A. For. Serv. Res. Note NC-283.

Read. D.J. 1968. Dieback disease of pines with special reference to Corsican pine (*Pinus nigra* var. *calabrica* Schn.) III. Mycological factors. Forestry 41:72-82.

Roll-Hansen, F. and H. Roll-Hansen. 1973. Microconidia formed by *Scleroderris lagerbergii*. 2nd International Congress of Plant Pathology. University of Minneapolis, MN.

Skilling, D.D. 1972. Epidemiology of *Scleroderris lagerbergii*. Eur. J. For. Pathol. 2:16-21.

Wendler, P.B., A.R. Gotlieb and D.R. Bergdahl. 1980. A newly discovered serotype of *Gremmeniella abietina*. Can. J. Bot. 58:1923-1928.

Zajchowski, S.J. and D.R. Bergdahl. 1982. Observations on a microconidiospore stage associated with *Gremmeniella abietina*. Phytopathology 72:267.

Zajchowski, S.J. and D.R. Bergdahl. 1983. Viability of Conidia of *Gremmeniella abietina* sprayed on new shoots of *Pinus resinosa*. Phytopathology 73: (In press).

INOCULUM POTENTIAL OF *Gremmeniella abietina* IN NEW YORK

Christopher J. Luley and Paul D. Manion[1]

ABSTRACT

In order to determine the combined effects of inoculum density and conditions favorable for infection by *Gremmeniella abietina*, potted red pine seedlings and spore traps were suspended by ropes and pulleys at various heights from the ground to the upper crown of trees. The seedlings and spore traps set in two red pine stands and a nearby hardwood stand were changed weekly. Spore densities were monitored with petroleum jelly coated microscope slides and a specially developed monofilament spore trap. Spore traps indicated that asexual spores of the pathogen were prevalent from ground level to 12 m in the upper crown in the red pine stands. Greatest spore densities were monitored in May and June and were significantly correlated with periods of high rainfall and 100% relative humidity. The monofilament spore trap was found to be a more efficient method to trap spores of the fungus under these conditions. Airborne conidia were collected at 9 m above the ground and at distances up to 25 m from the inoculum source. Combinations of moist, cool environmental conditions and high spore density in May and June favored the infection of the greatest number of seedlings and the infection of seedlings in the upper canopy of both red pine sites.

INTRODUCTION

Gremmeniella abietina (Lagerb.) Morelet (*Brunchorstia pinea* (Karst.) Hoehn., the asexual stage), the causal agent of Scleroderris canker, incited a dramatic increase in mortality of large red pines (*Pinus resinosa* Ait.) in northern New York in the mid-1970s. The most severe damage occurred in two major infection centers: one in the Boonville area and the other in the Lake Clear region of the Adirondack region (Skilling 1977). Since the factors responsible for this explosive epidemic are not completely understood, it is difficult for the forest pathologist to properly prescribe control tactics and assess the future potential impact of the disease.

Environmental effects on both the host and pathogen are known to exert a major influence on disease development of Scleroderris canker. This combined interaction of the host, pathogen and environment constitutes the inoculum potential of a disease (Van der Plank 1975). Specifically, inoculum potential

---

[1] Graduate student and professor, State University of New York College of Environmental Science and Forestry, Syracuse, NY 13210. Present address of senior author: Department of Plant Pathology, Iowa State University, Ames, IA 50011.

is a summation of the amount of susceptible host tissue, the amount and virulence of the inoculum, plus the combined effect of the environment on each (Horsfall and Diamond 1963).

The amount of disease produced by the pooled interaction of the host, pathogen and environmental factors have historically been a common measure of inoculum potential, but Van der Plank (1975) contends that inoculum potential should be viewed in terms of its separate components and not by the amount of disease produced.

Outbreaks of Scleroderris canker in Europe, Asia and North America have at various times been associated with changes in the host, pathogen and environment (Dorworth 1971). Increases in the inoculum potential of Scleroderris which favors the pathogen's ability to cause disease have been reported as resulting from predisposition of the host (Day 1945, Ohman 1966, Read 1966, 1968b); changes in environmental conditions which are conducive to infection and spread of the disease (Carvin 1980, Donaubauer 1972, Dorworth 1972, 1973, 1979a); and a new virulent race of the fungus (Skilling 1977, Dorworth 1981).

The increase in inoculum potential which led to the epidemic of Scleroderris canker in New York is speculated to have been associated with two major factors. The first was the appearance of a new virulent race of the fungus which is similar to European isolates of *G. abietina* (Skilling 1977, Dorworth et al. 1977). The second was a change in regional climatic conditions which favored rapid disease buildup (Carvin 1980). Presently, although disease increase rates have slowed, the potential for future outbreaks of the disease is still believed to exist.

Knowledge of the new physiologic race of *G. abietina* in New York presented by Skilling (1977), Dorworth et al. (1977), and the factors affecting the intensification of the disease in young red pine plantations studied by Carvin (1980) have aided in understanding this disease complex in the state. A study to clarify which factors influence the spread of *G. abietina* into the forest canopy would further aid in understanding the disease and its potential in New York. Therefore, the objectives of this study were as follows:

(1) To determine the inoculum density as vertically distributed in the forest canopy at different levels of disease severity in two red pine stands and a nearby hardwood stand. This entailed the development of a more efficient method to trap spores of the fungus.

(2) To monitor the microclimatic conditions within the forest canopy that could affect disease development.

(3) To determine infection periods (weekly) during the growing season at various heights in the forest canopy under the different inoculum densities as found in the two red pine stands and an adjacent hardwood stand.

MATERIALS AND METHODS

In the spring of 1981, three field sites, located northeast of Boonville, New York were selected for monitoring *G. abietina* inoculum levels and infectior periods.

Site Descriptions

Site A was a high infection area situated within a 50-year-old red pine plantation on the south facing slope of a slight topographic depression. This site could be considered a Scleroderris "infection pocket" since many of the trees had either been killed by the fungus or were living and contained multiple upper branch infections.

Site B was a low infection area located in a second red pine stand 50 m northeast of site A. The canopy of this stand was fully closed except for occasional missing trees. Upper branch infections were evident in some trees but the degree of disease incidence and mortality was significantly less than in site A.

Site C was established in a mixed northern hardwood stand 40 m north of site B. This site was 25 m from the nearest pine and was established to monitor drift distances of *G. abietina* conidia.

Sampling Design

Five sampling stations located at various heights in the canopy were established at each site. A pulley was attached to a branch in the upper crown of a dominant tree. Sampling stations were fastened and raised or lowered with a rope fed through the pulley. Each station contained two types of spore traps and potted red pine seedlings (Figure 1). In addition, sampling stations in site A held dew sensors. A seven day recording hygrothermograph was also housed at ground level in site A.

Spore traps and seedlings were changed weekly from early April until late August. Phenology of red pine in northern New York was noted relative to data collection dates.

Inoculum Densities

Inoculum densities were monitored with petroleum coated microscope slides and monofilament spore traps.

Petroleum coated microscope slides were held in place on the rope with rubber bands. Slides were collected weekly and scanned at 100X magnification for conidia and ascospores. The average number of spores resulting from five scans per slide was recorded. A single scan covered an area of 33 mm^2. Spore characteristics described by Skilling and O'Brien (1979) were used as a guideline for identifying spores. Higher magnification (400X) was used when the identity of individual spores was in question.

The monofilament spore trap was designed specifically to trap spores dispersed in fog or splashed rain water on the monofilament line. The trap consisted of a metal frame with 8 m of 290 um monofilament line strung over an area of 29 cm^2 (Figure 2). An aluminum wind vane was attached to the monofilament spore trap to orient the spore trap in the direction of prevailing winds. Water containing spores picked up by the monofilament line was collected in 35 mm film containers attached with parafilm at the base of the trap. The volume of water collected weekly was recorded, centrifuged for 20 minutes at 7000 rpm with a clinical table top centrifuge, the supernatant was drawn off to 0.25 ml, the pellet resuspended and then placed on a microscope slide for counting. Spores were counted in the same manner as petroleum coated slides.

Spore trap data was evaluted with respect to (1) influence of environmental variables on periods of spore release, (2) variation in spore density among heights within each stand, and (3) variation in spore density among different stands.

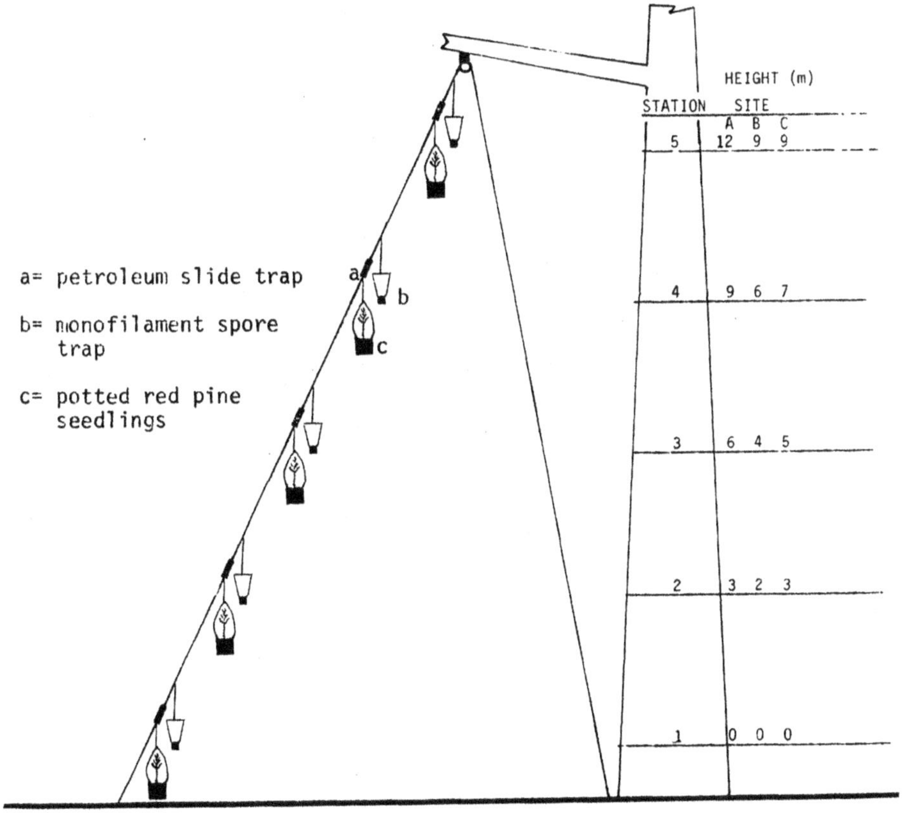

Figure 1. Sampling design and sampling heights within site A, site B, and site C.

Environmental Data

A seven day recording hygrothermograph and an Ag-Tech Dewdynamic (Ag-Tech Instruments Co., Savannah, GA 31404) system monitored temperature, relative humidity and periods of free water present in site A.

Figure 2. Monofilament spore trap designed for this study to trap conidia of B. pinea that were dispersed in free water (See text p. 11 for description).

The hygrothermograph continuously recorded temperature and relative humidity at site A. The Dewdynamics system consisted of five dew sensors that were individually attached to each of the five stations near the spore traps and potted red pine seedlings. Dew sensors were capable of detecting small amounts of surface wetness by changes in electrical resistance on gold plated sensing plates. Hourly changes in moisture resulting from fog, rain or dew were recorded with this system. Values recorded with the dew sensors ranged from 0 to 100 units, with values of less than 30 units indicating light dew; 30-69 units a moderate dew; and 70 units or above a heavy dew.

Weekly Infection Periods

Thirteen centimeter pots containing three 3-year old red pine seedlings served as an infection bioassay for sampling weekly infection periods. Each week uninfected potted seedlings from an infection free area near Lafayette, New York were moved into each station. The following week, seedlings were removed to a Scleroderris free area near Lowville, New York and another set of seedlings was placed in sampling stations. Ten pots of unexposed seedlings brought to Lowville in the beginning of the season were used as controls.

In November the seedlings were moved to Syracuse and exposed to winter conditions in a caged, screened storage area. In mid January seedlings were relocated to a greenhouse to allow symptoms to develop.

Infected seedlings were identified by the presence of a yellow-green discoloration beneath the bark and by confirming isolations from the necrotic tissues.

RESULTS

Inoculum Densities

No ascospores of G. abietina were collected in either the monofilament spore traps or the petroleum coated slide traps during the 17 week sampling period. Conidia of B. pinea, the asexual stage, were collected from the first week in April when data collection commenced. Conidial release continued throughout the sampling period and varied greatly with the time of collection (Figures 3 and 4). Greatest spore counts were recorded in early May through early June and coincided with the period of elongation and development of red pine shoots.

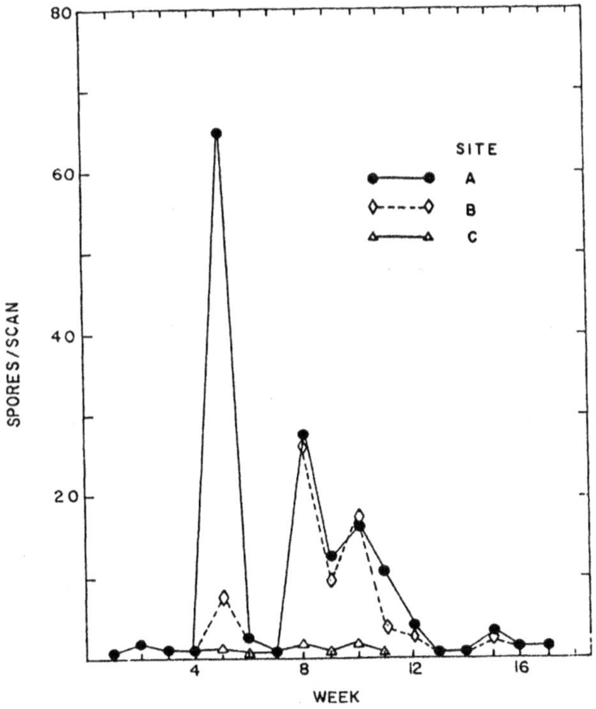

Figure 3. Weekly mean spore density monitored by the petroleum coated slide trap for site A, B, and C from April 10 to August 6, 1981. Means are an average of 5 stations at each site.

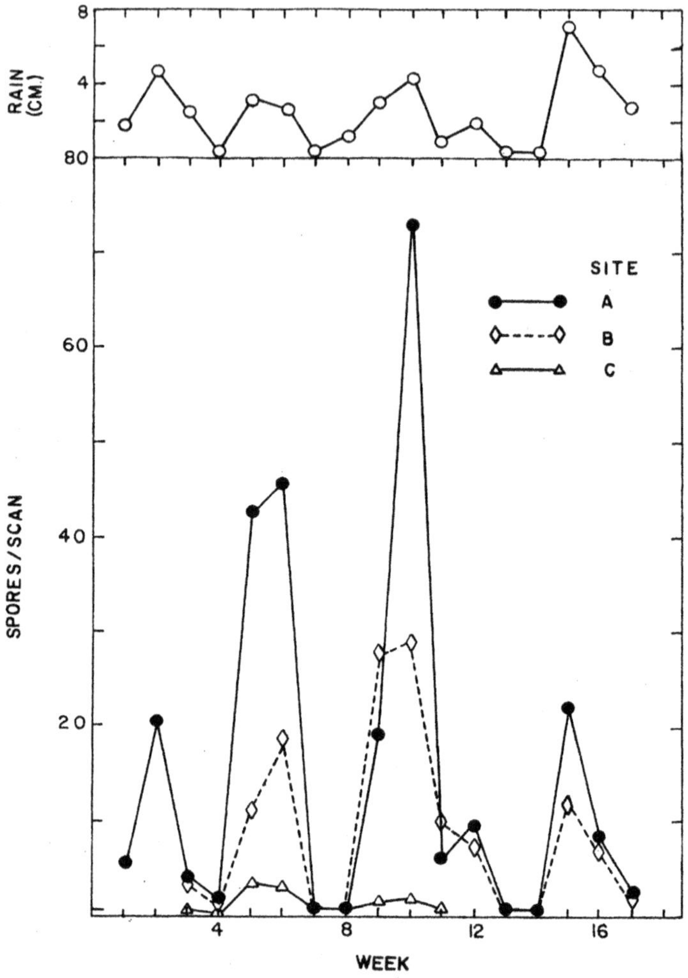

Figure 4. Weekly rainfall and weekly mean spore density monitored by the monofilament spore trap for site A, B and C from April 10 to August 6, 1981. Means are an average of five stations at each site.

Spore counts recorded by the monofilament spore trap were significantly correlated with weekly rainfall, total hours of 100% relative humdity for the week and the longest continuous period of 100% relative humidity for the week (Table 1). Spore counts recorded on the petroleum coated slide traps were not significantly correlated with any of the environmental variables measured (Table 1).

Correlations were not significant between amounts of free water monitored by dew sensors and spore counts recorded by either of the spore traps at the various sampling stations at site A.

Both the monofilament spore trap and the petroleum coated slide trap monitored uniformly low levels of conidia at all heights within site C (mixed hardwood) (Table 2).

Mean spore density for the year in the two red pine sites did not vary significantly with height (Table 2) or between stands (Figure 3) when monitored by the petroleum coated slide trap. Significant differences in spore density between site A and site B were detected with the monofilament spore trap. Spore density was generally found to be the greatest in site A during weeks of high rainfall (Figure 4). Analysis of variance indicated a plot-week interaction was present between site A and site B.

Spore density at the various heights within site A and site B varied significantly for the year when monitored by the monofilament spore trap. Within site A, greatest mean spore counts for the year were recorded near ground level (Table 2). Within site B, greatest mean spore counts were recorded within 2 m of ground level and at the highest station, 9 m in the forest canopy (Table 2).

Seedling Infection

Very few exposed seedlings develop typical symptoms when growth resumed in the greenhouse in January. Isolations were performed on all seedlings that had lateral or terminal buds that failed to expand or showed abnormal drying symptoms. Abnormal drying symptoms were observed in both the control and exposed seedlings and were not correlated with infection periods in the field. *G. abietina* was not isolated from any of the control seedlings.

Twenty three of the 515 seedlings exposed during the sampling period were identified as being infected based on symptoms and or culturing.

Table 1. Correlation coefficients of relationships between environmental variables and the number of spores collected with the monofilament spore trap (MST) and the petroleum coated slide trap (PST) for site A, site B, and site C combined.

Environment	Spore Trap	
	MST	PST
Rain[a]	.432*	.070
Maximum RH[b]	.412*	.069
Total RH[c]	.413*	.050
Average Temp.	-.094	-.109
Vol. Water MST	.441*	.041
MST	---	.459*

[a] cm per week
[b] longest continuous period of 100% relative humidity for the week
[c] total hours of 100% relative humidity for the week
* significant at P = .01

Table 2. Mean spore densities for the year at the various heights within site A, site B and site C for the monofilament spore trap (MST) and the petroleum coated slide trap (PST).

Site A			Site B			Site C		
Height (m)	MST[a]	PST[a]	Height (m)	MST	PST	Height (m)	MST	PST
0	28.4 A[b]	11.2[c]	0	10.0 AB	7.1[c]	0	2.2[c]	0.4[c]
3	12.5 B	7.2	2	9.1 AB	5.1	3	1.6	0.6
6	8.9 B	5.4	4	3.3 B	2.3	5	1.5	0.7
9	16.0 B	4.9	6	4.7 B	2.6	7	1.4	0.6
12	12.3 B	3.2	9	16.8 A	4.7	9	1.2	0.6

[a] expressed as spores per scan with a scan = 33 mm^2
[b] means with a common letter within a column are not significantly different at P = .05 according to Duncan's multiple range test (SAS Institute, Raleigh, NC 27605)
[c] no significant difference among heights indicated by analysis of variance at P = .05

Nine seedlings were infected at the various heights from site A (Table 3). Infections at site A occured on seedlings exposed at ground level on week 2, 6 and 9; 3 m from the ground on week 15; 6 m from the ground on week 10; 9 m from the ground on week 6; and 12 m from the ground on week 10.

Fourteen seedlings were found to be infected at the various heights in site B (Table 4). Infections were identified from seedlings exposed at this site at ground level on week 2, 5, 6, 8 and 10; 2 m from the ground on week 6, 8 and 15; 6 m from the ground on week 5; and 9 m from the ground on week 6. No infections were found to occur from seedlings exposed 4 m from the ground in site B.

No infections were found from seedlings exposed in site C.

In relation to the phenology of red pine, 19 out of 23 seedlings infected came during weeks when candles were elongating and needles had begun to emerge and develop. Two infections occurred when red pine was dormant in the spring and two infections occurred when growth had ceased and terminal buds were being formed.

An interaction of inoculum density, environmental conditions and the developmental stage of red pine meristematic tissue affect the inoculum potential of Scleroderris canker at a particular time and location. The highest level of infection for the two red pine sites combined (31%) occurred during a period of high spore density and cool, moist environmental conditions in mid-May. These conditions were conducive enough to 43% of the seedlings exposed in the closed canopy in one of the sites and infection of seedlings at upper positions in the canopy of both sites.

Table 3. Number of seedlings infected from exposure to natural inoculum and environmental conditions at each height and week within site A.

	Number of Seedlings Infected					
	Week[a]					
Height (m)	2	6	9	10	15	Total
0	1/2	1/3	1/2	0/3	0/2	3/12
3	0/3	0/3	0/3	0/3	1/3	1/15
6	0/3	0/3	0/3	2/3	0/2	2/14
9	0/2	2/3	0/2	0/3	0/2	2/12
12	0/3	0/3	0/3	1/3	ND	1/12
Total	1/13	3/15	1/14	3/15	1/9	9/65

[a] seedlings exposed during other weeks did not become infected
ND no data

Table 4. Number of seedlings infected from exposure to natural inoculum and environmental conditions at each height and week within site B.

	Number of Seedlings Infected						
	Week[a]						
Height (m)	2	5	6	8	10	15	Total
0	1/16	1/3	3/3	1/3	2/2	0/2	8/29
2	ND	0/3	2/2	1/3	0/3	1/3	4/14
4	ND	0/3	0/3	0/3	0/3	0/1	0/13
6	ND	1/2	0/3	0/3	0/3	0/2	1/13
9	ND	0/3	1/3	0/3	0/3	0/1	1/13
Total	1/16	2/14	6/14	2/15	2/14	1/9	14/82

[a] seedlings exposed during other weeks did not become infected
ND no data

DISCUSSION

The monofilament spore traps and the petroleum coated slide traps monitored ambient levels of conidia present, since the spore traps were not placed in close proximity to B. pinea pycnidia. Spore counts are significantly lower than those monitored by Skilling (1969) but are comparable to spore densities monitored by Selliff and Manion (1980) in the Boonville, New York area.

Spore density monitored from April to August in this study support the data of Skilling (1969, 1972, 197) that spore dispersal can occur over an extended time period whenever favorable environmental conditions exist. Periods of high rainfall and relative humidity were significantly correlated with spore release as monitored by the monofilament spore trap. The influence of these environmental variables on spore release have been reported previously (Skilling 1969, 1972; Dorworth 1972; Smerlis 1968).

Efficiency of both spore traps was affected by environmental conditions. It was evident that washing of the petroleum jelly from the petroleum coated slide occurred during periods of high rainfall. Washing of the petroleum surface by direct rainfall and the hydrophobic reaction of the petroleum jelly to free water probably greatly reduced the efficiency of this spore trap during rainy periods. This could account for the lack of significant correlation between the slide trap and environmental variables.

The monofilament spore trap was specifically designed to collect water containing spores, but was ineffective during dry periods. During week 8, 1.24 cm of rain fell and high spore counts were recorded by the petroleum coated slide trap. Evaporation dried all water from the monofilament spore traps. No spores were recorded for that week even though water was added to the collecting vials from this week and then centrifuged as before. It was later determined that spores strongly adhered to the wall of the plastic film canister and required scrubbing to be removed. In weeks of increased rainfall, the monofilament spore trap recorded an increase in spore density but the petroleum coated slide trap recorded a decrease in spore density.

Low levels of conidia were monitored at all heights in site C. Conidia from inoculum sources in the canopy of adjacent pines apparently became airborne and were transported laterally short distances. Lateral movement of airborne conidia for distances up to 100 m has been shown to occur for Dothistroma pini Hulberry (Gibson et al., 1964). Conidia of D. pini are believed to become airborne during periods of rainfall and high relative humidity by a splash "takeoff" method. Conidia of D. pini are of similar size and shape to that of B. pinea and are also primarily dispersed in splashed water. This would suggest that a similar mechanism of spore dispersal is possible for B. pinea.

High spore counts 12 m in the canopy indicates a movement and concentration of spores or a possible source of inoculum present at these heights. High spore counts also near ground level in this study indicates conidia are being concentrated away from their initial source since a lack of pine regeneration in the understory and the natural pruning of red pine branches had eliminated most of the low sources of inoculum.

Movement of airborne conidia would be favored during periods of high relative humidity and rainfall. Lateral movement of conidia that become airborne under such conditions is known to occur for D. pini (Gibson et al., 1964). Location of site A in a topographic depression would favor air movement that would carry airborne spores into this site. Dorworth (1972) showed air drainage into a topographic depression "amplified" ascospore levels in comparison to area outside the depression. Dorworth believed drainage of air into the depression carried windborne ascospores that were released during periods of high relative humidity and rainfall. A similar movement of conidia into the depression in site A during periods of high relative humidity and rainfall could account for the amplified levels of inoculum near the ground.

Data suggests that this type of lateral dispersion of airborne conidia in Site B was decreased by environmental and site conditions in this stand. Lower spore counts were generally recorded near the gound than were recorded at the spore trap 9 m in the canopy. Location of this site on level ground and the presence of understory foliage would decrease lateral movement of conidia in this stand.

The low number of seedlings infected by exposure to weekly inoculum densities and environmental conditions could represent actual potential because Setliff and Manion (1980) also observed low numbers of infection for seedlings exposed similarly. On the other hand, low numbers of infected seedlings may be a conservative estimate of actual infection potential in the field in 1981, because early warning of seedlings in the greenhouse in January may have decreased fungal invasion of stems and needles from initial infection courts. Infection experiments have shown that trees acquire resistance to invasion by G. abietina when growth resumes in the spring (Read, 1968b; Sletten, 1971). Early growth of seedlings in the greenhouse may have limited fungal invasion into stems and have in part contributed to the low number of infections observed. This could also have contributed to the lack of typical symptom development in most seedlings. Infection of seedlings in a particular week and location should, therefore, be interpreted as indicating that conditions are highly conducive to infection.

Lack of infections in site C under the low inoculum density monitored in this stand and by the lack of infections in site A and site B during weeks of low inoculum density suggests a certain level of inoculum must be present for a successful infection to occur. Dorworth (1979) found that 2.5×10^6 conidia per seedling was required to kill 50 percent of seedlings in field trials. Inoculum levels of this magnitude were never present during these experiments. It is not possible to extrapolate these trap space counts to spores per seedling, but the inoculum density collected in this line spore trap probably never exceeded 1000 spores per week.

Most of the infection of seedlings occurred in May and June coincident with weeks of maximum spore release by the fungus and shoot elongation and development of red pine. Infections in the upper canopy in site A or site B also occurred in this period during the three weeks of greatest spore release as monitored by the monofilament spore trap. Dorworth (1971) believed this would represent the period of maximum infection by G. abietina. Although infection can occur in the early spring and fall (Skilling 1979), spring is probably the period of maximum susceptibility of red pine and the period of rapid intensification of the disease. It is difficult to determine if high inoculum density or the presence of susceptible pine tissue is responsible primarily for the high degree of infection in this period.

Moist environmental conditions are often associated with infection in the canopies of large pines (Donaubauer, 1972; Read, 1968a). Infection of seedlings in the canopy in this study, however, cannot be attributed only to moist environmental conditions. Infection of these seedlings in the canopy did not occur during periods of greatest rainfall or of longest continuous 100% relative humidity.

It may be important that infection in the canopy occurred in both sites during week 6. This was also the week the greatest number of seedlings became infected (31%). This week stands out from all the others by having lower temperatures during periods of rainfall and high relative humidity. This combination of cool temperatures and high relative humidity and rainfall was conducive for infection of a large number of seedlings in the canopy in both sites.

Infection during other weeks, of lower inoculum density, was restricted to within 3 m of the ground in both site A and site B. This is similar to situations in the Lake States and Canada where the pathogen rarely infects branches in the canopies of large pines (Skilling and Cordell 1966). Infection at these lower heights in this study could be due to more favorable environmental conditions or generally higher inoculum density monitored near the ground. Dew sensors in site A indicate similar amounts of free water were present near the ground and in the canopy. This suggests infection near the ground in these cases may be due to the higher inoculum density at this point. In contrast, site B with the highest inoculum density at 9 m resulted in the infection of only one seedling at this height. This suggests that infection is not strictly related to inoculum density.

Results presented herein show that variable interactions among host, pathogen and environment affect the inoculum potential of B. pinea at a particular time and location. Skilling (1977) suggested that a more virulent race of G. abietina was also involved in the inoculum potential of Scleroderris canker in New York. Infection in the canopy of large pines and the great inoculum potential of Scleroderris canker in New York was clearly not due to a single factor, but must be interpreted as resulting from a combination of factors which when occurring together in proper sequence were responsible for the disease flare up in the state.

LITERATURE CITED

Carvin, R.P. 1980. Factors affecting the intensification of Scleroderris canker in young red pine plantations. MS Thesis, SUNY College of Environmental Science and Forestry. Syracuse, N.Y. 43 pp.

Day, W.R. 1945. A discussion of causes of dying back of Corsican pine, with special reference to frost damage. Forestry 19:4-26.

Donaubauer, E. 1972. Environmental factors influencing outbreaks of Scleroderris lagerbergii Gr. Eur. J. For. Pathol. 2:21-25.

Dorworth, C.E. 1971. Disease of conifers incited by Scleroderris lagerbergii Gr.: A review and analysis. Can. For. Serv. Dept. of Fisheries and Forestry Pub. No. 1289.

Dorworth, C.E. 1972. Epidemiology of Scleroderris lagerbergii in Central Ontario. Can. J. Bot. 50:751-765.

Dorworth, C.E. 1973. Epiphytology of Scleroderris lagerbergii in a kettle frost pocket. Eur. J. For. Pathol. 3:232-242.

Dorworth, C.E. 1979. Influence of inoculum concentration on infection of red pine seedlings by Gremmeniella abietina. Phytopathology 69:298-300.

Dorworth, C.E. 1981. Status of pathogenic and physiologic races of Gremmeniella abietina. Plant Dis. 65:927-932.

Dorworth, C.E., J. Krywienczyk, and D.D. Skilling. 1977. New York isolates of Gremmeniella abietina (Scleroderris lagerbergii) identical in immunogenic reaction to European isolates. Plant Dis. Rep. 61:887-890.

Gibson, I.A.S., P.S. Christensen and F.N. Munga. 1964. First observations in Kenya on a foliage disease of pines, caused by Dothistroma pini Hulberry. Comm. For. Rev. 43:31-39.

Horsfall, J.G. and A.E. Diamond. A perspective on inoculum potential. J. Indian Bot. Soc. 42A:46-57.

Ohman, J.H. 1966. Scleroderris lagerbergii Gr.: The cause of dieback and mortality of red and jack pines in Upper Michigan plantations. Plant. Dis. Rep. 50:402-405.

Read, D.J. 1966. Dieback disease of pines with special reference to Corsican pine, Pinus nigra var. calibrica Schn. I. The nature of the disease symptoms and their development in relation to the crown and aspect. Forestry 39:151-161.

Read, D.J. 1968a. Some aspects of the relationship between shade and fungal pathogenicity in an epidemic disease of pines. New Phytol. 67:39-48.

Read, D.J. 1968b. Dieback disease of pines with special reference to corsican pine, Pinus nigra var calibrica Schn. III. Mycological factors. Forestry 41:72-82.

Setliff, E.C. and P.D. Manion. 1980. Forest pathology and acid rain. Chap. 6 in: Effects of acid precipitation on a forest ecosystem. New York Energy Research and Development Authority, Albany, N.Y.

Sletten, A. 1971. Infection biology and chemical control of Scleroderris lagerbergii Gremmen on Pinus sylvestris L. Nor. For. Res. Inst. Vollebeck, Norway. pp. 118-134.

Skilling, D.D. 1969. Spore dispersal by Scleroderris lagerbergii under nursery and plantation conditions. Plant Dis. Rep. 53:291-295.

Skilling, D.D. 1972. Epidemiology of Scleroderris lagerbergii Gr. Eur. J. For. Pathol. 2:16-21.

Skilling, D.D. 1977. The development of a more virulent strain of Scleroderris lagerbergii in New York State. Eur. J. For. Pathol. 7:297-302.

Skilling, D.D. 1979. Scleroderris Canker Study Progress Report No. 3. Unpublished.

Skilling, D.D. and C.E. Cordell. 1966. Scleroderris canker on national forests in Upper Michigan and Northern Wisconsin. USDA Forest Service North Central Exp. Sta. Res. Pap. NC-3. 10 pp.

Smerlis, E. 1968. Spore discharge of Scleroderris lagerbergii in Quebec. Can. Dep. For. Rural Develop. Bi-mon. Res. Note 24:10.

Van der Plank, J.E. 1975. Principles of plant infection. Academic Press, New York, New York. 216 pp.

THE INFECTION PROCESS IN PINES BY *Gremmeniella abietina*

R.F. Patton, R.N. Spear, and P.V. Blenis[1]

ABSTRACT

Gremmeniella abietina infects shoots of pines through stomata numbering about 3-20 in each of two short irregular rows on the abaxial surface of bracts that subtend short shoots. The germ tube penetrates between guard cells, sparsely colonizes bract tissue, and reaches the periderm that separates bract from stem tissue by late summer or fall. The fungus crosses this periderm only after about early February and then rapidly colonizes the short shoot and surrounding cortical tissue. This invasion produces a resinous brown necrotic area of cortical parenchyma and phloem beneath the bract as the first visible symptom of infection.

INTRODUCTION

Agreement exists among most workers on *Gremmeniella abietina* (Lager.) Morelet that infection occurs through the buds or current season's shoots. The exact infection court and the manner of entry into the stems of pine hosts, however, has not been described. This paper is a report of entry of the fungus through stomata on bracts that subtend the short shoots, which grow from the current season's long shoots, and of early stages of colonization of stem tissues of the host.

The early visible symptoms of the disease are indicators of the presence of infection in the buds or shoots. The most obvious early symptom is the orange discoloration of the needle base in spring prior to shoot elongation (Skilling et al. 1979). Read (1968) stated that resin on bud scales was the earliest symptom of the dieback disease on Corsican pine in England, caused by this fungus, although he considered the dieback to result largely from an extension of necrosis from the bud.

The most critical attention to the infection process was given in the work reported by Siepmann (1972, 1976) and Lang and Schutt (1974). Siepmann (1972) reported that dieback started from scales of the long shoot. Necrotic spots

---

1 Professor, Specialist, and former Research Assistant, respectively, Department of Plant Pathology, University of Wisconsin-Madison, WI 53706. Research supported by the College of Agricultural and Life Sciences, University of Wisconsin Madison and by Research Agreement No. 13-591 with the U.S.D.A. Forest Service, and by the Wisconsin Department of Natural Resources.

were observed in cortical tissue under the scales, occasionally as early as January and regularly by March. Penetration of fungus hyphae into the epidermis and hypodermis of long shoots of Pinus nigra was seen to have occurred by August (Siepmann, 1976), and colonization of chlorophyll- containing cells of the cortex had occurred by the end of January of the following year. In terminal buds, the fungus first colonized dead cells in blade margins of the covering scales. Lang and Schutt (1974) confirmed Siepmann's report of the presence of the fungus in the epidermis and subepidermal cells and considered that the scales of shoots must be considered sites of the infection process, in conformance with Siepmann's observations (1972) of necrosis underneath the scales. They assumed, however, that penetration did not occur until autumn when the fungus grew from the axils of scales into the shoot by penetration of the epidermis. They believed that colonization of the phloem occurred during early winter.

We directed our attention to determining the specific infection court, the mode of entry into tissue of the stem or buds of pine, and the early stages of colonization of the short shoot or tissue in its immediate vicinity.

METHODS

Emphasis was placed on microscopic examination of pine shoots bearing infections of various ages. Through dissection and serial sectioning and working back from the oldest to the youngest infections we hoped to find the site of original infection.

Specimens for examination included red pine (P. resinosa) and Scots pine (P. sylvestris). These came from seedlings about 2 or 3 years old and plantation trees from about 5-15 years old.

Seedling pines were planted in rows in beds approximately one meter wide in forest areas near Loretta, Wisconsin and Boonville, New York. In the Wisconsin plot, established May 1978, trees were exposed (mid-June to mid-September) to inoculum borne on infected twigs distributed over vinyl coated wide-mesh wire fencing which was stretched over the seedlings and supported on stakes several centimeters above their tops. Collections of 1978 shoots were made at intervals from June 1978 to April 1979. In the New York plots seedlings were inoculated in the same manner in the springs of 1979, 1980, and 1981. Collections of the last season's terminal shoots were made at approximately 2- to 3-week intervals from October through March or early April, all trees being used up in the process each year.

Segments for microscopy were taken from the upper 5 cm of the last season's shoot, and since there were no visible symptoms, either were selected at random or after dissection in a search for necrotic areas. Other segments from shoots collected in infected red pine plantations in northern Wisconsin or Upper Michigan represented samples selected by a "pull test," whereby infected needle fascicles could be detected by a gentle pull stress sufficient to detach them from the long shoot. Serial sections for light microscopy were prepared from material embedded in either paraffin or Spurr's™ resin.

Growth of the fungus on the surface of bracts was observed by fluorescence microscopy and scanning electron miscroscopy (SEM). Spore germination and surface growth of germ tubes were observed by fluorescence microscopy of spores treated with a fluorescent brightener (Setliff and Patton 1974). Greater detail of stomatal entry was obtained by standard SEM methods (Patton and Spear 1978), which were applied to bracts selected from stems of seedlings that had been sprayed or brushed with suspensions of conidia produced in culture in concentrations that ranged from 1 x 10^5 to 10^8 spores/ml, depending upon the nature of the experiment. After inoculation trees were incubated up to 3 days in a dew chamber at 20C.

RESULTS

Behavior of the fungus was the same in both red and Scots pines, and information obtained from microscopic examination has been summarized without regard to species.

Working first with cross sections or longitudinal sections examined by light microscopy, and later supplementing these observations with SEM examination, we discovered that the bract that subtends the needle base bears two short irregular rows of 3-20 stomata each in a localized region of its abaxial surface. These stomata serve as the initial infection court for the fungus (Fig. 1).

Entry of the fungus is made through a stoma on the bract (Figure 2). A germ tube from a germinated spore on the bract surface enters the stomatal antechamber, penetrates between the guard cells to the substomatal space below, and then grows sparsely through the parenchyma cells of the bract (Figure 3). This mode of infection was confirmed with both naturally and artificially inoculated material and with both light microscopy and SEM. No evidence of direct penetration of epidermal cells was ever seen and no other infection courts have been observed to date.

Confirmation of the mode of entry on material inoculated naturally under screens bearing infected twigs or under the overstory of infected pines was obtained by examination of bracts inoculated artificially with spore suspensions

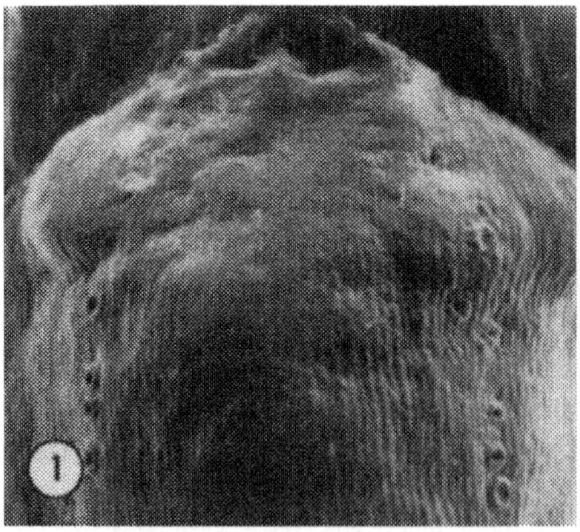

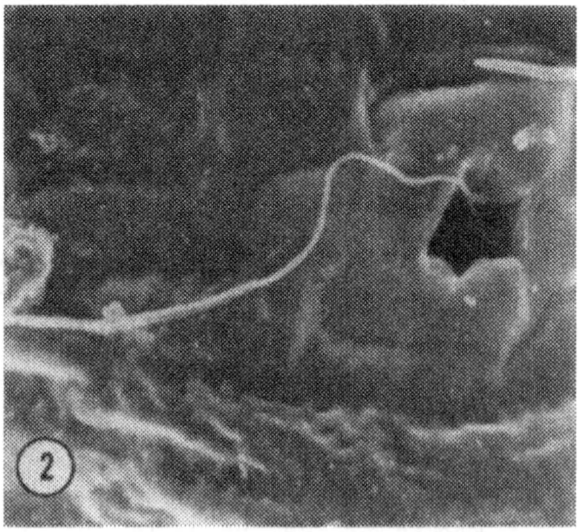

Figure 1. The two irregular rows of stomata on the abaxial face of a bract.

Figure 2. Entry of the germ tube of a conidium into the antechamber of a stoma on a bract.

in the laboratory. After observation of stomatal entries on inoculated bracts by SEM the bracts were sectioned and examined by light microscopy. Some correlation of SEM and light microscopy observations were made, although, because position of the fungus relative to the site and plane of sectioning could not be controlled, evidence had to be compiled from a series of sections, and no good photographic correlations were obtained. We were able to confirm, however, that entry, penetration between guard cells, and early stages of colonization were similar in naturally infected and artifically inoculated pine shoots.

Epifluorescence microscopy of Calcofluor-stained bracts showed that a high percentage of conidia germinated after incubations for 2 or 3 days, but not after 1 day. Hyphal overgrowth and anastomoses of hyphae prevented determination of an accurate germination percentage. A number of stomatal entries by germ tubes were seen, and when several of the bracts were subsequently observed by SEM, penetrations of the germ tubes between the guard cells were visible.

In a separate experiment on the effect of temperature on stomatal penetration, bracts on terminal shoots of 16 red pine seedlings were brushed with a suspension of 1 x 10^5 conidia/ml and the trees were incubated in groups of four each at approximately 11, 15, 20, or 22C for 3 days. Bracts were removed, stained with Calcofluor for 20 minutes, and examined by epifluorescence microscopy until 100 stomata were counted for each temperature. Germination on water-agar plates after 3 days was 50% at 12C and 90-94% at 15-22C. Most stomatal penetrations occurred on trees incubated at 15 and 20C (2.0 and 1.8% per tree, respectively) and they were less frequent at 12 and 22C (0.5 and 0.8% per tree, respectively). Although the relationship between temperature and stomatal penetration was similar to published curves relating germ tube development to temperature (Skilling 1968), standard deviations were high (ranging from 0.6 to 2.5) and variation between temperature treatments was less than variation within treatments, probably because of variability among the trees which led to variation in spore density on the host surface.

The internal tissue of a bract (which is the area colonized by the fungus after penetration through stomata) is separated from the phloem or cortical tissue of the stem by a periderm layer, at the base of the bract which the bract joins the stem, and immediately beneath the region of emergence from the stem by the short shoot (Figure 4). Colonization of the internal tissue of the bract begins immediately after stomatal penetration, although there is not intensive colonization of the bract tissue at any time. In bracts from pine shoots artificially inoculated in the laboratory spread of the fungus through the bract and down to the basal periderm layer was completed relatively soon after penetration (within about 2-4 weeks). In samples from natural infections in both the Wisconsin and New York collections the fungus reached the periderm layer at the base of the bract by October. The fungus apparently remains localized in the bract and only after about mid-January does the fungus cross the periderm to infect tissues of the main stem. After the periderm has been crossed, the phloem and short shoot immediately beneath the bract are rapidly colonized, become necrotic, and incipient symptoms become evident in late February and March.

Since the periderm that separates bract from stem tissue is a major barrier to the colonization of the stem immediately after infection, it has been difficult to determine the exact manner and site of crossing this barrier, as well as the time of this activity. Hyphal fragments have been seen interspersed among cell layers of the periderm in occasional samples from February and March collections. No massive breakthrough has ever been seen and presumably the fungus can proceed gradually by direct penetration of cells until it reaches the living cells of the cortex. Once the cortical tissue is reached the fungus rapidly invades parenchyma and phloem tissue in the immediate vicinity of the bract and favors particularly the tissues of the needle trace, which it follows in its invasion of the short shoot of the needle fascicle.

The fungus is often closely associated with the spirally-thickened tracheids which extend as a trace of primary vascular tissue into the bract. Although this tissue seemed like a good pathway or channel for the fungus to follow across the periderm, we found that formation of the periderm across the base of the bract in late June or July disrupts these tracheids so that continuity is broken, and the periderm becomes a solid wall separating proximal portions of the tracheids in the stem from the distal portions in the bract. Thus, at the junction of the basal portion of the bract with the stem the periderm is unbroken, and there is no channel into the stem cortex through which the fungus might easily grow.

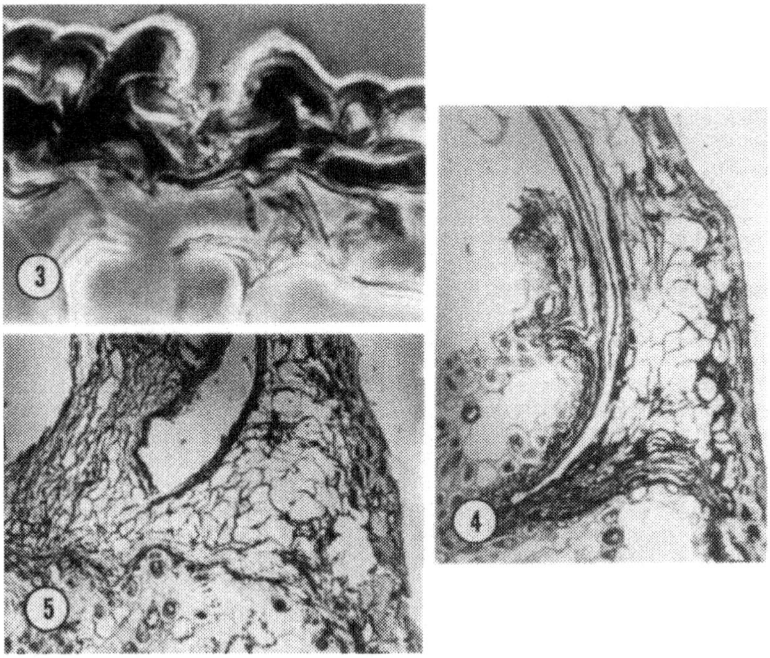

Fig. 3. **Penetration of a germ tube between** guard cells of a stoma into the substomatal chamber.

Fig. 4. Longitudinal section of bract (right) and portion of short shoot (left) showing periderm at base which separates bract from short shoot and stem cortex.

Fig. 5. Longitudinal section through bract (right) and short shoot (left). Along this plane parenchyma tissues of bract and short shoot were not separated by the periderm.

At the junction of the adaxial face of the bract with the short shoot, the periderm, which is continuous with the epidermal layers of the inner face of the bract, separates the parenchyma cells of the bract from those of the short shoot. At this junction, however, in a few longitudinal sections of some samples, the periderm was very thin and in some appeared to be lacking entirely (Figure 5). Hyphae of G. abietina were seen in this region both in short shoot tissue and parenchyma of the bract, and it appeared that the fungus could have grown from bract to short shoot without having to breach a periderm. Similarly, in some cross sections a U-shaped indentation of the periderm was seen at the outer margins of attachment of bract to stem at the base of the short shoot (Figure 6). At the sharp curve of the "U" the periderm appeared to blend into parenchyma, so that parenchymatous tissue of bract and stem were not separated. Hyphae of the fungus were traced from bract to short shoot in such a region (Figure 7).

Once the fungus has invaded the short shoot the cells are killed and the tissue becomes brown and resinous and tends to lose its integrity. At this time no discoloration is evident in the needles beyond the sheath, but beneath the sheath some discoloration may be seen and the needle fascicle and surrounding basal sheath is easily dislodged at the point of attachment of short shoot with the long shoot. This is the stage when the "pull test" is helpful in locating sites of incipient stem lesions.

DISCUSSION

Our examinations of infected shoots of red pine from both Wisconsin and New York enabled us to compare the mode of entry and subsequent colonization by the North American and European races of G. abietina. We could observe no differences in these early stages of infection between the two.

Indications from the literature are that buds and current season's shoots are the most likely sites of infection on pines. Siepmann (1976), for example, believed that hyphae of G. abietina pentrated into buds from dead cells of the

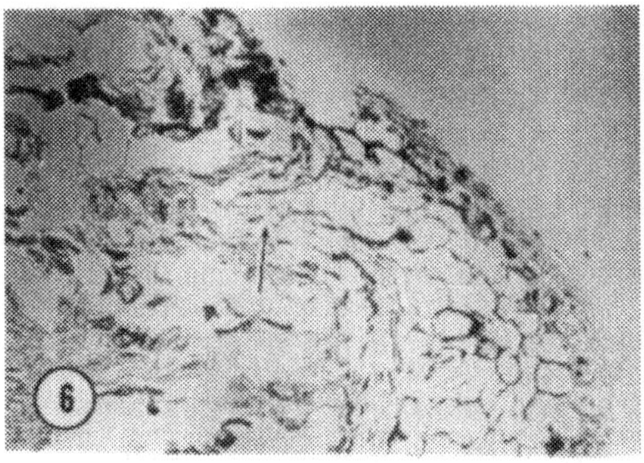

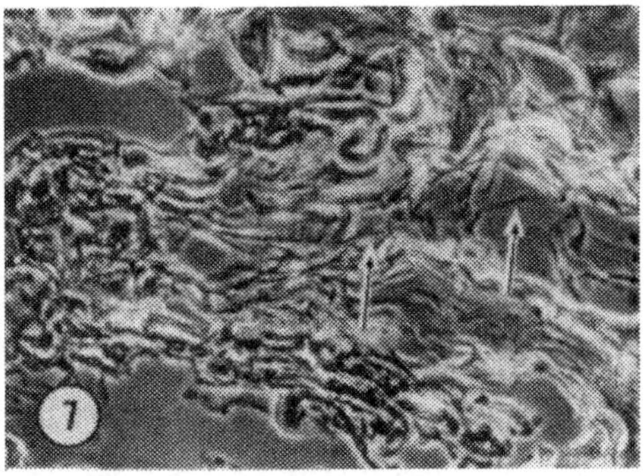

Figure 6. **Cross section** of bract and main stem near base of short shoot. In "channel" provided by indentation of periderm (arrow) hyphae were traced in several sections from bract (right) to stem cortex (left).

Figure 7. Hyphae growing through "break" in periderm at indentation seen in Figure 6.

blades of scales that covered the bud. From our observations of the infection process in red pine, however, we conclude that the buds of red pine are not often infected directly.

Examination of SEM of red pine buds from shoots collected in late summer show that overlapping of the bracts (bud scales) is such that stomata are not exposed. Also, the bracts are so tightly appressed to one another that spores probably would not infiltrate readily between bracts nor would germ tubes of germinated spores be likely to encounter stomata. Buds, of course, are readily infected by growth of the fungus from the shoot, and the common appearance of symptoms in the needles at the tip of the shoot indicates that infection often occurs through bracts near shoot tip. This seems to be the common situation in nursery stock, where the terminal shoot is usually upright and the whorl of needles surrounding the bud forms a cone or funnel which might help direct or influence the final deposition of spores on the bracts of the needles near the shoot tip. On the other hand, the "pull test" and observation of early needle discoloration on shoots of lateral branches in the lower whorls of plantation red pines about 1 to 2 meters in height indicated that infection might occur at any point along the length of the shoot and was not necessarily restricted to the tip of the shoot nor associated with the bud. Since stomata are present on the bracts that form the bud scales, it seems certain that they could serve as infection courts were germ tubes of G. abietina to reach them. Differences in the tightness or arrangement of overlapping bud scales among different species of pines might make buds of some species more suitable infection sites than others, although we have made no such comparisons.

Colonization of the bract tissue is very sparse and the fungus is difficult to locate in thin sections. Presumably the thin-walled parenchyma cells in the interior of the bract become non-functional in the summer after the disruption of the vascular connection to the shoot. Most of the cells are probably dead and may not provide enough nutrients to the fungus to support more than a minimal development of sparingly-branched hyphae in their growth down toward the periderm at the base of the bract.

The long latent period in the development of this disease poses problems in experimental work on its etiology and epidemiology. The usual delay of several months between the time when infection takes place and the appearance of obvious symptoms causes difficulty in evaluating the success of artificial inoculations and differentiating the significance of the influence of various environmental factors on disease development. The peak periods of spore dispersal are reached in May through July in North American conditions (Skilling 1972, 1981), although spores can be released any time during the year if the temperature is above freezing (Skilling 1981). Usually symptoms develop in the spring of the year following infection, but Skilling (1977) recorded the rather unusual appearance of symptoms in July 1976 after inoculations of several different conifer species in May 1976 in New York. The usual situation, however, seems to be that the fungus reaches the periderm at the base of the bract long before it crosses it to colonize the short shoot and cortical parenchyma. The reasons for the long delay are unknown, but we have conjectured that the relatively rapid advance into the living cortical tissues may be triggered by some physiological change in the host toward the end of the dormant period, which favors intensive mycelial colonization of the tissues accompanied by marked structural deterioration of the host cells resulting from action of the extracellular enzymes produced by the fungus. If such a change does occur and if it could be controlled or manipulated, then rapid progress might be made in a variety of investigations such as, for example, the use of artificial inoculations for progeny testing or in comparative studies of virulence of various isolates or races.

LITERATURE CITED

Lang, K.J. and P. Schutt. 1974. Anatomische Untersuchungen zur Infections-biologie von Scleroderris lagerbergii Gr. [Brunchorstia pinea (Karst.) von Hohn.]. Eur. J. For. Pathol. 4:166-164.

Patton, R.F. and R.N. Spear. 1978. Scanning electron microscopy of infection of Scotch pine needles by Scirrhia acicola. Phytopathology 68:1700-1704.

Read, D.J. 1968. Dieback disease of pines with special reference to Corsican pine Pinus nigra var. Calabrica Schn. III. Mycological factors. Forestry 41:72-82.

Setliff, E.C. and R.F. Patton. 1974. Germination behavior of Scirrhia acicola conidia on pine needles. Phytopathology 64:1462-1464.

Siepmann, R. 1972. Zur Fruchtkörperbilding und zum Infektionsverlauf bei Scleroderris lagerbergii-Befall an Schwarzkiefer (Pinus nigra Arnold). Forstwiss Centralbl. 91:153-160.

Siepmann, R. 1976. Ein Beitrag zur Infektionsbiologie des durch Scleroderris lagerbergii verursachten Schwarzkieferntriebsterbens. Eur. J. For. Pathol. 6:103-109.

Skilling, D.D. 1968. Biology of Scleroderris canker in the Lake States. Ph.D. thesis, Univ. of Minnesota, St. Paul, Minnesota. 90 p.

Skilling, D.D. 1972. Epidemiology of Scleroderris lagerbergii. Eur. J. For. Pathol. 2:16-21.

Skilling, D.D. 1977. The development of a more virulent strain of Scleroderris lagerbergii in New York State. Eur. J. For. Pathol. 7:297-302.

Skilling D.D. 1981. Scleroderris canker - The situation in 1980. J. For. 79:-95-97.

Skilling, D.D., J.T. O'Brien and J.A. Beil. 1979. Scleroderris canker of northern conifers. U.S.D.A. Forest Service, For. Insect and Dis. Leaflet 130. 7 p.

EFFECT OF ENVIRONMENTAL FACTORS ON THE POST-INFECTION BEHAVIOR OF *Gremmeniella abietina*

Peter V. Blenis, Robert F. Patton and Russell N. Spear[1]

ABSTRACT

Conidial production by *Gremmeniella abietina* was enhanced by growing the fungus in 8-oz. jars containing V-8 agar and a red pine twig. Survival of *G. abietina* at temperatures of 12 to 34C was closely related to average temperature and duration of exposure. The fungus survived 27 months in vivo without causing extensive necrosis of the host. Symptom development was more frequent among trees placed outdoors than among trees placed in a coldroom or greenhouse.

INTRODUCTION

Gremmeniella abietina (Lagerb.) Morelet can infect pines by growing through stomata located on the abaxial surface of bracts subtending needles. Following spring or early summer infection, the fungus survives inside these bracts until late winter or early spring when it grows into the shoot cortex.

Disease severity presumably could be determined by the ability of *G. abietina* to penetrate the host, its ability to survive during the summer or its ability to rapidly colonize host tissues. *G. abietina* pentrated red pine shoots at temperatures between 11 and 22.5C (Blenis, 1982) thus suggesting that the fungus has rather broad temperature requirements for penetration and that temperature during this stage in the disease cycle may not be critical in determining disease severity. Evidence suggests, however, that the fungus is quite sensitive to high temperatures and thus potentially limited by its ability to survive through the summer. *G. abietina* was killed by a 24-hour exposure to 40C (Skilling, 1968) and by a one-week exposure to 33C (Slettin, 1971). It reportedly has a thermal death point between 25 and 30C (Dorworth, 1972). Even if the fungus survives the summer, disease severity may depend on the effects of subsequent environmental conditions on its ability to grow from the bract to the shoot cortex or its ability to extensively colonize shoot tissue. In the Searchmount Management Unit in Ontario, for example, greater mortality of trees was found inside depressions than was found on the surrounding areas

---

[1] Research Assistant, Professor and Specialist, respectively, Department of Plant Pathology, University of Wisconsin, Madison, Wisconsin, 53706. Present address of senior author: Department of Plant Science, University of Alberta, Edmonton, Alberta T6G 2P5

The authors thank S.A. Vicen for technical assistance in producing the figures.

even through 90-100% of the trees on the latter areas were infected (Dorworth, 1971).

The primary objective of the research reported here was to determine the effect of temperature on the survival and colonization phases of G. abietina. Before experimentation could begin in earnest, however, it was necessary to develop a technique for obtaining large numbers of conidia in culture.

METHODS

Obtaining Conidia In Culture

Four culture systems were compared for their effectiveness in stimulating conidia production by G. abietina: 1) petri dishes containing 20 ml of V-8 agar (200 ml Campbell's V-8 juice, 800 ml water, 30 g Difco Bacto-agar), 2) 8-oz. jars containing 75 ml of V-8 agar, 3) 8-oz. jars containing 75 ml of V-8 agar and a red pine twig, 4) French square bottles containing an apricot juice-wheat seed medium. This medium was prepared by blending 900 ml of apricot nectar, 150 g of wheat seeds and 150 g of chopped red pine twigs in a Waring blender for 15 min. One hundred ml of the resultant slurry was then poured into each bottle. Isolates 113 and 310 from Wisconsin and isolate MN1 from Minnesota were grown under each of the four culture systems and seven replicates were made of each isolate/culture system combination. Conidial production was determined by adding 15 ml of water to each jar or dish and measuring four aliquots of the resultant suspension with a hemocytometer. Spores were harvested from the French square bottles after 52 days; conidia were harvested from the other three culture systems after 24 days. Samples of conidia produced under each of the four culture systems were assayed for germination.

Survival of G. abietina in vitro

A conidial suspension was made from an 18-day-old culture of a single-spore, Wisconsin isolate of G. abietina. Approximately 0.2-0.3 ml of this suspension was pipetted into each of 336 V-8 agar slants. The slants were then placed on a lab bench for one week to permit the spores to germinate. Next, the tubes were divided into 28 groups of 12 tubes each. Each of the 28 groups received one of 28 different temperature combinations generated by placing the cultures in unlighted incubators at either 24, 28, 32, or 34C for 9 hours and in unlighted incubators at either 12, 15.5, 20, 24, 28, 32, or 34C for 15 hours each day. Because the cultures were not moved between the incubators at the same time each day, the time at which the cultures were moved was recorded and the average temperature was determined from that data. One week after the temperature treatments had begun, the viability of four cultures from each of the 28 groups of cultures was determined by plating out two pieces of mycelium from each of the four cultures, on a V-8 agar plate. The cultures were incubated at room temperature and were examined one month later. At this time, the viability of the 28 groups of four cultures was rated as follows: 1) all four cultures dead, 2) all four cultures alive, 3) some of the four cultures alive and some of the four cultures dead. The latter category was considered to indicate the combination of temperature and duration of exposure at which cultures were beginning to be killed. Viability of the remaining cultures was determined in the same way, 2 and 4 weeks after the temperature treatment had begun.

Survival of G. abietina in vivo

Nine groups of trees were inoculated between 1977 and 1979 by inserting mycelium into stem incisions. The trees were then placed outdoors or in a greenhouse. At various times following inoculation, trees were dissected and attempts were made to reisolate the fungus.

Effect of Incubation Temperatures and Conditions on Colonization of Host Tissues by G. abietina

Three groups of trees were inoculated. Each group consisted of 45 trees that were inoculated with approximately 2 ml of a spore suspension of a Wisconsin isolate containing approximately 4×10^6 spores/ml and six trees that were inoculated with distilled water. Following inoculation, the trees were sprayed with atomized water, partially covered with two or three wet paper towels, covered with a plastic bag which was suspended either by pot stakes or by the tree itself, and placed in a controlled-temperature dew chamber for 3 days at 17-20C. The first group of trees, which was inoculated on 4 December 1980, consisted of 3- and 4-year-old trees that had been outdoors prior to inoculation. A hypodermic syringe with a needle was used to inject spores into the terminal bud. The second group of trees, which was inoculated on 8 December, consisted of 2-year-old trees that had been in a greenhouse. Spores were applied, with a hypodermic syringe to surfaces between the bud scales. The needle was covered with a plastic tip to minimize wounding. This was done by inserting the needle into the bore of a piece of polyethylene tubing with an inside diameter of 1.2 mm. The tubing was cut so that 3 mm of plastic extended beyond the end of the needle and was then sharpened, with a razor blade, to a fine point. The third group of trees, which was inoculated on 11 December, consisted of 2-year-old trees which had been in the greenhouse. A bare (untipped) syringe was used to inject spores into the terminal buds of the trees. Following inoculation, each of the three groups was divided into three subgroups which were incubated either in a greenhouse, a coldroom, or outdoors in an uncovered coldframe. The coldroom was equipped with four 200 W incadescent lights, placed 0.6 m above the trees, which provided a 12-hour photoperiod. Although temperature was not recorded, previous experience had indicated that the temperature normally fluctuated between 1 and 9C in time with the lights. The greenhouse was equipped with fluorescent lights which provided 14 hours of light per day. Attempts to isolate the fungus from these trees were made between 13 January 1981 and 6 June 1981. The earlier isolations were attempted at times prior to the appearance of external symptoms. Later, isolations were made from those trees showing typical syptoms of G. abietina infection as well as from some that apeared to be suffering from root disease and/or water stress.

RESULTS

Obtaining Conidia in Culture

More conidia were produced by the V-8 agar/pine twig culture system than under any of the other systems (Figure 1). This difference was found to be significant at P =.05. Spores produced in the French square bottles had a germination frequency of 67%; the frequency of germination of conidia produced under the other three systems ranged from 88-100% (average 94%).

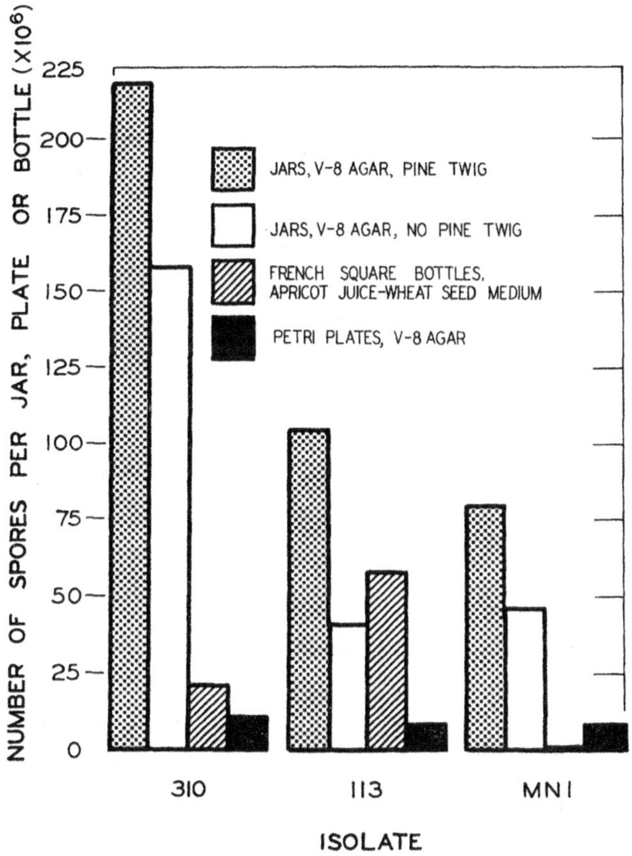

Figure 1. Comparison of four culture systems for promoting sporulation of G. abietina.

Survival of G. abietina in vitro

Survival of cultures at average temperatures between 30 and 34C was determined by the duration of exposure to these temperatures (Figure 2). Survival of the cultures was more closely related to average daily temperature than to extreme daily temperature. For example, all four cultures were killed by a 4-week exposure to a constant temperature of 32C. In contrast, all cultures maintained at a lower average temperature but a higher extreme temperature (low = 24C, high = 34C, average = 27.7C) were alive after 4 weeks.

Survival of G. abietina in vivo

Typical symptoms of G. abietina infection failed to appear on any of the trees that had been inoculated by placing mycelium into stem incisions. Necrosis rarely extended more than a few mm from the wound. The number of asymptomatic trees from which G. abietina was recovered at various times after inoculation was as follows: after 6-9 months, 17 trees; 9-12 months, four trees;

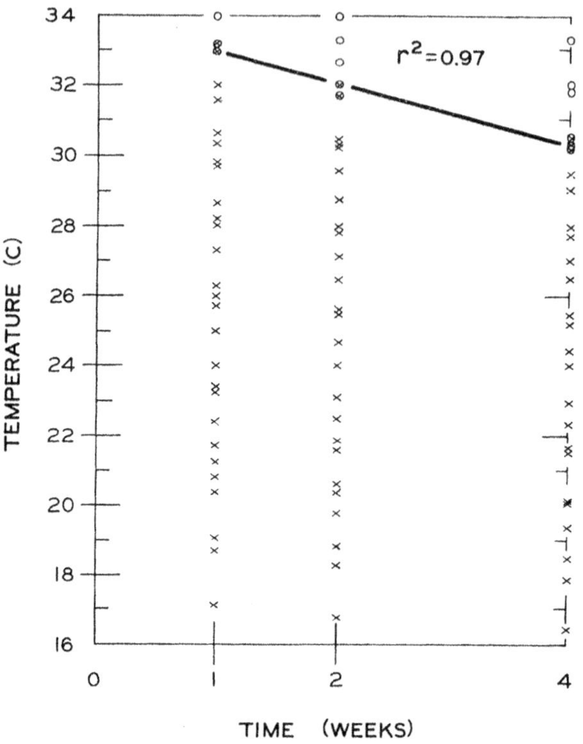

Figure 2. Viability of four cultures placed at each of 24 temperatures for 1, 2, or 4 weeks. X- all four culture viable; O-none of the four cultures viable; X - one, two, or three of the cultures viable.

trees; 13 months, three trees; 23 months, three trees; and 27 months, one tree.

Effect of Incubation Temperature and Conditions on Colonization of Host Tissues by G. abietina

The pattern of disease development among trees inoculated with a tipped needle was different from that observed among trees inoculated with a bare needle. The general pattern of disease development in trees inoculated with a bare needle and incubated in the coldframe was as follows. In mid- and late-January there were not external indications of G. abietina infection. Necrotic areas were observed on the needle primordia of terminal buds, however, when the bud scales were removed. Typically, these necrotic areas were superficial and encompassed fewer than ten pine needle primordia. The lesions were vertically elongate, presumably as a result of the vertical orientation of the syringe needle that was used to wound the buds. On 28 February, G. abietina was recovered from two trees. The terminal buds of both trees were loosely attached, but otherwise the trees appeared healthy. The fungus was reisolated from necrotic tissue which extended less than 4 mm downward from the base of the buds. On 30 April, three G. abietina pycnidia were found beneath a bract near the terminal bud of one tree On 4 May, the fungus was reisolated from 6 trees, all of which had necrotic stem tissue extending 1-2 cm down from the bud. Two of the trees bore needles with the red basal discoloration that is typical of *G. abietina* infection. Pycnidia were present on one of these trees. Needles near the buds on the other four trees were pale green, grey, or yellowish at the base.

The results of the tipped-needle inoculation differed from those of the bare-needle inoculation in that from the former group of trees the fungus was not recovered before April. *G. abietina* was reisolated from one of those trees in April and from four more in May. There was a higher frequency of disease development among trees incubated in the coldframe than among trees in the greenhouse or coldroom (Table 1).

Table 1. Number of trees showing symptoms of *G. abietina* infection, and from which the fungus was reisolated, after incubation in different locations.

	Incubation Location		
Date of Inoculation[a]	Coldframe	Coldroom	Greenhouse
4 December	7	0	0
8 December	4	1	0
11 December	2	1	0

[a] Trees inoculated on 4 and 11 December were inoculated with an untipped syringe, whereas those inoculated on 8 December were inoculated with a plastic-tipped syringe. Trees inoculated on 4 December had been in the coldrame prior to inoculation, wherease those inoculated on 8 and 11 December had been in the greenhouse.

DISCUSSION

G. abietina was able to survive daily exposure to 34C for 4 weeks as long as there were interceding temperatures such that the average temperature was below 30C. There is no guarantee, however, that the fungus would be as resistant to high temperatures inside the bract as it was inside the culture tubes. First, there was greater fungal biomass in the tubes than would be present inside a red pine bract during the summer. Second, under field conditions, direct insolation could raise bract temperatures higher than 34C. Third, the rapid changes in temperature that would likely occur as bracts are alternately shaded and reexposed to sunlight could be more damaging to the fungus than a relatively constant high temperature. Fourth, the physiological status of both the fungus and the substrate are different under field and laboratory conditions. On the other hand, the persistence of *G. abietina* inside several trees for over a year indicates that the fungus is able to survive inside the host for a prolonged period of time.

Frequency of symptom development was greater among trees overwintered in the coldframe than among those placed in a coldroom or greenhouse. Temperature was not recorded in the coldroom, so there was no assurance that the temperature inside this facility remained above freezing. The experiment was repeated the following year. Trees that had been inoculated with a bare-needle were placed in the coldframe, placed in the coldroom or regularly moved from one location to the other according to three different schedules. An average infection rate of 79% was attained, with no significant differences being found between treatments. However, a chronic malfunction in the coldroom caused the temperature to drop below freezing on a number of occasions, thus preventing a comparison between chilling and freezing temperatures.

CONCLUSIONS

Large amounts of conidia can be obtained by culturing G. abietina in jars containing V-8 agar and a red pine twig. Circumstantial evidence suggests that exposure of infected trees to conditions of freezing, or freezing and thawing, favors colonization of the host by the fungus. However, since such conditions will be present during the winter in the locations with which we are concerned, it is unlikely that such conditions are closely related to disease severity within the geographical range of this fungus. In order to determine the precise role of environmental conditions on disease severity, it will be necessary to develop a technique for inoculating trees under controlled conditions and to develop a protocol for maintaining trees following inoculation, such that symptoms will appear. The results presented in this paper represent some progress towards that ultimate goal.

LITERATURE CITED

Blenis, P.V. 1982. Influence of environmental factors on the disease cycle of Scleroderris canker. Ph.D. thesis. University of Wisconsin, Madison, Wisconsin. 102 pp.

Dorworth, C.E. 1971. Diseases of conifers incited by Scleroderris lagerbergii Gremmen: a review and analysis. Can. Dep. Fish. For. Can. For. Serv. Publ. 1289. 42 pp.

Dorworth, C.E. 1972. Epidemiology of Scleroderris lagerbergii in Central Ontario. Can. J. Bot. 50:751-765.

Skilling, D.D. 1968. Biology of Scleroderris canker in the Lake States. Ph.D. thesis. University of Minnesota, St. Paul, Minnesota. 90 pp.

Slettin, A. 1971. Infection biology and chemical control of Scleroderris lagerbergii Gremmen on Pinus sylvestris L. Medd. Nor. Skogforsoksves. 29:117-134.

Gremmeniella abietina IN NORWAY SPRUCE, LATENT INFECTION, SUDDEN OUTBREAKS, ACID RAIN, PREDISPOSITION

Pia Barklund, Gunnar Axelsson and Torgny Unestam[1]

ABSTRACT

The pathogen *Gremmeniella abietina* often grows endophytically in twigs without causing symptoms. Stressing conditions may predispose the tree to infection and/or break the latent infection. Inoculation of spruce plants, kept in a climate chamber, resulted in typical disease symptoms. Experimental treatment with acid rain increased leakage of ions, enhanced the germination of conidia and changed the phylloplane flora. Detergents can be used to simulate the effects of acid rain treatment.

INTRODUCTION

A sudden and widespread die-back of young pure stands of Norway spruce (*Picea abies*) occurred in southern Sweden and Norway in the spring 1977. The cause of the disease was suggested to be a combination of climatic stress on the trees and fungal infection by *Gremmeniella abietina* (Barklund & Rowe 1981).

The geographic distribution of the disease was not entirely related to that of the severe drought occurring in southern Sweden 1975-76 but rather to the distribution (both in Sweden and Norway) of acid precipitation (Oden 1976), which could be an additional factor facilitating fungal infection (Shriner 1978).

Gremmeniella on Norway Spruce

In Sweden *Gremmeniella* was initially considered a threat only to Scots pine (*Pinus sylvestris*). The damage to spruce was minor and occurred usually in mixed stands, where the spruce was an understory to pine.

In West Germany attacks of *Gremmeniella* on Norway spruce were reported in 1978 (Hartmann) and in 1983 (Schutt personal communication). Previously the fungus was only rarely found on Norway spruce in central Europe.

The relations between *Gremmeniella* and Norway spruce seem to be altered or more unstable than before. The possibility of new fungal strains being the reason has not been investigated.

---

1 Research leader, Research assistant and Professor, The Swedish University of Agricultural Sciences, Department of Forest Mycology and Pathology, Box 7026, S-750 07 Uppsala, Sweden.

SYMPTOMS ON NORWAY SPRUCE

It was impossible to establish whether all spruce, with die-back in 1977, were diseased by Gremmeniella, but the more serious the disease was the more prevalent was the infection of the fungus. In those cases where the trees recovered fast, the role of the fungus was minor.

Symptoms of the natural infection were seen before the spring flush. Internal necrosis was found at the base of the annual shoot and the last season's needles became red beginning at the necrotic tissues. Such shoots were often killed and, where they survived, the flushing was sometimes postponed one year or the new shoots were stunted.

From branches with dead apical shoots latent shoots developed the following year. Such latent shoots had a high degree of infection of Gremmeniella, but showed no symptoms. A latent infection like this may make the trees more vulnerable to environmental stress (Schoeneweiss 1981).

Infection trial

Three year old plants with fully elongated current shoots were inoculated with Gremmeniella. A suspension of conidia was sprayed over the whole plant, and the plant was kept in autumn conditions (+10C and light 8 hours/day). After 7 months a widespread latent infection was found, most prevalent in the base of the annual shoot. Necrotic symptoms occurred in a few plants in the same place. This is in coincidence with natural infection.

ACID RAIN TREATMENT

Direct effect on the fungus

Conidia of Gremmeniella germinated in vitro over a wide range of pH, 3-7, and in a medium containing artificial acid rain, germination was best at pH 3.5. The conidia also survived acid treatment down to pH 2.8 for at least 24 hours without loosing germination capacity. This may give Gremmeniella an advantage in the competition with more sensitive epiphytic organisms.

Effect on the surface microflora

Three year old plants of Norway spruce were grown under controlled conditions in a climate chamber. Once a day the plants were sprayed for a few seconds with artificial acid rain of pH 2.8 or 3.5. After one month we made a study of the epiphytic flora on the annual shoots. At pH 3.5 bacteria were totally absent, yeast fungi declined and the filamentous fungi were virtually eliminated. Yeast fungi are known to be able to play a role in controlling pathogens on cereals (Fokkema 1981). In nature a situation with reduced epiphytic flora may occur periodically and will, at least temporarily increase the risques for fungal attack.

Predisposition of the shoot tissue

Treatment with artificial acid rain in climate chamber caused leakage of ions and other cell contents of annual shoots. Leaching of nutrients caused by acid rain has been suggested to change the leaf surface as a habitat for microorganisms (Lepp and Fairfax 1976).

In one experiment acid rain of pH 3.3 was sprayed once a day, with an additional weekly dose of pH 2.8 in order to imitate a common situation in many forest regions. After only three weeks leakage of ions were doubled compared with untreated control. This indicates that the cells of the shoot surface were damaged.

We could cause the same type of leakage even easier by treating the shoots with detergents or organic acids. The leachate was used as a substrate for Gremmeniella conidia. A very dilute leachate stimulated spore germination.

Different clones of Norway spruce showed varying sensitivity to acid rain treatment. Inoculation of plants from different clones treated with detergent resulted in more severe symptoms in the most readily leached clones.

CONCLUSION

We have some evidence that Gremmeniella is favored by acid rain but still more experiments are needed for the confirmation of the role of the acid rain in the infection biology of Gremmeniella and other pathogens in the forest tree canopy.

As acid precipitation solely or together with natural stress factors may affect the interaction between the host and the parasite at different stages of the pathogenesis, it is necessary in this kind of disease to investigate each step under controlled conditions.

LITERATURE CITED

Barklund, P. and J. Rowe. 1981. Gremmeniella abietina (Scleroderris lagerbergii), a primary parasite in a Norway spruce dieback. Eur. J. For. Pathol. 11:97-108.

Fokkema, N.J. 1981. Fungal leaf saprophytes, beneficial or detrimental? In "Microbial Ecology of the Phylloplane". (J.P. Blakeman ed). pp. 433-454. Academic Press, London.

Hartmann, G. 1978. An epidemic attack of Scleroderris lagerbergii Gremmen in Norway spruce in Northern Germany. Proc. Second Session Working Canker Dis. (Scleroderris), Subject Group S. 2.06.02. Division II, I.U.F.R.O., Munich, Aug. 1978.

Lepp, N.W. and J.A.W. Fairfax. 1976. The role of acid rain as a regulator of foliar nutrient uptake and loss. In "Microbiology of Aerial Plant Surfaces". (C.H. Dickinson and T.F. Preece, eds). pp 107-118. Academic Press, London.

Oden, S. 1976. The acidity problem - an outline of concepts. Water, Air, and Soil Pollut. 6:136-166.

Schoeneweiss, D.F. 1981. The Role of Environmental Stress in Diseases of Woody Plants. Plant Dis. 65:308-314.

Shriner, D.S. 1978. Effects of simulated acidic rain on host parasite interactions in plant diseases. Phytopathology 68:213-218.

THE ROLE OF INSECTS IN THE DYNAMICS OF SCLERODERRIS CANKER IN NEW YORK

G.N. Lanier, L.P. Abrahamson and H. Schoeneck[1]

ABSTRACT

The twig-infesting scolytid beetles Pityophthorous pulicarius and P. puberulus were exonerated from the role of suspected vectors of Scleroderris canker but a significantly reduced rate of infections in red pines treated with the systemic insecticide carbofuran (Furadan[R]) indicated that insects may nonetheless exacerbate the disease outbreak by providing infection courts for windborn and rain-splashed conidia.

INTRODUCTION

Explosive increases of the European strain of Scleroderris canker in New York between 1974 and 1977, despite the virtual absence of wind born ascospores (Skilling 1977), suggested that insects might be vectors of asexual propagules (conidia) of the causal fungus, Gremmeniella abietina (Lagerb.). Twig beetles of the genus Pityophthorous were especially suspect because they were generally found breeding in twigs killed by this fungus. Teneral adults emerging from Scleroderris-infected twigs in the Great Lakes Region were contaminated with spores of G. abietina (North American strain) (Smerlis and Finnegan, 1981) and were able to apparently infect wounded healthy saplings to which they were exposed in the laboratory (Frederick et al. 1976).

We report an investigation of the role that twig beetles and other insects play in the disease dynamics of the European strain of Scleroderris canker in New York. Our specific objectives were: 1, identify the insects associated with Scleroderris canker-infected pines; 2, study the life histories of putative vectors; 3, determine whether putative vectors have specialized fungal-transport organs (mycangia); 4, demonstrate the ability (or lack of ability) of putative vectors to infect healthy twigs; 5, determine whether reducing insect feeding on trees can effect rates of Scleroderris infection.

---

[1] Professor, Sr. Research Associate and Graduate Research Assistant, State University of New York, College of Environmental Science and Forestry, Syracuse, NY 13210.

Research was supported by Cooperative Aid Agreement 13-563, US Forest Service, North Central Experiment Station.

METHODS

Basic Approaches and Localities

The above objectives were approached in the following five ways: 1, examining the insect fauna associated with diseased and healthy red and Scots pine; 2, studying the biology of some of these insect species; 3, discerning spore loads on insects by microscopic observation and culturing techniques; 4, caging contaminated insects on healthy material; and 5, comparing infection in healthy trees treated, or not treated, with the systemic insecticide, carbofuran, (Furadan[R], FMC Co.).

Field studies were concentrated in two New York locations: pine plantations owned by Oneida County, near Boonville, and several state owned roadside plantations near Saranac Lake. Experiments were also conducted in the laboratory at Syracuse. All pine material transported to Syracuse was in sealed polyethylene bags which, along with pine material, were later incinerated. The laboratory was quarantined and windows were sealed with duct tape. Insect specimens were preserved in alcohol or incinerated.

Survey of Insect Fauna in Infected Pine Plantations

During the late summer of 1977 and beginning the spring of 1978, we made extensive observations on, and collections of, insects making wounds, visiting cankers, or breeding in diseased twigs and dying trees. Those species that appeared to be incidental and those that were definitely associated only with the lower boles of trees or dead and dying material were not further considered. In addition to collecting twig samples and sweeping foliage with an insect net, we sampled insects associated with pines by placing sticky bands around branches and hanging sticky hardware-cloth panels in tree crowns.

Life History Studies

Life histories of *Pityophthorous* species were studied by making periodic observations of uninfested diseased twigs and of infested material. Emergence of overwintering adults from the forest floor was monitored by establishing 1.6m X 1.6m X 0.5m tents of black polythelene sheeting with a mason jar at the apex. Emerging beetles, attracted by light passing through the jar, bored into pine twigs placed therein.

Isolation of Microbes (G. abietina) From Beetles

All beetles used for attempted isolations of *G. abietina* were excised from galleries in the field or caught as they emerged from branch tips or from overwintering sites on the forest floor.

Adults were processed through a biological fractional sterilization method (Abrahamson 1967) prior to isolation studies. The adult beetles were held for 12 hours on wet sterile filter paper in a sterile petri dish then were transferred to a sterile dry filter paper and dish for 12 hours. This alternation was repeated three times so that most loosely adhering spores of microbes would germinate under the moist conditions, then perish in the dry environment; microbes more intimately associated with the beetles (e.g. in mycangia) would survive. The beetles were then anesthetized with ethyl acetate, asceptically divided into head, prothorox, meso-metathorax and abdomen sections. Each insect part was individually fragmented and the total of each part introduced into V-8 juice agar medium in a petri dish. Individual live, sequentially sterilized beetles were also placed on sterile V-8 juice agar medium in a petri dish.

Effects of Insecticide Prophylaxis on Insect Feeding

Two previously thinned stands of juvenile red pines 10-30 cm in diameter were used for insecticide prophylaxis tests. At Mud Shanty Corners and on Moose River, near Boonville, we selected and numbered 20 and 40 trees, respectively. At each site, half of the trees were randomly designated for treatment; the remaining trees served as experimental controls. Both sites were at the margin of the Scleroderris epidemic when the experiment was initiated.

On April 15, 1978, granular carbofuran (10% AI) was applied at the rate of 90 gm/cm diameter in a shallow trench cut at the drip line of each treated tree. Treatment was repeated on April 24, 1979.

The activity of the carbofuran as a prophylactic against insect feeding was monitored in twigs clipped from the top, middle, and lower thirds of the crown by confining 2 or 4 Ips pini (Scolytidae) in petri dishes with two twigs from one of the three crown levels. The trees monitored in this manner were subsets of the treatment and check groups, except for the July 31, 1977, bioassay which tested all trees. At each sampling period all discolored branch tips on the plot trees were clipped and taken to the laboratory for diagnosis.

RESULTS

Survey of Insect Fauna in Infected Pine Plantations

Many insect species were observed or collected in pine plantations infested with G. abietina. Those that breed in cankered limbs and/or cause wounds in needles and twigs of healthy hosts are listed in Table 1.

During our initial inspection of the outbreak of Scleroderris canker in the Boonville area in July 1977, we noted that adults of the pine spittle bug, Aphrophora parallela, were so abundant that their liquid excretion fell as fine rain. We found a twig beetle, Pityophthorous pulicarius colonizing diseased twigs wherever we made inspections. We also occasionally encountered another twig beetle, P. puberlus, and larvae of Magdalis weevils in dead twigs. In addition to these insects, at Saranac Lake we found a few galls on red pine limbs caused by the spindle gall weevil, Podapion gallicola. The northern pine weevil, Pissodes approximatus and the engraver beetles, Ips pini and Orthotomicus caelatus, were common in the boles and larger limbs of dying trees, but we never found these species in diseased branch tips.

Initially we concentrated on the Pityophthorous species because they were the most common insects associated with cankered limbs and teneral adults emerging from these limbs had been shown to bear G. abietina contamination (Frederick et al. 1976; Smerlis and Finnegan 1981).

We had previously observed that small (0.3 - 0.4 mm) punctures were common on cankered and uncankered limbs of red pine. Beneath the aperature, the puncture expanded into an urn-shaped cavity in the inner bark. We recognized these punctures as similar to feeding cavities made by Pissodes approximatus, al-

Table 1. Insects associated with Scleroderris canker in New York: A list of potential vectors or infection facilitators.

Order and Family	Species	Remarks
Homoptera		
Cercopidae	Aphrophora parallela (Pine spittle bug)	Very abundant at Furman Road in 1977. Necrotic spots caused by stylets may be infection courts. Spittle of immatures may protect fungal spores.
Coleoptera		
Scolytidae	Pityophthorous pulicarius (Pine tip beetle)	Common in cankered twigs. Has been mentioned as a possible vector.
	Pityophthorous puberlus (a twig beetle)	Smaller than P. pulicarius and less common.
	Ips pini (Pine engraver beetle)	Ips and Orthotomicus were common in boles of dying trees and sometimes in cankered limbs but are unlikely vectors.
	Orthotomicus caelatus (an engraver beetle)	
Curculionidae	Pissodes approximatus (Northern pine weevil)	Found breeding in cankered limbs as well as boles of dying trees. Adults feed on healthy limbs.
	Podapion gallicola (Spindle gall weevil)	Not common in study plots but reported to be causing extensive branch mortality in Lewis County.
	Magdalis perforatus	Breeds in infected limbs and twigs. Adults feed by making punctures in pine shoots.
Cerambycidae	Monochamus scutellatus (White spotted sawyer)	Breeds in large limbs and boles of dying trees. Adults make ragged wounds when they feed on needles & bark of twigs. Injuries were too infrequently found to be a major factor in disease spread.

though some of the holes seemed more likely to have been made by a smaller weevil, probably Magdalis perforatus and perhaps Podapion gallicola. Extensive searching, foliage beating and sticky band trapping from red pine plantations during 1979 and 1980 at Boonville resulted in the collection of a few Pissodes approximatus and Magdalis perforatus and several additional small unidentified (and probably incidental) weevils. We found one P. approximatus specimen in the act of feeding on the tip of a red pine branch about 2 m above the ground level. However, most of the injuries that we saw on the samples of our plot trees conform more closely to injuries of Magdalis perforatus, as described in Martin (1962). We did not find evidence of feeding at needle fascicles that would be attributed to Magdalis austera or M. hisoides, as described by Plumb (1950).

Investigations of the Life Histories of Pityophthorous Species

Pityophthorous pulicarius and P. puberlus attacked only injured or dying twigs; even under forced conditions these beetles would not attempt to bore into healthy twigs and buds. Both species of Pityophthorous abandoned infested twigs in the fall to overwinter in the duff. These adults emerged from May 22 to June 26 at Boonville in 1978. Two days after we collected the first beetles emerging from traps on the soil we found beetles established in twigs that had been infected by G. abietina during the previous year.

Attacks continued throughout the summer. Apparently these were made by a combination of adults emerging from the duff through June and second attacks by the parent adults. In mid-July, new adults began to initiate their own attacks which continued until the onset of freezing temperatures in the fall. Shortly after the first cold snap, parent and brood adults emerged from the twigs, leaving any remaining immature stages to perish during the winter. The Pityophthorous we studied never inflicted wounds in healthy tissue.

Pityophthorous publicarius Confined On Pine Seedlings and Cultured For G. abietina

Two red pine seedlings about .5m tall on which 50 P. pulicarius were placed did not develop any evidence of the disease. Beetles traversed branches and needles of the trees but no feeding injuries nor attempts to bore into the branch tips were observed.

Although Smerlis and Finnegan (1981) cultured the American strain of G. abietina from Pityophthorous species of O. caelatus that had emerged from infected branches, we observed no G. abietina growing in V-8 juice agar plates (Abramhamson 1967) that were seeded with parts of 8 surface sterilized P. pulicarius. Our results indicated that P. pulicarius does not carry fungal spores in a mycangium-like structure. However, beetles emerging from fungus-infected twigs must frequently be surface-contaminated. Nevertheless, contamination alone does not establish the beetle as a vector of the fungus.

Prophylactic Treatment, Feeding Punctures and Scleroderris Canker

If insects are important vectors of Scleroderris canker, or if they indirectly facilitate infection by providing entrance courts, rates of disease should be reduced by treating trees with insecticides. Bioassays for prophylaxis conducted by confining pine engraver beetles. Ips pini, with cut twigs indicated that carbofuran-treated trees became potently insecticidal. Although our first tests (05/27/78) revealed no effect on feeding deterrency or beetle mortality, on 08/31/78 the effectiveness of the treatment was obvious. Treated trees retained feeding deterence and insecticidal activity through the last bioassay, 06/11/80 (Table 2).

Table 2. Bioassays of feeding and survival by Ips pini on carbofuran-treated and untreated red pines, Boonville, NY.

Samples	Trees	Twigs[a]	Beetles per sample	Feeding injuries	Feeding[b] scores/ beetle	Survival rate (48 h)
First carbofuran treatment 15 April, 1979						
May 27, 1978						
Treated	9	54	2	84	0.777	0.834
Check	9	54	2	106	0.983	0.955
August 31, 1978						
Treated	8	48	2	49	0.510*	0.522*
Check	8	48	2	83	0.865*	0.916
Ocotber 12, 1978						
Treated	10	60	2	11	0.092*	0.708
Check	10	60	2	31	0.258*	0.791
April 24, 1979						
Treated	5	10	2	1	0.053*	0.369*
Check	4	8	2	11	0.687*	0.875*
Second carbofuran treatment 15 April, 1979						
May 18, 1979						
Treated	10	60	4	43	0.180*	0.200*
Check	10	60	4	122	0.510*	0.646*
July 31, 1979						
Treated	26	152	2	33	0.107*	0.254*
Check	28	164	2	140	0.407*	0.514*

	Trees	Twigs	Beetles per sample	Beetle feeding (injuries/beetle) Biting	Beetle feeding (injuries/beetle) Boring	Survival rate (48 h)
October 12, 1979						
Treated	10	60	2	20(0.16)	0(0.00)*	0.70
Check	10	60	2	16(0.13)	13(0.11)*	0.80
June 11, 1980						
Treated	27	54	2	22(0.20)	1(0.01)*	0.54*
Check	30	60	2	8(0.06)	21(0.17)*	0.87*

* $p < .05$, Chi-square analysis

[a] Except 4/24/79 two twigs taken from top, mid and lower crown 1/3 for a total of 6 sample trees.

[b] Bioassays 5/27/78 and 8/31/78 recorded number of beetles feeding. Later bioassays recorded feeding scores. Two or more beetles may feed at the same site; thus beetles feeding $\geq$ feeding scores.

In parallel with our laboratory feeding assays, we observed that the number of weevil feeding punctures in the randomly sampled twigs of carbofuran-treated trees was about 50% of the number of samples from untreated trees (Table 3). We also noted that feeding in treated tree branch tips was less contagious than in untreated trees; rather than occurring in clusters as on untreated trees, almost all feeding punctures on treated trees occurred singly. These results, with the laboratory bioassays of feeding by Ips pini (Table 2), indicate that insect wounds should be fewer and less extensive throughout the treated trees, compared to those untreated trees.

On three occasions in 1979, and once in 1980, all of the plot trees were carefully inspected and all flagging branch tips were removed by a pole pruner. Clipped tips were examined in Syracuse for Scleroderris canker symptoms and/or G. abietina conidia.

At each observation the aggregate number of new cankers was less on the treated than on the untreated trees (Table 4). This is strong circumstantial evidence that insects are a factor in epidemiology of Scleroderris canker.

The contcomitant reduction in insect feeding and depression in Scleroderris canker incidence could reflect either a vector or facultative relationship between the insects and the fungus. After the onset of this project, spore traps demonstrated the virtual omnipresence of conidia of G. abietina in the general area of heavily infected stands (Luley 1981). A vector simply is not necessary to account for the observed spread of the disease and no such role has been demonstrated.

Two types of parallelism between insect population and Scleroderris were noted in this study. The first was an outbreak of the pine spittle bug which apparently was coincidental with the Scleroderris build-up in the mid 1970's. Initially we considered spittle bugs to be of little potential importance in the disease because they have no particular association with cankers. The evidence that insect feeding wounds and new infections move together raises the possibility that multiple spittle bug feeding punctures intensified or facilitated the Scleroderris outbreak, at least in Scots pine. The second type of association was an increase in populations of twig and bark beetles and weevils resulting from the availability of canker-weakened host material. Potentially most important of these are Pissodes and Magdalis weevils which, as emergent adults, make urn-shaped feeding punctures in the bark of healthy twigs.

Propagules of fungal plant pathogens have a tenuous existence. To successfully infect a new host the spores must be released when temperature and humidity promote germination. Once germinated, the vulnerable hyphae must quickly find a means of penetrating the protective layers of the plant. Fortunately, the vast majority of the new fungi perish without entering a host. Those that chance to germinate on or near a breech in the plant surface probably enjoy a monumental increase in survival rate.

We conclude that insects are probably important in exacerbating Scleroderris epidemics by providing entrance courts for wind-blown and rain-splashed spores. When trees are weakened by cankers, they are invaded by insects that breed in the inner bark. Increase in abundance of certain of these species (e.g. Pissodes approximatus) due to the expansion of breeding space, in turn, leads to greater wounding on healthy trees and probably to enhancement of opportunities for infection by G. abietina.

Table 3. Feeding punctures in carbofuran-treated and untreated red pine trees, Boonville, NY.

	No. of Tips Sampled	Injured	Punctures/ Injured Tip	No. of Tips Sampled	Injured	Punctures/ Injured Tip	2-yr. Total Punctures Tip
Treated	73	8*	1.00	54	14*	1.15	0.17
Check	78	27*	2.74	60	26*	2.13	0.38

Table 4. New Scleroderris cankers in carbofuran-treated and untreated red pine trees.

	1979			1980	
	June 6	June 29	July 27	July 8	Total
Treated	16	19	5*	13*	53*
Untreated	23	29	21*	26*	99*

* Data for treated and untreated trees are significantly different ($P < .05$; chi-square test).

LITERATURE CITED

Abrahamson, L.P. 1967. Mycangia and symbiotic microbes of *Xyloterinus politus* (Say) and *Trypodendron betulae* Swaine. M.S. Thesis, University of Wisconsin.

Frederick, D.J., N.F. Sloan, and W.S. Skowron, Jr. 1976. Potential Insect Transmission of *Scleroderris lagerbergii* Scolytid beetles, Plant Dis. Rep. 60:411-413.

Luley, C.J. 1982. Inoculum potential of Scleroderris canker in New York. M.S. Thesis, SUNY College of Environmental Science and Forestry, Syracuse, N.Y.

Martin, J.L. 1962. *Magdais perforatus* Horn on *Pinus resinosa* Ait. in Ontario. Can. Dept. For. Bi-Month. Prog. Rep. 18: No. 4.

Plumb, G.H. 1950. The adult feeding habit of some conifer-infesting weevils. Canad. Entomo. 82:53-57.

Skilling, D.D. 1977. The development of a more virulent strain of *Scleroderris lagerbergii* in New York State. Eur. J. For. Pathol. 7:297-302.

Smerlis, E. and R.J. Finnegan. 1981. Bark beetle carriers of *Gremmeniella abietina* and other pathogenic microfungi. Can. For. Serv. Research Notes 1:2-4.

EFFECTS OF SULFUR DIOXIDE ON INFECTION OF RED PINE BY *Gremmeniella abietina*

J.A. Laurence, K.L. Reynolds, D.C. MacLean, Jr., G.W. Hudler, and L.S. Dochinger[1]

ABSTRACT

Relatively low doses of SO_2 were found to significantly reduce the growth of red pine and inhibit infection of trees by *G. abietina* in field plots that received low concentrations of inoculum. Under conditions of high inoculum density, SO_2 did not affect infection or disease development.

INTRODUCTION

Scleroderris canker (caused by *Gremmeniella abietina*) is an important disease of red pine in the northeast. This species of pine along with others which are susceptible to the pathogen, have been used extensively for reforestation and are important commercial species.

The areas where Scleroderris canker is currently a probelm are mostly rural forested regions. These regions are not isolated, however, from the effects of air pollutants which are transported over long distances or from point sources of SO_2 . In addition, significant concentrations of ozone can be present in these areas.

The objectives of this study were to determine the effects of SO_2 on the growth of red pine and on infection and disease development by *G. abietina*.

---

[1] Plant Pathologist, Research Assistant, and Summer Research Assistant, Boyce Thompson Institute, Ithaca, NY 14853, Assistant Professor, Department of Plant Pathology, Cornell University, Ithaca, NY 14853, and Supervisory Research Plant Pathologist, USDA-Forest Service, Northeastern Forest Experiment Station, Forestry Sciences Laboratory, P.O. Box 365, Delaware, OH 43015.

This work was funded in part by Cooperative Agreement No. 23-247 between USFS and Boyce Thompson Institute and was performed under work unit FS-NE-2208. It would not have been possible to conduct these experiments without the extraordinary cooperation of the New York State Department of Environmental Conservation, and in particular Mr. Bruce Schneider, NY DEC, Lowville, N.Y. The technical assistance of Guy Knudsen and Mary Ann Beall is also acknowledged. We also thank John Evans and Jeff Morris of the Biometrics Unit, Cornell University for assistance with statistical analyses and Colleen Sloan for manuscript preparation.

MATERIALS AND METHODS

Plant Culture

Eighteen-hundred 2-2 seedlings of *Pinus resinosa* (red pine) were obtained from Western Maine Forest Nursery, Freyburg, ME. The trees were planted in 18 plots of four rows of 25 trees each in a field near Harrisville, NY in May of 1979. The field soil was a sandy loam, suitable for the culture of red pine.

Measurements of the height of each tree were taken at the time of planting, and periodically during the course of the study.

Inoculation

Spores of *G. abietina* were washed from cultures of the fungus with deionized water. The concentration of spores in solution was determined using a calibrated counting cell and the inoculum solution was adjusted to contain 10^5 spores per ml.

After dark (approximately 2100 EDT) on 27 June 1979 and 28 July 1980, trees in half of each plot were inoculated by spraying to run-off with a pressurized garden sprayer. Aliquots of inoculum were collected on water agar periodically during each inoculation. After a suitable incubation period, spore germination was checked to assure that viable inoculum was used. In both cases, spore germination was high (90%).

SO_2 Exposures

Trees were exposed using linear gradient exposure devices (Laurence et al., 1982; Shinn et al., 1977). Exposures were carried out at four times during the course of the experiment (July and August, 1979 and July and August, 1980). Each exposed plot was treated with SO_2 during one 4 week period during 1979, 1980, or 1979 and 1980.

SO_2 was supplied from tanks of 20% SO_2 in N_2. Concentrations of SO_2 were monitored at 12 points in each of the 4 plots that were being fumigated with pulsed fluorescence analyzers modified for rapid response (Thermoelectron Corp., Model 43, Hopkinton, MA). Monitors were time-shared by use of 24 point sequential samplers, resulting in about 1 measurement per hour. Data were acquired automatically with an A.D. Data systems data logger. Exposures were conducted daily from 0700 to 1900 EDT, seven days per week.

Disease Measurement

Two measurements of disease severity were taken in both 1980 and 1981. 1980 was a typical year, with symptom development beginning in April and progressing until June. In 1981, an extended warm spell in February caused symptom development to begin early, hence data in 1981 were of a categorical nature (i.e., healthy, infected, alive, dead).

Statistical Analyses

In addition to growth measurements, 10 other variables were calculated to assess infection, infection rates, and death rates. These variables were discussed in detail in the next section.

The factorial experiment was conducted in a split-split plot design with SO_2 applied to whole plots, inoculation in 1979 as split-plots and inoculations in 1980 as split-split plots. Whole plot treatments consisted of exposure in 1979, exposure in 1980, exposure in both years, and non-exposed controls. Inoculations resulted in split plots inoculated before exposure, after exposure, or before and after exposure. This design allowed the evaluation of repeated SO_2 exposures and repeated inoculations, at various times with regard to SO_2 treatment, on disease development.

RESULTS

SO_2 Exposure Concentrations

Concentrations during exposure were generally less than 0.25 ppm. Occasional peaks were measured, primarily during times when adjustments were being made in the exposure equipment. SO_2 doses (ppm x hr) ranged from 0.73 to about 25 ppm x hr over the course of the study. Concentrations within plots and between plots were more variable than we expected based on our past experience with the linear gradient exposure system (Laurence et al., 1982). We feel this was due to the open canopy that existed in the field.

Visible Response of Plants to SO_2

Visible reponses of red pine to SO_2 were observed on three occasions during the course of the study. In each case, only two or three trees responded. The injury consisted of necrosis of the terminal 2-3 cm of most of the needles on the tree. The injured needles did not cast prematurely. Chronic injury due to SO_2, which would be expected to consist of a general chlorosis of the needles, or premature casting, did not occur on any trees. Red pine has been reported to be visibly sensitive to SO_2 (Applied Science Associate, 1976), but the results of this study do not confirm that classification. The pines did not respond on many occasions when other vegetation (i.e., ragweed) growing in the plots did show visible symptoms.

Effects of SO_2 on the Growth of Red Pine

1979 tree growth. Following planting in 1979, the initial heights of all trees were measured. Subsequent analysis showed that the initial heights of trees did vary significantly from plot to plot. These differences were unrelated to any treatment imposed on the trees. Subsequent analyses of tree growth included initial height as a co-variate. Measures of growth in 1979 that also should have been independent of treatment (elongation was basically complete before inoculation or fumigation took place) were found to be uniform across plots. Based on these analyses, we concluded that the trees were growing at the same rate in all plots.

Regression of growth in 1979 vs SO_2 dose were performed for each row of trees in each plot. Slopes and intercepts of these regressions were subjected to both univariate and multivariate analyses of variance. No significant effects of SO_2 dose on tree growth in 1979 were found.

1980 tree growth. Both SO_2 and inoculation with *G. abietina* caused significant effects on growth. Inoculated trees showed an average growth of -3.3 (due to tip dieback) while non-inoculated trees grew an average of 2.7 cm. The SO_2 X inoculation interaction was not significant.

There was a signficant difference due to time of fumigation; SO_2 reduced growth of red pine when applied in July, but not when the fumigation took place in August.

When non-fumigated plots were removed from the analysis, there were no signficant differences among the growth in fumigated plots; the effect of fumigation was the same from plot to plot. When fumigation took place in both years, the effects were additive. Carry over effects of fumigations were not detected. A summary of these data are presented in Table 1.

Table 1. Effects of fumigation with SO_2 on 1980 growth of red pines.

Fumigated in	Growth (cm)[1]
Control	-0.27
July	-1.53
August	0.89
Year 1	1.10
Year 2	-1.29

[1] Negative growth indicates the effect of disease as death of terminals and tips of branches.

Effects of SO_2 on Disease Development

Measurements of disease severity were made periodically during the course of the experiment. No differences were found in the severity of disease with respect to SO_2 dose. The data were then categorized to allow analysis of infection rates, death rates, etc. The variables created are defined below.

$$\text{Percent infection spring 1980} = \frac{\text{No. infected spring 1980}}{\text{No. alive spring 1980}}$$

At the first disease measurement time, 70% of inoculated trees were infected, while only 5.6% of trees that were not inoculated became infected (due to natural inoculation or contamination during the inoculation).

$$\text{Change in infection 1979 to spring 1980} = \frac{\text{No. new infections + No. died}}{\text{No. healthy at end of 1979}}$$

A significant effect of inoculation was found (as above) and a marginally significant fumigation X inoculation interaction was present ($p = 0.09$). The infection rate was similar in plots that were inoculated, regardless of fumigation (0.85-0.88), but natural infection was considerably lower in fumigated plots (0.28) when compared to non-fumigated plots (0.41).

$$\underline{\text{Proportion dead spring 1980}} = \frac{\text{No. dead spring 1980}}{\text{No. alive at end of 1979}}$$

A significantly greater proportion of trees died in plots that were not fumigated (0.38), when compared to trees in plots that did receive SO_2 (0.28). This is likely due to the increased natural G. abietina infection in non-fumigated plots.

$$\underline{\text{Percent infection summer 1980}} = \frac{\text{No. infected summer 1980}}{\text{No. alive summer 1980}}$$

A large effect of inoculation was found, with 95% of inoculated trees infected vs 24% in non-inoculated plots. Significant effects of fumigation were not observed.

$$\underline{\text{Proportion dead summer 1980}} = \frac{\text{No. dead from spring to summer 1980}}{\text{No. alive spring 1980}}$$

No signficant effects of fumigation were observed.

$$\underline{\text{Change in infection spring 1980 to summer 1980}} =$$

$$\frac{\text{No. infections spring to summer + No. died spring to summer}}{\text{No. healthy spring 1980}}$$

A large effect of inoculation was observed, but effects of SO_2 fumigation were not found.

$$\underline{\text{Percent infected spring 1981}} = \frac{\text{No. infected spring 1981}}{\text{No. alive spring 1981}}$$

Differences due to inoculation, but not fumigation were observed. Thirty-three percent of inoculated trees died, compared to only 5% of non-inoculated trees. There was an indication of a fumigation X inoculation interaction, but it was not significant (Table 2).

$$\underline{\text{Proportion dead spring 1981}} = \frac{\text{No. dead fall 1980 to spring 1981}}{\text{No. alive fall 1980}}$$

Differences due to inoculation, but not fumigation were observed. Thirty-three percent of inoculated trees died, compared to only 5% of non-inoculated trees. There was an indication of a fumigation X inoculation interaction, but it was not significant (Table 2).

$$\underline{\text{Change in infection summer 1980 to spring 1981}} =$$

$$\frac{\text{No. infected summer 1980 to spring 1981 + No. died summer to spring}}{\text{No. healthy summer 1980}}$$

A similar trend to those described above was found for this variable. There was a large effect of inoculation, but none for fumigation. At this point, infection rates are negative because a large number of diseased trees died, removing them from the infected pool. Results for the fumigation X inoculation interaction are given in Table 5, and once again, suggest a protective effect of SO_2.

$$\text{Percent infected summer 1981} = \frac{\text{No. infected summer 1981}}{\text{No. alive summer 1981}}$$

Percent infection summer 1981. At this measurement, one year after the last fumigation, inoculation is the only identifiable effect. The suggested protective effect of SO_2 is no longer apparent in the interaction means, where no significant effects occur and contribution of the trend towards protection is not apparent.

Table 2. Effect of SO_2 fumigation and inoculation with *G. abietina* on measured infection and death of red pine in spring 1981.

Fumigation	Inoculation			
	Yes		No	
	% infected	% dead	% infected	% dead
Yes	83	26	25	4
No	79	39	19	7

Table 3. Effects of SO_2 fumigation and inoculation with *G. abietina* on the change in infection from summer 1980 to spring 1981.

Fumigation	Inoculation	
	Yes	No
Yes	-2.71[1]	0.04
No	-2.11	-0.20

[1] Negative rates are due to large numbers of diseased trees that died and are removed from the infected pool, thus the larger the number, the fewer dead trees.

DISCUSSION

It is obvious from the results of this study that Scleroderris canker is a serious disease of red pine. It not only causes growth effects through tip dieback, but the pathogen spreads rapidly and in most cases infections are fatal. By the end of the experiment, all trees were dead. SO_2 did reduce growth of trees, but the effect was not nearly as significant as that of the disease. More importantly, the two stresses did not interact to speed the mortality of trees.

In several cases, it did appear that SO_2 inhibited infection by G. abietina. This was apparent in non-inoculated plots, where the natural rate of infection of SO_2 -treated trees was somewhat lower than the corresponding rate in non-fumigated trees. In inoculated plots, this effect was not detected, probably due to the high disease intensity resulting from a heavy inoculum load. The protective effect of SO_2 was transitory, however, as it was much less apparent in the summer of 1981, one year after the last fumigation. This is similar to responses observed in other pollutant-pathogen interactions (Laurence, 1981; Laurence and Reynolds, 1982).

Scleroderris canker did not seem to alter the susceptibility of red pine to SO_2. Visible symptoms of SO_2 injury were not observed for infected trees any more often than they were on non-infected trees. Symptoms typical of chronic SO_2 injury were not observed during the course of the study in either inoculated or non-inoculated plots. Red pine has been reported to be visibly sensitive to SO_2 (Applied Science Associates, 1976), but our results would indicate that it is at least intermediate in tolerance to SO_2. This is probably not the case when growth is considered.

Based on the results of this study, it does not appear that the recent emergence of Scleroderris canker as an important problem in the northeast U.S. is related to the increased presence of sulfur oxides in the air. These experiments were conducted in the presence of ambient ozone and acidic precipitation, so the effects of those air quality variables cannot be ruled out, however, increasing the pollutant load with SO_2 did not cause an intensification of the disease.

This study does indicate possible significant effects of SO_2 on tree growth however. The detection of such growth effects at relatively low concentrations should be a cause for concern, as the concentrations we used (generally less than 0.25 ppm) can occur near point sources of SO_2. It must be pointed out, however, that the duration and frequency of exposures in this study were much greater than would normally be observed in the vicinity of a source of SO_2.

In light of the effects of SO_2 on growth, the apparent inhibition of infection by the fungus in SO_2 -treated trees is likely to be of minimal importance; any gain from protection against disease is likely to be lost in growth. Furthermore, once the inoculum level builds to a certain point, the protective effect is lost (as seen in plots inoculated and fumigated). As a result, one would only expect a delay in the epidemic.

LITERATURE CITED

Applied Science Associates. 1976. Diagnosising Vegetation Injury Caused by Air Pollution. Office of Air and Waste Management, Office of Air Quality Planning and Standards, Control Programs Development Division, U.S. Environmental Protection Agency.

Laurence, J.A. 1981. Effects of air pollutants on plant-pathogen interactions. Z. PFL. Krankht. Pfl. Schutz 88:156-172.

Laurence J.A., and K.L. Reynolds. 1982. Effects of concentration of SO_2 and other characteristics of exposure on lesion development by Xanthomonas phaseoli in Red Kidney bean. Phytopathology 72:1243-1246.

Laurence, J.A., D.C. MacLean, R.H. Mandl, R.E. Schneider, and K.S. Hansen. 1982. Field tests of a linear gradient system for exposure of row crops to SO_2 and HF. Water, Air, and Soil, Pollut. 17:399-407.

Shinn, J.A., B.R. Clegg, and M.L. Stuart. 1977. A minimum field fumigation method for exposing plants to controlled gradients of air pollution levels. Preprint of paper presented at "Department of Energy, Division of Biomedical and Environmental Research Workshop, Ecological and Agricultural Effects of Caol Combustion, Oak Ridge, TN.

EVALUATION OF POSSIBLE EFFECTS OF ACID RAIN ON SCLERODERRIS CANKER OF RED PINE IN NEW YORK

Robert J. Bragg and Paul D. Manion[1]

ABSTRACT

The possible affects of acid rain on Scleroderris canker of red pine were evaluated through two procedures. In the first, red pine seedling field plots, protected and unprotected from natural rain, were exposed to natural inoculum and treated with either deionized distilled water (pH 5.6), artificial acid rain (pH 3.5), or lime. Acid rain treatments did not increase disease infection levels over distilled water or lime treatments. Soil pH was slightly lowered by acid rain treatments. The second procedure evaluated pH and buffering capacity of red pine bark and buds from twelve plantations in the Adirondack region of New York. In addition, soil pH and soil cation content of the stands were determined and evaluated in relation to bark pH and natural Scleroderris canker disease increase rates. The effect of acid rain on red pine bark is probably minimal since bark surfaces were generally more acidic than acid rain. Buffering capacity against acid was small. Soil pH and spring bark surface pH were found to be inversely correlated with disease increase. These two procedures generated contrasting interpretations: (1) Scleroderris canker disease did not increase in response to artificial acid rain treatment of seedlings; (2) Scleroderris canker disease increase rate in red pine stands was somewhat related to soil pH. None of these relationships are very strong, and, therefore, it would appear that the various interacting factors of acid rain on the system generate a neutral minor impact on the Scleroderris disease problem in New York.

INTRODUCTION

Scleroderris canker, caused by the fungus *Gremmeniella abietina* (Lagerb.) Morelet, has caused major dieback and cankering of planted pines and spruces in Europe and Great Britain for almost 100 years (Donaubauer 1972a, Dorworth 1970). The disease was first noticed in North America in northern Michigan in 1951, and Ontario, Canada, in 1954. The pathogen caused cankering and mortality of young red pine (*Pinus resinosa* Ait.) and jack pine (*P. banksiana* Lamb.) plantations (Ohman 1966).

---

[1] Graduate Student and Professor, State University of New York, College of Environmental Science and Forestry, Syracuse, N.Y. 13210. Present address of senior author: Cary Arboretum, Box AB, Millbrook, N.Y. 12545

This research was supported by the USDA Forest Service, North Central Forest Experiment Station.

Scleroderris canker was first reported in the Adriondack Mountains of New York in 1964 (French and Silverborg 1967). In 1975 G. abietina was recovered from large dying red and Scots pine (P. sylvestris L.) in northern New York plantations (Setliff et al., 1975). This recent extensive mortality of large trees was associated, presumably, with a serological race of the fungus which had the ability to infect tissues in both upper and lower crowns of mature trees (Dorworth 1981, Dorworth et al., 1977; Skilling 1977). No one has tested the pathogenecity difference among the serological races (Dorworth 1981).

Another possible interpretation of the recent disease build up in New York could be associated with a shift toward more favorable environmental conditions. The intensification of Scleroderris canker depends heavily on climatic and other ecological conditions (Donaubauer 1972b, Dorworth 1972).

Current interpretation of previously collected data suggests to some that the precipitation in the northeastern United States has become more acidic in the past few decades (Likens and Bormann 1974) others question the change and suggest that the percent pH is a long time characteristic of the region. Precipitation in the Adirondack region of New York today averages about pH 4.2 with individual events recorded to 3.2 (Raynal et al., 1980). The rapid development of the serious Scleroderris disease problem in the northeast coincides in position and time with the acid rain problem. The possibility that acid rain may be an environmental factor which has altered recent G. abietina interactions with red and Scots pine in New York is an intriguing question.

The objectives of this study were to assess the possible interactions of Scleroderris canker and acid rain through two approaches. The first approach involved experimental treatments of red pine seedlings in the field with natural rain, artificially produced acid rain, deionized distilled water rain, and lime soil application and evaluation of effects on natural infection by Scleroderris canker. The rain treatments were to test primarily direct effects. The lime treatment was to evaluate possible indirect soil, plant nutrition, and plant disease defense interactions.

The second approach tested the hypothesis that acid rain of the Adirondack region directly or indirectly affects the pH and buffering capacity of buds and bark of red pine and that the change is more favorable for disease development. Bark pH has been shown to respond to sulfur dioxide concentration (Grodzinska, 1971, 1976, Staxang 1969). Grodzinska found that buffering capacity increased with increasing incidence of air pollution. Stephan and Sholz (1979) evaluated pH and buffering capacity of Austrian pine (Pinus nigra Arnold) clones in relation to Scleroderris canker resistance. They found a significant correlation between susceptiblity to the disease and bark buffering capacity.

Carvin (1980), in 1978, established permanent plots in red pine plantations in the northern and western Adirondack region of New York to evaluate disease increase over time. The plots, visited yearly, show spatial and temporal differences in rates of disease increase. The pH and buffering capacity of spring and fall collections of bud and branch samples from these permanant plots were evaluated statistically for correlation with soil pH and the disease increase rate of the stands.

METHODS

Artificial and Natural Rain Treated Red Pine Plots

Plots of red pine seedlings were established in April 1981 on New York State forest land, 5.5 km northeast of Boonville, N.Y. This same location used by Setliff and Manion (1980) was a mixed *Pinus resinosa*, *P. rigida*, *P. banksiana* stand established in the 1930's to stabilize a blow sand site. Although the Scleroderris disease is well established in the stand, mortality due to the disease is rare. Ten plots (one meter square) were established in each of three 17 m^2 or larger openings in the stand.

Each of the thirty plots consisted of 48 three-year-old seedlings planted 7 cm apart in rows of 10 cm apart. Half of each plot was planted with seedlings from the New York State Department of Environmental Conservation nursery at Saratoga, N.Y. and the other half from the SUNY College of Environmental Science and Forestry nursery at Lafayette, N.Y. At each of the three sites, the seedlings of a given seed source were randomized among the ten half plots. This was accomplished by sequential individual row planting of all ten plots at the same time.

At each of three replicate sites the ten plots were separated by a 1-3 m unplanted buffer zone. Two plots at each site were treated with 1500 g/m^2 soil applications of lime. Four of the other eight plots were covered with 1 m square Filon type 550 greenhouse panels (A.H. Hummer Seed Co.) (Figure 1).

Figure 1. Acid rain treatment plots showing covered and uncovered red pine seedlings.

An artificial rain system similar to Shriner's (1974, 1978) was established using two nozzles (Bete Fog Nozzles, Inc., Greenfield, MA) attached 0.25 m from the ground on a pine stake in the center of each plots. The nozzles delivered droplets from 0.59 to 1.16 mm diameter (Shriner 1974). A pump (Flotec, Inc., model B251-1100, Norwalk, CA) and Tygon tubing delivered approximately 5 liters (0.5 cm) of rain per m^2 plot in 15-20 minutes from a Nalgene carboy. Deionized distilled water, which had been allowed to equilibrate with air, served as the artificial rain of pH 5.6. An artificial rain of pH 3.5 was prepared using a 6:3:1 ratio of 0.5 m solutions of H_2SO_4, HNO_3, and HCl, respectively (Cogbill 1976). Rain treatments were applied at 11-weekly intervals from April 24 to July 1.

About 5 l of (0.5 cm) deionized distilled water was applied directly to the soil of each covered plot weekly. This was done to supplement natural moisture levels which averaged 2.5 cm/week on uncovered plots.

Weekly rain samples were collected with a plastic funnel rain gauge. The amount and pH of the rain was measured with an Orion model 601A digital pH meter. Weekly spore samples were collected on two vaseline coated slides placed horizontally in covered and uncovered plots at each site. Soil samples were collected in June. Three random soil samples from the upper 10 cm of each plot were mixed and the pH was determined in a 1:2 soil to water suspension (Wilde et al., 1979). The 1981 and 1982 heights of the planted seedlings were measured in late summer.

Disease incidence was determined in the spring of 1982 by calculating the percentage of seedlings with Scleroderris canker on each plot. Seedlngs were considered to be diseased if they showed typical browning from the base of the needles outward. Isolations from a sample of seedlings with typical symptoms verified _G. abietina_ as the cause.

Bark and Bud pH and Buffering Capacity and Soil pH in Relation to Disease Increase Rates in Red Pine Plantations

Disease development in twelve red pine plantations in the Adirondack region of New York, was determined by Carvin (1980) based on numbers of trees and numbers of branches infected in an approximately 0.1 hectare permanent plots in 1978, 1979, 1980, and 1981. Three trees in each of these red pine plantations were sampled for lower (less than 1 m) and upper (2.5-3.5 m) twig bark and bud tissue in the spring and fall of 1981. Samples were collected in plastic bags and stored up to two weeks at 15C. Three pH readings of each current-year bark sample were measured with a Corning surface electrode and averaged. Tissue pH and buffering capacity were determined using the modified methods of Skye (1968) and others (Grodzinska 1971, Staxang 1969, Stephan and Schultz 1979). One gram of current-year bud or bark tissue was ground in 10 ml of distilled water for one minute using a Tissumizer (Tekmar Co.). Solids were removed by suction filtration through Whatman #1 filter paper, and the pH of the filtrate was measured with an Orion pH meter 24 hours later.

Titrations were performed on the ground bark and bud tissue filtrate using 0.01 ml increments of 1N HCl or NaOH.

In the spring of 1981, five random soil samples were collected and mixed from each 0.1 hectare plot. The plot soil pH was determined as before. The magnesium, calcium, potassium and sodium content of the soil was determined by netural ammonium acetate extraction and atomic absorption spectroscopy (Wilde et al. 1979).

RESULTS

Artificial and Natural Rain Treated Red Pine Plots

Average disease incidence based on six replicates of five plot treatments for Lafayette and Saratoga seedlings is summarized in Table 1. Disease incidence was slightly less but not significantly different between acid rain and deionized distilled water rain plots. Acid rain covered and uncovered Saratoga seedlings had statistically different disease levels.

Table 1. Average percent disease among plot treatments.

Treatments	Percent Disease: Lafayette Seedlings	Percent Disease: Saratoga Seedlings
pH 3.5 Rain Covered	62[1]	63[2]
pH 3.5 Rain Uncovered	68	86[2]
pH 5.6 Rain Covered	70	70
pH 5.6 Rain Uncovered	79	83
Lime added to Soil	71	83

[1] Number of diseased seedlings divided by the total number of seedlings (29-90) times 100.

[2] Significantly different at the .001 level using a paired G-test (Sokal and Rohlf 1969). All other paired comparisons were not significantly different at the .05 level.

The 1981 height growth of Saratoga Seedlings averaged 9.8 cm and Lafayette seedlings averaged 10.2 cm. The values were not significantly different. Growth of Saratoga and Lafayette seedlings was not affected by the various plot treatments ($p > .05$).

Inoculum levels varied by week, site and plot. The number of spores per five scans (about 169 mm^2) of the vaseline-coated slide varied from 0 to 79. The highest inoculum levels occurred between May 27-June 24. No consistent differences in spore densities were recorded between plots. Inoculum levels by week were not significantly correlated with rainfall ($R = .15$).

The acidity of natural rain ranged from a low of 2.95 for June 17-24 to a high of pH 5.34 for June 24-July 1 (Table 2). The average rain pH for the entire period was 4.0.

The composite soil pH of treatment plots is summarized in Table 3. Lime treatments substantially increased soil pH. The covered plots had a slightly higher soil pH when compared to the uncovered plots in all cases. The pH 5.6 treated plots had a slightly higher pH than the pH 3.5 treated plots in all but one case. Soil samples from non-treated areas of each site ranged from pH 4.65 to 4.71.

Table 2. Total weekly rainfall at Boonville, NY for 1981.

Date	Amount (cm)	pH
5/6 - 13	3.0	4.49
5/13 - 20	2.9	4.09
5/20 - 27	.1	3.71
5/27 - 6/3	1.4	3.69
6/3 - 10	2.8	3.74
6/10 - 17	4.8	4.02
6/17 - 24	1.4	2.95
6/24 - 7/1	1.6	5.34
7/1 - 22[1]	7.2	4.28
7/22 - 29	4.8	4.76
7/29 - 8/6	2.9	3.84
8/6 - 27[1]	7.0	3.20

[1] Three-week period

Table 3. Soil pH[1] of treated and untreated plots at three sites at Boonville, NY

			Treatment		
Site	Limed	pH 5.4 Covered	pH 5.6 Uncovered	pH 3.5 Covered	pH 3.5 Uncovered
1	6.34	4.87	4.64	4.76	4.32
2	5.56	4.35	4.33	4.44	4.23
3	6.12	4.40	4.32	4.26	4.08

[1] Each value is one measurement from a mixture of three samples taken in June, 1981.

Bark and Bud pH and Buffering Capacity and Soil pH in Relation to Disease Increase Rates in Red Pine Plantations

Bark surface pH from red pine plantations of the northern and western Adirondacks was significantly less than ($p < .05$) bark tissue and bud tissue pH (Table 4).

Tissue pH varied by season. Spring and fall bark and bud tissue samples were significantly different ($p < .05$) for upper and lower sample. Fall tissue was about 0.5 pH units lower than spring tissue for upper or lower samples. Spring surface pH, however, was not significantly different from fall ($p > .05$ for upper and lower sample).

The upper samples were more acidic than lower samples in all three sample types, but only upper and lower samples of bark surface pH were significantly different.

Disease increase rates (van der Plank 1963) were available or calculated from additional data on the Carvin (1980) Adirondack region red pine plots. Rates (r) based on numbers of trees infected and also for numbers of branches infected were calculated for one-, two- and three-year intervals from 1978 to 1981 for each plot. The significant relationships among disease increase rates and bark surface pH are shown in Table 5. Low bark surface pH in the spring and high bark surface pH in the fall were correlated to 1978-1980 disease increase rates. Bark surface pH ranged from 2.4 to 5.0. None of the correlations with bark or bud tissue and disease increase rates were significant.

As shown in Table 6, soil pH on the red pine plots range from 4.28-6.10. Soil cation levels varied considerably among plots.

Table 4. Average[1] pH values of red pine bark and bud samples collected in the spring and fall from upper (2.5 - 3.5 m) and lower (<1 m) portion of the crown.

	pH Spring		pH Fall	
	Upper Sample	Lower Sample	Upper Sample	Lower Sample
Bark Surface	3.20[a] [2]	3.40[b]	3.17[a]	3.52[b]
Bark Tissue	4.22[c]	4.26[c]	3.68[d]	3.79[d]
Bud Tissue	4.04[c]	4.11[c]	3.55[d]	3.67[d]

[1] Each number is an average of 3 samples from each of 11 red pine stands.

[2] Values followed by a common letter are not significantly different at the 0.05 level with a Duncan multiple range test.

Table 5. Correlation coefficients of spring and fall bark surface pH[1] with disease increase rates.

	Spring		Fall	
Disease Rate	Upper Sample[a]	Lower Sample	Upper Sample	Lower Sample
1978-80 Trees	-.41[b] [2]	-.51[c]	.22[b]	ns[3]
1978-80 Branches	-.53[c]	-.55[d]	.50[d]	.36[d]
1980-81 Trees	ns	ns	.28[c]	ns
1980-81 Branches	ns	ns	.42[d]	.41[d]

[1] Average of three readings from three samples.

[2] b,c, d - significant at the .05, 0.1 and .001 levels, respectively.

[3] Not significant.

A correlation analysis was performed between surface pH and soil pH and cations. In the fall, upper and lower surface pH's were inversely related to soil pH (Table 7).

Disease increase rates based on the numbers of diseased branches were negatively correlated with soil pH (R = -.41, P $<$.05). Disease increase rates based on number of trees infected were directly correlated with soil potassium levels (R = .34, P $<$.05).

Table 6. Soil pH and nutrient characteristics[1] of permanent red pine plots (5) in New York.

Plot	pH	MG^{++}	Ca^{++}	K^{+}	Na^{+}
		(ppm)			
1	6.10	46.2	274.5	14.6	.8
2	5.06	.9	1.1	7.1	2.2
3	4.73	4.2	34.8	12.9	.6
4	4.46	3.0	19.1	20.9	.5
5	4.58	4.0	44.9	31.5	.3
6	4.48	6.9	106.6	34.2	6.6
7	4.39	17.5	156.4	41.9	3.8
8	4.38	18.9	220.7	36.5	1.6
9	4.93	70.1	482.5	15.7	1.5
10	3.82	6.8	72.6	11.5	.5
11	3.76	1.3	3.8	6.4	.4
12	nd [2]	3.2	13.4	13.9	2.8

[1] Each value is one measurement from a mixture of five samples.

[2] No data.

Table 7. Correlation coefficients of spring and fall bark surface pH[1] with soil pH.

	Spring		Fall	
	Upper Sample	Lower Sample	Upper Sample	Lower Sample
Soil pH	.61[c] [2]	.55[c]	-.48[c]	-.30[b]

[1] Average of three readings from three samples.

[2] b, c, - significant at the .05 and .01 levels, respectively.

Buffering capacity of the upper and lower samples were approximately the same. An average titration curve, for spring bud tissue is shown in Figure 2. Bark tissues have somewhat less buffering capacity against the base than do bud tissues. Bark and bud tissues have similar and very small buffering capacities against the acid. Both tissue samples have a greater buffering capacity in the spring as compared to the fall.

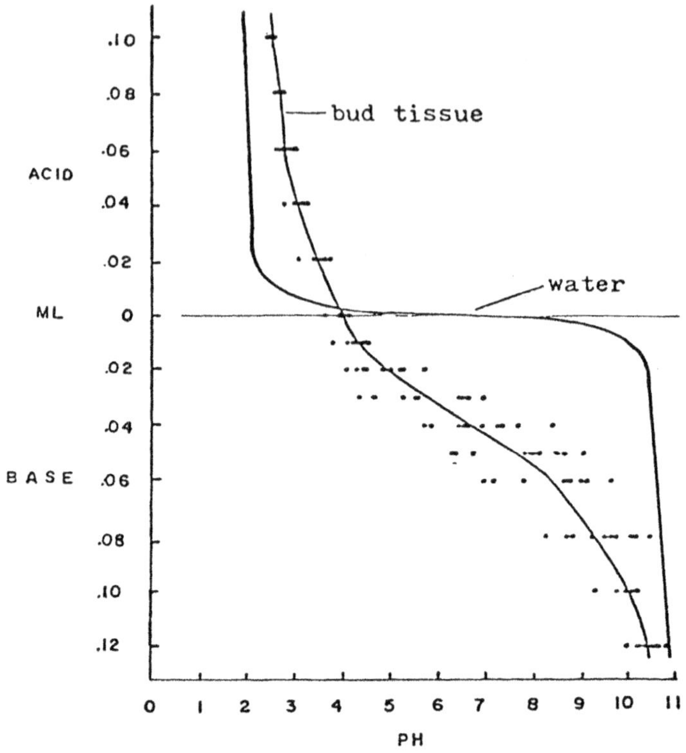

Figure 2. Average bud tissue titration for spring 1981. Each point represents one of three samples from each plot.

DISCUSSION

Shriner (1978) and others (Bruck et al., 1981, Lacy et al., 1981, Leben 1954) reported that acid rain may have both positive and negative effects on specific disease systems. Acid rain may alter soil or foliar nutrient levels, spore germination levels or tissue susceptibilities. Our eleven weekly pH 3.5 acid rain treatments of red pine seedlings slightly, but not significantly, reduced the Scleroderris canker infection. These data are consistent with Setliff and Manion's (1980) preliminary results.

Whether or not acid rain influences the incidence of diseases on plants may depend on when the treatments are applied: (1) before inoculum is present (pre-treatment), (2) at the time when inoculum first appears, or (3) after inoculum is present (post-treatment). Shriner's (1978) acid rain treatments of bean plants increased or decreased incidence of halo blight depending on when the acid rain was applied. Leven (1954) produced reduced disease levels when the pre-treated tomato with acid rain prior to infecting the tomato plants with early blight.

We do not know precisely when inoculation and infection took place in relation to the rain treatments in our experiment. Data from another experiment performed by Luley (1982) during the same period, in the same area, identified at least six of the 11 treatment weeks as having conditions sufficient for infection. Additional weeks for infection occurred after the treatment period. Therefore, our experiments are best identified as a pre or during infection treatments.

Seedling growth was not affected by plot treatments. Growth was probably influenced more by nursery conditions of the previous year than the various acid or lime treatments.

When covered or uncovered plots were treated with 3.5 pH acid rain, soil pH was usually slightly lowered. When treatment plots were covered, soil pH became slightly higher. Although the changes are slight, the data agrees with Setliff and Manion's (1980) earlier results. This suggests that soil pH of this sandy site may be readily affected by rain pH. This change in pH may be the indirect effect of changes in moisture levels, changes in microflora population levels or leaching of nutrients from the soil.

The slight increase of soil pH with plot covering and slight disease reduction suggests that lower soil pH might favor higher disease levels. The differences are not significant and can be likewise attributed to inoculum levels, water levels and other environmental variables associated with covering plots.

Tree bark persists in the ecosystem for long periods of time and undergoes pH changes (Grodzinska 1971, 1976; Skye 1968; Staxang 1969). Red pine bark surfaces, on which G. abietina germinates, were found to be more acidic than internal tissues. Grodzinska (1976) found this to be true for several tree species studies. The average tree bark surface in spring, when infection takes place, was pH 3.20 to 3.40. Spore germination on the media adjusted to these levels remains high (Setliff and Manion 1980, Bragg 1982). Germ tube length, however, may be lightly affected (Setliff and Manion 1980). The average bark pH is lower than weekly natural rain pH except for one week. Therefore, it is unlikely that bark surface pH of red pine is directly affected by acid rain.

Disease infection is more likely to occur on the lower branches in the spring of the year (Dorworth 1972). Red pine bark samples are somewhat less acidic when collected from the lower portion of the crown and in the spring of the year. Therefore, it is unlikely that a more acidic bark pH, resulting from intense acid rain should favor Scleroderris disease development. In contrast, correlation analysis suggests that low bark surface pH in the spring is related to high disease increase rates, whereas low bark surface pH in the fall correlated with low disease increase rates.

Buffering capacity against the base is greatest in the spring. Grodzinska (1971) and Skye (1968) also found that tree bark buffering capacity against the base is highest in spring. Buffering capacity against the acid is small and remains so throughout the year. Since bark pH is generally more acidic than acid rain itself, very little, if any, buffering capacity may be needed to maintain bark pH. The effect of acid rain on red pine bark buffering capacity is probably minimal.

Disease increase rates based on numbers of disease branches was negatively correlated to soil pH suggesting that low soil pH, which may be leached of soil nutrients or alter in microflora, may favor disease intensification. Soil pH itself was positively correlated with bark surface pH in the spring of the year.

In summary, analysis of bark pH and buffering capacity and field acid rain treatments suggested that acid rain is not an environmental factor contributing to Scleroderris canker disease. On the other hand, lower soil pH does seem to be somewhat related to higher disease increase rates. None of these relationships are very strong, suggesting that acid rain is not a major environmental factor in the Scleroderris canker disease. The subtle positive and negative aspects of acid rain on microflora, soil nutrients, host condition, spore germination, and fungus growth, may interact in ways to produce a neutral condition.

LITERATURE CITED

Carvin, R.P. 1980. Factors affecting intensification of Scleroderris canker in young red pine plantations. M.S. Thesis, SUNY College Environmental Science and Forestry, Syracuse, N.Y. 43 pp.

Cogbill, C.V. 1976. The history and character of acid precipitation in eastern North America. In Dochinger, L.S. and Seliga, T.A. (eds.). Proceedings of the First International Symposium on Acid Precipitation and the Forest Ecosystem. Ohio State University.

Donaubauer, E. 1972a. Distribution and hosts of Scleroderris lagerbergii in Europe and North America. Eur. J. For. Pathol. 2:6-10.

Donaubauer, E. 1972b. Environmental factors influencing outbreaks of Scleroderris lagerbergii Gremmen. Eur. J. For. Path. 2:21-25

Dorworth, C.E. 1970. Scleroderris lagerbergii Gremmen and the replant problem in Central Ontario. Forest Res. Lab., Ontario Region Info. Rept. O-X-139.

Dorworth, C.E. 1972. Epidemology of Scleroderris lagerbergii in Central Ontario. Can. J. Bot. 50:751-765.

Dorworth, C.E. 1981. Status of pathogenic and physiologic races of Gremmeniella abietina. Plant Dis. 65:927-931.

Dorworth, C.E. J. Krywiewczyk and D.D. Skilling. 1977. New York isolates of Gremmeniella abietina (Scleroderris lagerbergii) identical in immunogenic reaction to European isolates. Plant Dis. Rept. 51:887-890.

French, W.J. and S.B. Silverborg. 1967. Scleroderris canker on red pine in New York State plantations. Plant Dis. Rep. 51:108-109.

Grodzinska, K. 1971. Acidification of tree bark as a measure of air pollution in southern Poland. Bull. Acad. Sc. Pol., 19:189-195.

Grodzinska, D. 1976. Acidity of tree bark as a bioindicator of forest pollution in southern Poland. Proceedings of the First International Symposium on Acid Precipitation and the Forest Ecosystem. Ohio State University.

Lebin, C. 1954. Influence of acidic buffer sprays on infection of tomato leaves by Alternaria solani. Phytopathology 44:101-106.

Likens, G.C. and F.H. Bormann. 1974. Acid rain: A serious regional environmental problem. Science 184:1176-1179.

Luley, C.J. 1982. Inoculum potential of Scleroderris canker in New York. M.S. Thesis, SUNY College Envir. Sci. For., Syracuse, N.Y.

Ohman, J.H. 1966. Scleroderris lagerbergii Gremmen: the cause of dieback and mortality of red and jack pines in upper Michigan plantations. Plant Dis. Rep. 50:402-405.

Raynal, D.J., A.L. Lenf, P.D. Manion. and C.J.K. Wang. 1980. Effects of acid precipitation on a forest ecosystem. New York State Energy Research Development Authority. Albany, N.Y.

Setliff, E.C. and P.D. Manion. 1980. Forest pathology and acid rain. In effects of Acid Precipitation on a Forest Ecosystem. New York State Energy Research and Development Authority, Albany, N.Y.

Setliff, E.C., J.A. Sullivan, and J.H. Thompson. 1975. Scleroderris lagerbergii in large red and Scots pine trees in New York. Plant Dis. Rep. 59:380-81.

Shriner, D.S. 1974. Effects of simulated rain acidified with sulfuric acid on host-parasite interactions. Ph.D. Thesis. North Carolina State University, Raleigh.

Shriner, D.S. 1978. Effects of simulated acidified rain on host-parasite interactions in plant diseases. Phytopathology 68:213-218.

Skilling, D.D. 1977. The development of a more virulent strain of Scleroderris lagerbergii in New York State. Eur. J. For. Pathol. 7:297-302.

Skye, E. 1968. Lichens and air pollution. ACTA Phytogeogr. Suec. 52:1-123.

Staxang, B. 1969. Acidification of bark of some deciduous trees. OIKOS 20: 224-230.

Stephan, Von B.R. and F. Scholz. 1979. Weitere untersuchungen sur unterschiedlichen antalligkeit on Pinus nigra Kloren gegenuber Scleroderris lagerbergii. Eur. J. For. Pathol. 9:46-51.

Van der Plank, J.E. 1963. Plant diseases: epidemics and control. Academic Press, Inc., New York. 349 pp.

Wilde, S.A., R.B. Corey, J.G. Iyer and G.K. Voigt. 1979. Soil and plant analysis for tree culture. Oxford and IBH Publishing Company, New Delhi.

FIVE-YEAR SURVEY OF SCLERODERRIS CANKER IN YOUNG RED PINE PLANTATIONS

Ronald P. Carvin and Paul D. Manion[1]

ABSTRACT

Thirteen permanent plots containing 5,245 red pine trees (1520 years) were surveyed each spring from 1978 to 1982 for Scleroderris canker. During this time, four of the plots had negative disease increase rates (-.04 to -.48; based upon percent trees infected) and three plots had negative rates (-.21 to -.39; based upon percent branches infected). Associated vegetational data were also collected and correlated with disease intensification. Positive correlations were found between amount of *Spiraea* cover and percent trees infected ($R = .65$ to $.78$; $P < .01$) and percent branches infected ($R = .41$ to $.74$; $P \leq .08$). Low correlations were found between disease rates and other associated vegetation. These data show the intensification of *Gremmeniella abietina* in these plantations has markedly decreased and some plantations are showing signs of recovery.

INTRODUCTION

Scleroderris canker, caused by the fungus *Gremmeniella abietina* (Lagerb.) Morelet (=*Scleroderris lagerbergii* Gremmen), is considered to be a serious and potentially the most dangerous conifer tree disease in North America (Skilling, 1972). The fungus has been demonstrated to infect and kill every species of pine tested (Skilling, 1980) including red pine (*Pinus resinosa*) which is the most susceptible. Spruce, fir, larch, hemlock and Douglas-fir are affected but damage on these species is minor.

In New York, Scleroderris canker has been present since 1959 (French and Silverborg, 1967) typically affecting young trees or the lower branches of older ones. In 1975, Scleroderris was reported to be present in the upper crowns of red and Scots pines (*Pinus sylvestris*) 45-60 feet tall (Setliff et al., 1975) and was causing widespread losses in northern New York State. Skilling (1977) characterized a new and more virulent strain of Scleroderris which is similar to the European strain in culture and immunogenic reaction (Dorworth and Krywienczyk, 1975; Skilling et al., 1979).

---

[1] Graduate Student and Professor, State University of New York, College of Environmental Science and Forestry, Syracuse, N.Y. 13210.

This study was supported in part by the USDA Forest Service North Central Experiment Station.

The purpose of this study was to document Scleroderris intensification in permanent plots established in young red pine plantations in New York. Also to evaluate the possible interrelationship between disease intensification and environmental factors.

METHODS

Permanent Plots

During the spring and early summer of 1978, red pine plantations, within the range of Scleroderris canker in New York, were visited and preliminary surveys were performed to determine the extent of *Gremmniella* infection. Thirteen permanent plots in eight plantations were selected for intensive study of disease spread (Carvin 1980). Plantations were similar in that they consisted of young red pine trees (between 15-20 years old, 3.66-5.88 m tall), and they contained at least one source of *Gremmeniella* inoculum within each plot. Young red pines were selected, to facilitate data collection within the upper crown. Plots were also established on the basis of amount of initial infection (between 5 and 50%, amount of initial mortality (less than 1%), topographic features, and dominant vegetation type other than red pine. All plots contained 273 possible trees with the exceptions of plots 1, 7 and 10 (1444, 483 and 588 trees respectively) for a total of 5,245 trees.

Data Collection

Data were collected during late May to early June, for five consecutive seasons from 1978 to 1982. The condition of each tree within the plots was recorded as being either missing (presence of an opening in the row), healthy (no evidence of *Gremmeniella* infection) or infected. A tree was considered infected if at least one twig showed signs or symptoms of *Gremmeniella* including; bud resinosus and death, red-brown discoloration of needle bases, needles easily removed from twigs, yellow green cambial discoloration, and/or the presence of pycnidia on twigs, needles or buds (Skilling et al., 1979; Skilling and O'Brien, 1979). Additional data were collected for trees that were infected including, height of the highest infected branch, year of the youngest whorl infected (that year in which the whorl was produced), number of infected branches, and the total number of branches up to and including the youngest infected whorl. Trees that were initially classified as infected were reclassified as healthy if they went for two consecutive years without showing signs of further infection or progression of the fungus. Disease increase rates (r) using van der Plank's compound interest formula (1963) were calculated based upon percent trees infected and percent branches infected.

Climatological data was obtained from the U.S. Department of Commerce[2] for the major infection periods (May to September) from 1971 to 1982. Data was representative of the weather patterns for the range of Scleroderris canker in New York. Average temperature precipitation and number of degree days was recorded at the Boonville, Lowville, Constable, and Hinckley weather stations. Weather data were correlated with disease increase rates.

---

2 U.S. Department of Commerce, National Oceanic and Atmospheric Administration, Environmental Data Service.

Stand characteristics data including site-index, density of red pine and associated vegetation were also collected during the survey period. Site-index was determined by the 5-year growth intercept method (Day et al., 1960). Vegetational data were collected using the line intercept method. Two one-hundred foot measuring tapes were extended between random selected rows of red pine. Herb species, shrub species, shrub height and width to red pine branches were recorded at every two foot interval. Percents of herbs and shrubs were calculated for each plot and correlated with disease increase rates.

RESULTS

Table 1 and 2 show the calculated disease increase rates based upon percent trees (r-trees) and percent branches (r-branches) infected.

Table 1. Scleroderris disease increase rates from 1978 to 1982 in thirteen permanent plots based upon percent trees infected.

Plot No.	Town in New York	1978-79	Disease Increase Rate (r-trees) 1979-80	1980-81	1981-82	1978-82
1	Westville	0.23	0.06	-0.87	-1.34	-0.48
2	Constable	0.00	0.00	0.41	0.00	0.10
3	Constable	0.06	0.10	-0.96	-0.38	-0.30
4	Russia	0.68	0.14	-0.19	0.24	0.22
5	Russia	1.36	0.16	-0.89	0.29	0.23
6	Boonville	0.38	0.18	-1.38	-0.54	-0.33
7	Boonville	0.55	0.15	-0.68	0.27	0.07
8	Boonville	0.80	0.08	0.00	0.05	0.35
9	Diana	0.33	0.05	-0.48	0.50	0.10
10	Pitcairn	0.19	0.13	0.73	0.27	0.33
11	Watson	0.48	0.13	0.52	0.62	0.44
12	Watson	0.40	0.01	-0.04	0.42	0.20
13	Watson	0.41	0.04	-0.77	-0.05	-0.09

Table 2. Scleroderris disease increase rates from 1978 to 1982 based upon percent branches infected.

Plot No.	1978-79	Disease Increase Rate (r-branches) 1979-80	1980-81	1981-82	1978-82
1	0.53	0.17	-0.90	-1.23	-0.36
2	0.00	0.00	0.79	0.28	0.27
3	0.22	0.22	-0.88	-0.38	-0.21
4	1.07	0.12	0.19	0.54	0.48
5	1.86	0.16	-0.58	0.29	0.43
6	0.93	0.12	-1.14	-1.47	-0.39
7	0.90	0.20	-0.60	0.64	0.29
8	1.10	0.08	-0.19	0.61	0.40
9	0.62	0.09	-0.34	0.66	0.26
10	0.35	0.20	-0.70	0.65	0.12
11	0.15	0.17	0.67	0.99	0.50
12	0.58	0.16	0.48	0.45	0.42
13	0.98	0.11	-0.24	0.11	0.24

From 1978 to 1979, r-trees ranged from 0.00 (plot 2) to 1.36 (plot 5). Rates for branches ranged from 0.00 to 1.86 again with plot 5 at the upper end.

From 1979 to 1980, most of the plots showed significant decreases, in both r-trees and r-branches, from the previous period (range of r-trees = 0.00 to 0.18; r-branches = 0.00 to 0.22). From 1980 to 1981, nine plots were calculated to have negative disease rates based upon percent trees infected or percent branches infected. This is because previously infected trees were reclassified as being healthy during this period. Some of the plots, however, did show an increase in their rates from the previous period.

From 1981 to 1982 a reverse trend is observed where a majority of the plots show higher rates for both r-trees and r-branches, than in the previous 1980-81 period.

Generally the rates based upon percent branches infected were slightly higher than rates based upon percent trees infected for all periods with the possible exception of 1980-81.

Disease increase rates were also calculated for the entire survey period from 1978 to 1982 to look at long term trends. Four of the plots have negative rates based upon percent trees infected and three have negative rates based upon percent branches infected, which indicates a recovery may be occurring in these plantations.

Correlations

Low correlations existed between disease increase rates and climatological data. However, relationships between disease intensification and overall weather patterns is still being investigated and will be reported in a subsequent paper.

Low correlations were also found between disease rates and site index and density of red pine stocking.

The three major herb species recorded at each line intercept point included "no herbs" (bare ground was observed), members of the Graminae family and a moss (*Polytricum*) or moss/lichen (*Polytricum*/*Cladonia*) combination. The moss/lichen combination was most prevalent in six of the plots with a range of 24 to 75% incidence. Low correlations exist between herb incidence and disease increase rates.

Shrub species were limited almost exclusively to *Spiraea latifolia* with minor occurrences of *Prunus* sp., *Cornus* sp., and *Juniperis communis*. High positive correlations were found between amount of *Spiraea* cover and percent trees infected ($R = .65$ to $.78$; $P < .01$) and percent branches infected ($R = .41$ to $.74$; $P \leq .08$). Low correlations were found between other shrub species and disease rates.

DISCUSSION

During the first two years encompassed by this survey, Sclerroderis canker incidence increased as reflected by the positive disease rates, both for r-trees and r-branches. However there was a dramatic slowing of the disease during the 1979-80 period. The average r-trees for all thirteen plots for 1978-

79 was 0.45 which dropped to 0.09 for 1979-80. Average rates for r-branches dropped from 0.72 to 0.14 for the same time periods. In essence, the amount of disease, with respect to the numbers of trees and/or branches infected, increased from 1978 to 1980, but at a progressively slower rate.

During the 1980-81 period a majority of the plots were showing signs of recovery as reflected by the negative disease increase rates (average r-trees = -0.24; average r-branches = -0.26). Negative rates were calculated because the amount of infected material from one year to the next decreased. This was due largely because of the method of evaluating tree condition. If a tree that was initially classifed as infected, went for two consecutive years without showing progressive signs or symptoms of Scleroderris, then it was reclassifed as healthy. In other words a tree classified as infected in 1978 may have been reclassified as healthy in 1980. The reasoning here is that if the initial survey were conducted in 1980 instead of 1978, many trees would not show signs or symptoms specific for Gremmeniella infection. Therefore they would be classed as healthy.

During the 1981-82 period a majority of the plots show positive disease rates again (average r-trees = 0.06; ave. r-branches = 0.16). However, plots 1, 3 and 6 are still showing signs of recovery.

These fluctuations in disease rates suggests that region wide as well as local environmental conditions have a significant impact upon Scleroderris disease incidence. Donaubauer(1972) speculates that alternating periods of high and low Scleroderris attack may be attributed to environmental factors that may predispose the host or stimulate the pathogen. Dorworth (1972) also emphasizes the effect of terrain in intensifying inoculum into "fog pockets" where cool moist conditions would be conducive for spore germination and infection.

During the course of this study several environmental characteristics were looked at in relation to disease incidence and seem to have little effect. In addition, site-index, density, weather patterns and most associated vegetation had low correlations with rates of disease increase. The only signficant correlations were between percent Spiraea cover and percent infection. It was also seen that increase rates based upon percent branches infected were generally slightly higher than rates based upon percent trees infected. This suggests that new infections are occurring more on trees already infected than on previously healthy ones. The spread of the disease is occurring in localized foci within the plantation possibly because of favorable conditions created by the Spiraea. However, Spiraea cover alone probably has little effect on disease intensification. Other factors, for example temperature, moisture, soils, etc., in conjunction with each other, are more appropriately responsible for Scleroderris intensification, and need to be investigated.

CONCLUSIONS

Intensification of Scleroderris canker in young red pine plantations in New York seems to be progressively slowing down. Indications suggest that some plantations may even be recovering from infections. The data also suggests that environmental factors may have a significant effect on the rate of disease incidence both on a local and regional basis. Although the disease seems to be moving at a slow rate currently, there is always the potential for a return to the serious and rapid intensification of the mid 1970's.

LITERATURE CITED

Carvin, R.P. 1980. Factors affecting intensification of Scleroderris canker in young red pine plantations. M.S. Thesis, SUNY College of Environmental Science and Forestry, Syracuse, N.Y. 13210. 43 pp.

Day, M.W., C.F. Bey and V.J. Randolph. 1960. Site index for planted red pine by the 5-year growth intercept method. J. For. 58:198-202.

Donabauer, E. 1972. Environmental factors influencing outbreaks of *Scleroderris lagerbergii* Gremmen. Eur. J. For. Pathol. 2:21-25.

Dorworth, C.E. 1972. Epidemiology of *Scleroderris lagerbergii* in Central Ontario, Can. J. Bot. 50:751-765.

Dorworth C.E. and J. Krywienczyk. 1975. Comparisons among isolates of *Gremmeniella abietina* by means of growth rate, conidia measurement, and immunogenic reaction. Can. J. Bot. 53:2506-2525.

French, W.J. and J.B. Silverborg. 1977. Scleroderris canker of red pine in New York State plantations. Plant Dis. Rep. 51:109.

Setliff, E.C., J.A. Sullivan and J.H. Thompson. 1975. *Scleroderris lagerbergii* in large red and Scots pine trees in New York. Plant Dis. Rep. 59: 380-381.

Skilling, D.D. 1972. Epidemiology of *Scleroderris lagerbergii*. Eur. J. For. Pathol. 2:16-21.

Skilling, D.D. 1977. The development of a more virulent strain of *Scleroderris lagerbergii* in New York State. Eur. J. For. Pathol. 7:297-302.

Skilling, D.D. 1980. Scleroderris Canker Study Progress Report No. 4. Unpublished. USDA Forest Service, St. Paul, MN.

Skilling, D.D., and J.T. O'Brien. 1979. How To identify Scleroderris canker. USDA Forest Service, North Central Forest Experiment Station Pamphlet.

Skilling, D.D., J.T. O'Brien and J.A. Beil. 1979. Scleroderris canker of northern conifers. USDA Forest Service, Forest Insect and Disease Leaflet #130. 7 p.

Van Der Plank, J.E. 1963. Plant disease: Epidemics and control. Academic Press, New York. 349 p.

FACTORS AFFECTING THE DEVELOPMENT OF DISEASE EPIDEMICS BY *Gremmeniella abietina*

Timo Kurkela[1]

ABSTRACT

Serious *Gremmeniella* epidemics can develop only when a considerable amount of host stands are potentially susceptible. Under normal conditions the host (pine) is resistant and the pathogen survives in damaged trees. During cold vegetative periods resistance may be broken or fail to develop. Pines introduced from different climatic conditions or planted on unfavorable sites are most likely to become susceptible. In that sense a Gremmeniella epidemic in pine stands can be a man-made problem. With increased inoculum potential the pathogen is also able to attack more resistant local pines.

INTRODUCTION

The potential susceptibility of the host to an attack of *Gremmeniella abietina* is a primary requirement of the development of serious epidemics. *G. abietina* can attack many coniferous hosts, e.g. pine, spruce, fir, Douglas fir, and larch. The main hosts have been, however, certain pine species, and for that reason I restrict this discussion to pines. The susceptibility of pines has both a genetic and an environmental basis and may develop only under certain circumstances. Pine tissue seems to be susceptible only when it is dormant or physiologically distrubed during the growing season. An epidemic may develop when the susceptible period of the host coincides with the spore dispersal of the pathogen. Local existence of the pathogen is also required since a rapid long distance dispersal of this kind of ascomycete can hardly occur.

In this paper I discuss the factors and circumstances which seem to be crucial for the onset of epidemics, and also the role of the pathogen in the progress of the disease.

THE SUSCEPTIBILITY OF HOSTS

Gremmeniella abietina can attack several pine species. Donaubauer (1972a) mentions 18 species on which the fungus has been observed. Damage on an economically significant scale has occurred in at least *Pinus nigra* and *P. sylvestris* in Europe and *P. resinosa* and *P. banksiana* in North America. These four pines, excluding *P. nigra*, have a wide geographical distribution which means that they have adapted to a wide variation of climatic and site factors. The

---

[1] The Finnish Forest Research Institute, Unioninkatu 40 A, SF-00170 Helsinki 17, Finland.

adaptation in different provenances, however, is so much more restricted that one should carefully consider which provenances can be used in addition to local ones. The situation seems to be most difficult with P. nigra in Central Europe because there it is outside of its natural range.

In the northern forests, productivity, wood quality, and frost hardiness are the main criteria used in choice of tree species for practical forestry. Diseases and pests also restrict the growth of trees and may even destroy stands, but these appearing occasionally are often omitted in the consideration of species.

Provenances moved from different climate conditions are the most likely to be potentially susceptible to G. abietina (see Dietrichson 1968), since many climatic factors are correlated with susceptibility or disease incidence. Read (1966, 1968) observed that pines growing with less light or artifically shadowed were more susceptible. In his studies, the effect of low temperature was not clearly evident in the development of the disease (Read 1967). Some Finnish experiences with pines introduced from the south to the north, have shown that low temperature combined with cloudy weather during the growing season can inhibit the development of resistance, especially in certain introduced provenances. Gremmeniella disease was destructive in the pine reforestations in northern Finland established 8 to 15 years earlier with provenances from Central Finland. The southern provenances were used because of the lack of local pine seed and the good experiences during the earlier, warm decades. Almost the same process was repeated in the 1950s and 1960s in southern Finland. It was thought that pine provenances could be moved within an area without risk. Some southern seed lots from the Soviet Union and Poland were also used to fill the occasional shortage. The growing seasons from 1975 to 1981 were cold in Southern Finland; and many of the pine stands with a foreign seed origin began to deteriorate, often regardless of topography or soil fertility. Native pine was more resistant, and the stands which were of local origin were at first only damaged at unfavorable sites, e.g. narrow river basins, frost pockets, and drained peat lands. Also in southern Sweden G. abietina has damaged German provenances (Kohn 1964).

Topographic depressions have been shown to be sites where the risk to Gremmeniella infection is greatest (Donaubauer 1972b, Dorworth 1973, 1978). During cold growing seasons, climatic variation is most unfavorable in such depressions. Soil fertility, especially in the depressions, may also be important in the developing of susceptibility. Scots pine is modest in its requirements for soil fertility. When pine is growing in conditions of unbalanced mineral nutrition, it may also be potentially susceptible. Inoculation of naturally regenerated pines growing in sites with different fertility resulted in different lesion extention during the winter (Kurkela and Norokoppi 1979). In northern Finland G. abietina has been most destructive on pines growing on fine textured fertile soils (Lahde 1974). A high level of nutrients and the transfer of seed to the north may act in the same way; vegetative activity is increased and development of resistance (and frost hardiness) is disturbed. In clearcut areas, small lots of higher fertility are disregarded in reforestation; and pine is also preferred in depressions because of problems with frost in young spruces.

When considering the susceptibility in other pines, P. nigra, P. resinosa and perhaps P. banksiana, we easily come to the same conclusion as with Scots pine. Pine stands are often established in areas where the provenance used is possibly adapted to the average climate but not to the whole scale of variation of different climatic factors.

THE ROLE OF THE PATHOGEN

Since abiotic factors cause predisposition in the host, it might be possible to reach a limit wherein the pathogen is no longer necessary for the appearance of damage. This kind of direct frost damage was emphasized by Eiche (1966) in Sweden and by Pomerleau (1971) in Canada. Tissue necrosis in stem and branches, caused by frost and ice formation or by snow bending during the winter, may serve as entrances for the pathogen (Roll-Hansen and Roll-Hansen 1973, Kurkela 1981). We also know that visible damage is not always necessary for successful infection, and frost damage does not always alter the susceptibility of pines to *G. abietina* (Dorworth 1977). It seems that in extreme climatic conditions, *physiogenic* diseases can be as important as *G. abietina* in killing pines (see Roll-Hansen and Roll-Hansen 1973), but in milder climates the role of the fungus becomes more prominent.

The long distance dispersal of the pathogen occurs by airborne sexual ascospores. The association of cold growing periods and the appearance of the disease is often difficult to explain, because the pathogen has two years of sexual development, and the spore yields are not produced regularly at the same time every year. In the summer of 1982 *G. abietina* did not produce ascospores at all in southern Finland; but now, one year later, an abundant crop of apothecia are developing. Development of pycnidia was also delayed in 1982; they did not mature until the winter or spring of 1983.

The amount of inoculum is important for successful infection even in susceptible pines, see e.g. Dorworth (1979) and Hudler et al. (1983). The resistance of pines restricts the increase of the population of *G. abietina*. The pathogen can survive under "normal" conditions only in damaged trees, or in older stands in "naturally" dying suppressed trees or lower branches. In extreme climates, physiogenic injuries occur more frequently, which means there are more possibilities for the pathogen to survive over times when the pine is not susceptible. Where the "normal" population of the pathogen is high, no long distance dispersal is required, and milder stress is enough for the beginning of an epidemic when pine resistance is decreased. Once the pathogen has entered a potentially susceptible stand, it seldom disappears. The fungus has survived in the wood of eight-year old, completely occluded wounds (Kurkela 1981). Because the most susceptible stands of southern provenances may be repeatedly damaged by snow and frost, the population of *G. abietina* there, can constantly remain at a higher level than in the local pine stands in the surrounding region. During the last epidemics in Finland, potentially susceptible artificial pine stands have served as primary infection sites for *G. abietina* at the onset of epidemics; and the same stands have further been excellent inoculum sources for the spreading of the disease to the stands of local pine.

LITERATURE CITED

Dietrichson, J. 1968. Provenance and resistance to *Scleroderris lagerbergii* Gremmen. (*Crumenula abietina* Lagerb.). The international Scots pine provenance experiment of 1938 at Matrand. Medd. Nor. ske Skogforsoeksves 25:395-410.

Donaubauer, E. 1972a. Distribution and hosts of Scleroderris lagerbergii in Europe and North America. Eur. J. For. Pathol. 2:6-11.

Donaubuaer, E. 1972b. Environmental factors influencing outbreak of Scleroderris lagerbergii Gremmen. Eur. J. For. Pathol. 2:21-25.

Dorworth, C.E. 1973. Epiphytology of Scleroderris lagerbergii in a kettle frost pocket. Eur. J. For. Pathol. 3:232-242.

Dorworth, C.E. 1977. Relative susceptibility of red and jack pine to Gremmeniella abietina. Bi-Mon. Res. Notes, Can. For. Serv. Res. Notes 33:6.

Dorworth, C.E. 1978. Presence of a late pleistocene drainage system manifested through depredation of red pine by Gremmeniella abietina. Ecology 59:645-648.

Dorworth, C.E. 1979. Influence of inoculum concentration on infection of red pine seedlings by Gremmeniella abietina. Phytopathology 69:298-300.

Eiche, V. 1966. Cold damage and plant mortality in experimental provenance plantations in northern Sweden. Stud. For. Suec. 36:2-219.

Hudler, G.W., G.R. Knudsen, and M.A. Beale. 1983. Dose-response relationships of five conifers to infection by conidia of Gremmeniella abietina. Plant Dis. 67:192-194.

Kohn, E. 1964. Om tallens gren- och granens topptorka och dess bekämpning. Skogen 51:200-203.

Kurkela, T. 1981. Versosyöpä (Gremmeniella abietina) riukuasteen männiköissä. Folia For. 485:1-12.

Kurkela and Y. Norokorpi. 1979. Pathogenicity of Scleroderris lagerbergii, Lachnellula pini and L. flavovirens and their cankers on Scots pine. Commun. Inst. For. Fenn. 97, 1:1-16.

Lahde, E. 1974. The effect of grain size distribution on the condition of natural artificial sapling stands of Scots pine. Commun. Inst. For. Fenn. 84, 3:1-23.

Pomerleau, R. 1971. Considerations on the cause of conifer damage in plantations attributed to the Scleroderris canker. Eur. J. For. Pathol. 1:114-122.

Read, D.J. 1966. Dieback disease of pines with special reference to Corsican pine, Pinus nigra var. calabrica Schn. I. The nature of the disease symptoms and their development in relation to the crown and aspect. Forestry 39:151-161.

Read, D.J. 1967. Dieback disease of pines with special reference to Corsican pine Pinus nigra var. Calabrica Schn. II. The relationship between frost resistance, microclimate and disease. Forestry 40:83-97.

Read, D.J. 1968. Some aspects of the relationship between shade and fungal pathogenicity in an epidemic disease of pines. New Phytol. 67:39-48.

Roll-Hansen, F. and H. Roll-Hansen. 1973. Scleroderris lagerbergii in Norway. Hosts, distribution, perfect and imperfect state, and mode of attack. Medd. Nor. Skogforsoeksves. 30:441-459.

THE MOVEMENT OF *Gremmeniella abietina* FROM THE ALPS TO THE APPENINES

Paolo Capretti[1]

ABSTRACT

Brunchorstia pinea occurs in Italy from the Alps to the Appenines. On the Alps the natural vegetation is composed mostly of conifers. Here *B. pinea* seems acclimatized and damage is moderate. On the Appenines conifers (particularly *Pinus nigra*) were introduced in plantations. Damage is heavy but concentrated in small areas, and is strongly related to microclimate and to variable morphology of the territory.

INTRODUCTION

Italy is a peninsula oriented along the meridians with the Alps representing the northern limit. Most of the peninsula consists of the Appenines mountains. The Italian territory is characterized by 80% mountains and hills and 20% plains and coastal areas.

Attacks by *Gremmeniella abietina* have been observed in Italy since the sixties by Moriondo (1963), both on the Alps and Appenines, on different host species with different intensities of damage. This pathogen has also been described by Donaubauer (1976) in the Austrian Alps and by Ettinger (1945) in Switzerland.

DAMAGE OBSERVED ON THE ALPS

Climate and Silvicultural Management Considerations

The southern slope of the Alps (the Italian side) is a continuous mountain chain going from West to East that links France to Yugoslavia. The mean altitude of the peaks is about 3500 m. The precipitation is mostly concentrated in the growing season as in the Northern countries of Austria and Switzerland. The snow covering lasts for months.

The forest cover on the Italian slopes is maintained mainly by natural generation, sometimes integrated by plantations but only on a few hectars. In Switzerland and Austria, the governments supported several plantation programs at high altitudes mainly to prevent erosion and avalanches, while on the Italian slopes of the Alps these problems are less serious.

---

[1] Ricercatore, Istituto di Patologia e Zoologia Forestale e Agraria, Piazzale delle Cascine, 28 - 50144 Firenze, Italy.

Species cultivated. The native conifers in this area are Norway spruce (35%), European larch (25%), Scots pine (10%), Pinus cembra (7%) and Austrian pine (4%). The latter mentioned species grows on the lower parts of the mountains.

Norway spruce and European larch are cultivated for timber and the others are cultivated mainly to protect the mountains against erosion. Norway spruce and Larch spread all over the Alps. Pinus cembra grows mostly in the eastern area, generally from 1400 m to 2300 m. Scots pine is almost monophyletic in the west and grows mixed with larch or Norway spruce in the east from 1000 up to 1800 m. Austrian pine is found mainly in the areas along the border between Italy and Austria and Italy and Yugoslavia at 400-1500 m.

Damage by Gremmeniella abietina. Observations over several years show that G. abietina occurs from the east to the west of the Alps but the fungus has been studied mostly in the Dolomite area.

Damage consists of terminal shoot death and sometimes death of the entire tree. On Pinus cembra damage effects mostly the natural generation and the lower parts of the crown of adult trees. On Scots pine we can also observe damage in the upper part of the crown, especially when the trees grow outside their altitude limits. Occasionally Norway spruce natural generation suffers from damage on terminal shoots. Austrian pine planted trees were attacked in some gardens close to an area where B. pinea was widespread. Damage on this species was found also in certain valleys where the natural generation was too dense.

Damage was more evident in the bottom of valleys and along the slopes, varying each year depending on the weather changes.

The fungus strain prevalent in the above mentioned alpine areas is the one that has a conidial form mainly with 3-4 septa as described by Morelet (1980) as Brunchorstia pinea var. tipica. In some parts of the Alps the conidial form with 1-7 septa corresponding to Morelet's Brunchorstia pinea var. cembrae was also found.

Observations carried out in natural forests of the Dolomite area on Plose mountain 2300 m showed that the conidial form B. pinea var. cembrae occurred mainly on the top of the mountain and on its northern slopes, both on Norway spruce and on Pinus cembra, and at a lower altitude on Scots pine. On the southern slope only the three septated conidial forms were present, but none of the forms showed differences of behavior with regard to pathogenicity. Both the conidial forms are present on the natural generation and on the lower parts of the crowns of adult trees.

On the Italian slope of the Alps the perfect state was never found.

These observations suggest that B. pinea is widespread in the alpine region. The fact that it does not seem particularly dangerous on the southern slopes of the alps could be related to the silvicultural management based mainly on the natural generation.

This fungus can be considered an indigenous Alpine parasite or a naturalized parasite, in fact it is balanced or tolerated by natural generation.

DAMAGE OBSERVED ON THE APPENINES

Geography, Climate and Silvicultural Considerations. The Appenines form a long chain of mountains that link the Alps to the southern end of the Italian peninsula. The mean altitude of the highest peaks is about 1700 m. The morphology of the territory is not uniform, the valleys having dissimilar directions and different widths.

The Appenines are characterized by quite different climates because of their north - south orientation. For this reason, considering also the strong Mediterranean influence, there is a wide range of microclimates depending on the specific slope and altitude. In this area the climate is quite variable even in the same place from one year to another.

The precipitation rate is maximum in autumn or at the beginning of spring. The minimum begins in the late spring and often extends to the end of the growing season. This is a characteristic of all Mediterranean climates. The variability of the climatic conditions exposes the trees to damage by frost during the spring or by drought in the summer.

The snow covering is present for 3 - 4 months on the tops of the mountains but in the forests is present for only a few weeks, and in some years is completely absent even at 2000 m.

Species Cultivated. In this area the original genera of trees are mostly broadleaves (beech, chestnut, oaks). The silviculture practice is based mostly on coppices.

The original conifers are very few and are confined to a few areas, as relics of glaciation. Although the broadleaf forests represent the predominant vegetation, they have no marketable product. For this reason they are gradually being replaced by conifer plantations. Austrian pine, Silver fir and Douglas fir are the trees most planted, while the least planted ones are Scots pine in the north and *P. halepensis* in the south.

Damage by *Gremmeniella abietina*. The presence of *B. pinea* on the Appenines has been observed since the sixties. Attacks were always present in plantations. In many cases *B. pinea* damages appeared on single trees or in small groups of plants.

Heavy damage has been observed in the northern Appenines on Austrian pine, Scots pine and Norway spruce. In the central Appenines they have been observed on Aleppo pine and Austrian pine; along the Adriatic coast, between the Alps and the Appenines, on *Pinus pinea* and *P. pinaster*.

In the first case, on the Northern Appenines there was damage in a large 40 year old Austrian pine plantation where the trees were killed after attacks causing terminal shoot blight. The damage appeared first on the lower parts of the crowns and then moved upward.

Less heavy damage was found on Scots pine and Norway spruce, attacked only in a frost pocket plantation. The damage was associated with frost rings on the main stems. The Austrian pine outbreak appeared to be associated with plantation density and the climatic conditions.

The areas most severely affected were often exposed to spring frost damage and crossed by wet winds during the growing season. The pathogen was not present in some plantations close to the damaged ones. These disease free plantations were situated on different (mostly southern) exposures, which shows the microclimate differences.

A second severe attack problem was observed in *Pinus halepensis* 20-year-old-plantations. This mediterranean species requires warm sunny sites. Damage began in the lower parts of the crown and expanded to eventually kill the trees. Small cankers appeared on the stems and along the branches. Damage was more severe in small valleys where the pines suffered from frost damage and low light intensity.

A third important disease situation was observed along the Adriatic coast on *P. pinea* which grows in this site under suboptimum climatic conditions. The spreading of the pathogen and its damage was also associated with winter frost damage.

B. pinea attacks were evident everywhere in the Appenines where the conidial form with 1-3 septations were found. The presence of only the conidial producing strain indicates its great adaptability to the environment; in this case warmer and drier than the alpine one. Recently, the perfect state was found in the northern Appenines more than 20 years after the first observations of the damage.

These observations suggest that *B. pinea* spread to the Appenines from the Alps along the mountain chain covered with numerous artificial plantations made during the last 30 years.

The fungal strain present on the Appenines appears particularly dangerous since the weather conditions place the conifer plantations of the area under environmental stresses.

CONCLUSIONS

Climate and morphology of the territories are very different on the Alps and on the Appenines. Silvicultural and cultivated species are also different.

On the Alps the heaviest damage from *B. pinea* occurred in the high mountain plantations. On these sites the pathogen found suitable conditions for rapid spread because of the uniformity of the stands, and the favorable environmental conditions.

On the Italian slope of the Alps this pathogen was less damaging, perhaps because of the differences of climate and site, but mainly because large plantation areas do not exist at high altitudes. Therefore infection occurs on the natural generation and seldom causes severe damage.

This fungus found better conditions for development on the Appenines and in the Padana valley, where conifers are usually planted. Severe disease reported on Austrian pine, *P. halepensis*, *P. pinea* and Norway spruce reinforce this opinion. Here, as well as in the Alps, the fungus was particularly dangerous in homogeneous stands.

In the Appenines however, the attacks were severe only when the trees were debilitated by environmental stress. In this area the irregular morphology of the territory and its Mediterranean climate with dry springs limits the spreading of the fungus. The damage can be severe only where localized favorable conditions for fungus growth and stress conditions of the host species occur together.

LITERATURE CITED

Donaubauer, E. 1974. Über das Kieferntriebsterben in Österreich (Scleroderris lagerbergii) Gremmen und Cenangium ferruginosum Fr. In: 100 Jahre Forstlicke Bundesversuchsanstalt, Mitt Versuchsanst. Wien. 67-98.

Ettinger, L. 1945. Über die Gattung Cremenula sensu Rehm, mit besonderer Berucksichtigung der Cremenula - Triebsterbens der Pinus - Arten. Beitr. Kryptogamenflor Schweiz. 10, 1-75.

Morelet, M. 1980. La maladie a Brunchorstia 1. Eur. J. For. Pathol. 10:268-277.

Moriondo, F. 1963. Nuovi reperti Fitopathologici rei boschi italiani - Ann. Accad. Ital. Sci. For. 21:343-345.

EXPERIENCES WITH SCLERODERRIS CANKER ON *Pinus cembra* L. IN AFFORESTATIONS OF HIGH ALTITUDE

Edwin Donaubauer[1]

ABSTRACT

The government of Austria started an afforestation project of subalpine sites about 30 years ago as part of an avalanche control program. About 15 years later, the first epidemics of Scleroderris canker were observed. Hundreds of hectars have been killed by disease. Epidemics built up very slowly at first but the process has accelerated in the past 8-10 years. Correlations have been observed between severity of disease attack and altitude of provenance as well as to altitude of plantation site. No correlations were found with snow-cover or forest type. Differences in the formation of perennial cankers, morphology of fungus and frequences of occurrence of the ascofructification were observed as compared with the Scleroderris disease in lower altitudes and on other host species.

INTRODUCTION

The central Alps in Austria were covered by forests up to an altitude of about 2200-2300 m centuries ago. During the past two to three centuries the upper tree-line has moved down because of extensive clear-cuts in the course of mining activities and to obtain grazing grounds for sheep and cattle. (Fromme 1957.) Serious accidents by avalanches occurred about 30 years ago and pilot studies showed that two/thirds of all avalanches originate from the former forested areas. Therefore, the government started a project for controlling the avalanches by technical means as well as by afforestation of formerly forested sub-alpine sites. The afforestation program was started immediately and the species *Pinus cembra* L. and *Larix europaea* L. were used. The first set-backs were caused by an epidemic occurrence of *Phacidium infestans* Karst., a snow mold fungus (Donaubauer 1963).

Since 1967 infections and epidemic occurrences of Scleroderris canker have become numerous in such afforestations. Therefore a long-term research project was initiated.

METHODS

First attention was directed at morphological studies on the differences of Scleroderris on *P. cembra* and other hosts in lower altitudes. (Donaubauer 1974). Later interest concentrated on epidemiological questions. Observation plots were established in several existing plantations (different altitudes,

---

[1] Head, Institute of Forest Protection, Forstliche Bundesversuchsanstalt, A-1131 Vienna/Austria.

exposures, ages of plants). Great numbers of plants were marked and severity of attack by Scleroderris canker was checked annually since 1975.

Further, trials for chemical control were carried out. A young plantation (locality Hippach/Tirol, 2000 m altitude) was treated with aerial application of 5 kg Maneb per hectar in June 1978. An area of 500 hectars was treated (Aeroplan: Helicopter Typ Lama, Low Volume Application: 5 kgs per 20 liters of water/hectar). The treated plantation was observed during the following years.

Plants from defined seed sources (mother trees) were exposed to natural infection in existing plantations and have been under observation for several years.

RESULTS

Epidemiological Characteristics. The observation of specifically described individual plants has shown, that the development of disease has accelerated. The time between first infection of a plant and death became shorter and shorter. Initial epidemics in new plantations were first observed about 15 years after establishment and now they occur within about 5-6 years. It is unknown if this development is due to the increasing amount of inoculum or due to a change of virulence.

There was observed a strong correlation between plantation altitude and mortality. After many years of experience we may state, that afforestations above 2000 meters of altitude have no chance of survival. Epidemics are not known in altitudes of 1700-1800 meters and the chance of sufficient surviving plants is great between 1800 and 1900 meters of altitude (Table 1).

Table 1. Scleroderris canker mortality on planting sites of different altitudes in subalpine afforestations (examples).

Plot no.	Altitude m	Year 1982	1981	1980	1979	1978	1977
				(% mortality)			
1	2000-2300	97	83	55	28	0	0
2	2000-2200	92	60	26	0	0	0
3	1900-2100	73	60	33	0	0	0
4	1750-1850	22	11	11	11	0	0

The first plantations were established with plants originating from seed sources of a more southern latitude (northern Italy) and lower altitudes (between about 1400 m and 1600 m). The results imply that there is a narrower 'amplitude' of tolerance to site conditions. Higher altitude provenances are known to have a higher frost resistance, and better protected against transpiration losses. Flushing and shoot growth are linked with day-length (less with temperature). (Tranquillini 1979). Provenances from lower elevations have their shoot-growth during the short days of spring. These plants seem to have a longer period of cambium susceptibility. Sizes of stem cankers on such plants indicate such correlations. No stem cankers were observed on autochthonous plants of natural regeneration.

Ecological Aspects. The planting of inadequate provenances led us to think of a direct correlation between frost hardiness and susceptibility to the disease. Thousands of plants were checked each year. They have shown that early symptoms of attack an apparent in late summer or early in the autumn before frost damage could have occurred. Often symptoms can be seen from a distance. Discoloration at the basis of needle fascicles occurs and pycnidia fructification can be found in autumn on the basal part of these needles. From these experiences we do not suspect that Scleroderris canker is directly related to frost on *P. cembra*.

Further, no indications were found that exposure or snow cover (depth and duration) were important influences.

The perennial cankers (from natural as well as from aritificial infection) develop very slowly at first. They are confined to small areas, sometimes for more than a year, and are surrounded by wound periderm. The cankers expand millimeters and sometimes a centimeter in the axial direction, but much less in the lateral direction. Canker enlargement obviously takes place during the autumn and/or early spring. Each expansion-area is surrounded by callus formation. The older shoots are able to restrict the fungus during the growing season. Girdling takes place sometimes 3-4 years after infection.

The die-back of one-year old shoots occurs on trees of all ages. In young plants the leader is often infected in epidemic area, but there is no proven explanation for this phenomenon.

Phacidium infestans Karst. is a serious disease of *P. cembra* on sites with long lasting and deep snow cover. Usually the shoots are also infected by Scleroderris, and show fructification of the fungus in the spring after snow-mold attack. It seems that Scleroderris is present everywhere and is able to secondarily infect the shoot after snow-mold.

Air Pollution as a Stress Factor?

In coordination with another research project, attention was directed toward the question of air pollution as a stress factor, which predisposes the *P. cembra* to Scleroderris canker infection. The acidity and load of pollutants (Sulfur, Nitrogen, other heavy metals) were analyzed in snow and rain and the SO_2, NO_x and Ozone concentrations of air were studied at different elevations between the bottom of valleys and about 2000 m altitude. All results obtained up to the present time show no correlations. In general polluted air and precipitation are found up to a level of about 1200 m and therefore can not be considered as a contributing stress. The concentration of Ozone (O_3) is high in or near urban areas in the valleys, decrease with altitude, but increase again at 2000 m of altitude. The higher Ozone concentration may be natural due to the higher radiation at higher altitude. (Smidt 1982). The concentration found is probably not serious enough to cause damage, but further investigations will be necessary to determine if the concentration is sufficient to stress provenances not adapted to the site.

Control and Prevention

The application of the fungicide (MANEB) was not an effective control in spite of the very high concentration used. The large area sprayed showed no difference in mortality to unsprayed surrounding plantations. Since the slowly growing P. cembra seems to be endangered by Scleroderris during the first 20-30 years of age the repeated application of fungicides is considered too expensive. The total costs per hectare reached about 95 US $. Further, the application is unrealistic because of the necessary protection of the drinking water supply for the densely populated valleys.

Mechanical control was carried out on some larger trial areas to evaluate if the removal (burning) of infected plant material reduces the inoculum density and infection rate. Results can be expected after several years.

We expect to reduce losses by selection of proper provenances. Small scale trials with progenies of 33 mother trees from various locations (1600-1900 m of altitude) have shown wide variability in mortality. Two progenies were infected (14 and 21% of the plants) but no mortality occurred during the three years of observation, whereas others had mortality rates up to 41% (percentage of infected plants up to 57%). These preliminary results suggest further efforts be made in selection and breeding for resistance.

Due to the experiences of past decades, and until enough plants of more resistant provenances are available, no further plantations are being established in altitudes above 1900-2000 meters.

Remarks on Morphology and Fructification of the Fungus

As reported earlier, P. cembra is infected by the long-sporeform of the fungus (pycnidia spores with greater numbers of septa). (Donaubauer 1974). These long same spores are seldom seem on Pinus nigra and P. sylvestris and were never found on Picea abies in Austria. The formation of apothecia is abundant on perennial cankers and sometimes on older shoots of P. cembra. Apothecia were very seldom found on the other hosts, which grow in lower altitudes.

LITERATURE CITED

Donaubauer, E. 1963. Über die Schneeschütte-Krankheit (Phacidium infestans Karst.) der Zirbe (Pinus cembra L.) und einige Begleitpilze. Mitt. Forstl. Bundes-Versuchsanst. Mariabrunn, H. 60:575-600.

Donaubauer, E. 1974. Über das Kieferntriebsterben in Österreich (Scleroderris lagerbergii Gremmen und Cenangium ferruginosum Fr.). In: 100 Jahre Förstliche Bundesversuchsanstalt, Mitt Versuchsanst. Wien. 67-98.

Fromme, G. 1957. Der Waldrückgang im Oberinntal (Tirol). Mitt. Versuchsanst Forstl. Bundes-Vers.-Anst. Mariabrunn, H. 54:3-221.

Smidt, St. 1982. Untersuchungen über das Auftreten von Ozon, seine kombinierte Wirkung mit SO_2 auf Fichte und das Vorkommen von sauren Niederschiagen in Osterreich.

Tranquillini, W. 1979. Physiological Ecology of the Alpine Timberline. Ecol. Stud. Berlin, 31:137 S.

DISEASE INCREASE RATES FOR SCLERODERRIS CANKER IN LEWIS COUNTY, NEW YORK

Paul D. Manion and Bruce Schneider[1]

ABSTRACT

Annual inspection, since 1975, of four permanent red pine plots provided a data base for calculating disease increase rates. Initial establishment of infection was estimated by projecting rate values back to first branch or first tree infected. Initial infection occurred within five years of plantation establishment and disease progressed to 90% infection within four to seven years of disease onset. These data show the extremely explosive potential for the Scleroderris canker pathogen.

INTRODUCTION

The mortality of large red and Scots pine (*Pinus silvestris* L.) in New York in the mid 1970's was of sufficient magnitude to be of concern. It was also important to quantatively document the epidemic nature of the disease and the potential for spread of the pathogen over time. This paper quantifies the rapid disease intensification in Lewis County, New York.

MATERIALS AND METHODS

In 1975, 0.37 hectare permanent plots were established in four red pine plantations of Lewis County, New York. Dead branches and branches with dead growing tips were examined for the typical yellowish green discoloration in the cambium layer. Branches with the yellowish green discoloration were recorded as being infected with Scleroderris canker. The location of each tree was mapped and the position of each infected branch recorded. Re-examination of the trees each year provided a record of numbers and position of branches infected, trees infected, and trees dead. Microscopic examination of pycnidia on some of the infected branches verified *G. abietina*. The fungus was isolated from branch samples in each of the tree plantations and identified as European strain based on serological tests (serological tests were performed by the U.S. Forest Service North Central Forest Experiment Station). Rates of disease increase were calculated based on the percent of trees infected and number of branches infected, and percent mortality in each plot.

---

[1] Professor, State University of New York, College of Environmental Science and Forestry, Syracuse, NY 13210 and Principal Forestry Technician, New York State Department of Environmental Conservation, Lowville, NY 13367, respectively.

RESULTS

Disease intensification was dramatic in all four plantations. In 1975, 8.8 to 39.4% of the trees were infected. By 1978, 77.5 to 99.3% of the trees were infected (Table 1). The rates of disease increase, (Van der Plank 1963) values, were very high (Table 1).

Logarithmic and logit plots of numbers of branches infected and percent of infected trees respectively follow relatively straight lines (Figures 1 and 2).

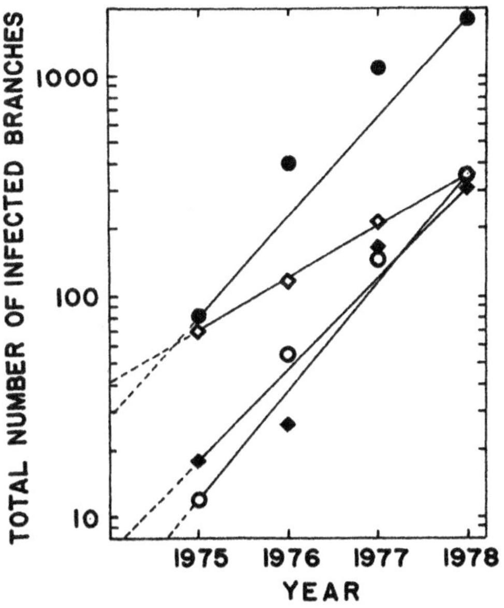

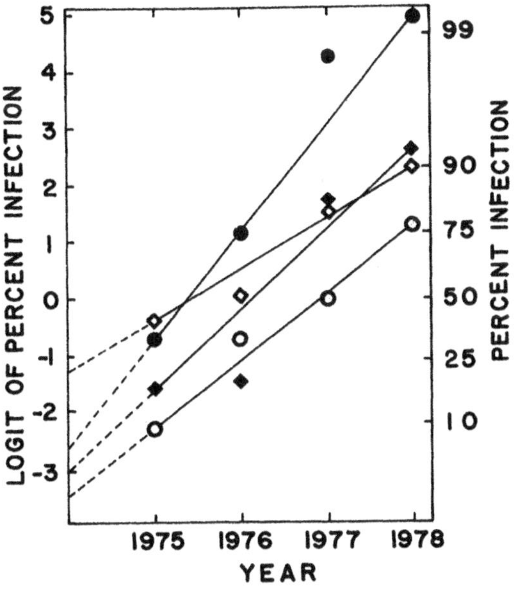

Figure 1. Scleroderris canker disease increase over time based on numbers of branches infected.

Figure 2. Scleroderris canker disease increase over time based on percent of the trees infected.

Year of first infections, as calcualted by projecting rates of increase back to one branch or one tree infected, were no more than three years apart in three of the four plots. Infection probably began in these plantations between 1969 and 1973.

Extensive mortality occurred in plantation two and four; however, mortality was not related to numbers of branches infected in previous years. Those trees that died had approximately the same number of branches infected as those that survived. Nine percent of the mortality occurred on trees with no previous branch infection. Eighteen percent of the mortality occurred with only a single branch infected in the previous year.

Table 1. Summary of Scleroderris canker intensification in four New York red pine plantations based upon percent disease, number of infected branches, and percent mortality between 1975 and 1978.

Plot	1	2	3	4
Date planted	1964	1971	1967	1967
Density (trees/ha)	583	1005	1317	2422
Average height (m)	4.6	1.1	3.3	4.1
Percent Diseased Trees				
1975 (%)	39.5	16.5	8.8	31.2
1978 (%)	90.0	93.4	77.5	99.3
Rate of increase (r_1)*	0.91	1.42	1.19	1.90
Year of first infected tree**	1971	1973	1973	1971
Number of Branches Infected				
1975 (#)	71	18	12	80
1978 (#)	359	311	351	1880
Rate of increase (r_1)*	0.54	0.95	1.13	1.05
Year of first infected branch**	1966	1972	1972	1969
Percent Motality				
1975 (%)	0	0	1.3	2.0
1978 (%)	0	45.0	2.5	38.7
Rate of increase (r_1)*	0	1.30		1.14

* Rate of increase (per unit per year) based on Van der Plank (1963).
** Estimated by projecting the disease progress curves back to one tree or one branch infected.

DISCUSSION

The rates of Scleroderris canker intensification in these plots are very high. They are comparable to the previously reported rates of increase for chestnut blight in Pennsylvania and West Virginia and Dutch elm disease in Illinois (Merrill 1967).

The epidemic nature of the disease is shown by the 90 percent distribution of the disease in Plot 1 in less that seven years from first infection. This rapid rate of disease increase would not allow much time for management options.

The slight variation in rates of increase from year to year do not reflect any consistent environmental relationships. Apparently, conditions for rapid intensification of Scleroderris canker prevailed over the four-year time period in this study.

Although the rates of increase vary somewhat, depending on whether percent infection, branches infected, or mortality were used, these do not seem to markedly affect the projection to first infection except in Plot 1. Rate of increase based on number of branches for Plot 1 indicate initial infection occurred in 1966. Plot 1 has a low density of susceptible plants interspersed with white pine and various hardwood species. If one assumes that these competitive plants affect the microclimate or availability of host tissue for local spread, then a different infection rate would be expected.

The year of first infections calculated by projecting back were relatively close together. They do not necessarily correspond in time with plantation establishment date. This suggests that, although some of these plantations might have been established with infected nursery stock (nursery records strongly suggest that they were not) that these plantations all become infected about the same time. Favorable weather conditions may have caused wide scale dissemination of a virulent form of the pathogen. This hypothesis is supported by the spatial analysis of disease trees over time by Carven (1980). He showed that up to about 10% infection was randomly distributed in the plantation. Increased infection incidence beyond 10% occurred in aggregated pockets.

Mortality caused by G. abietina, is generally assumed to occur when the host is seriously weakened through large numbers of branch infections, but mortality in these plantations was not related to numbers of branches infected in the previous year. Bark beetles may be a contributing factor in plantation four whereas smaller tree size may be a factor in plantation two. Examination of disease development in a larger number of planatations may be necessary to better interpret the factors affecting the overall impact of Scleroderris canker.

LITERATURE CITED

Carven, R.P. 1980. Factors affecting intensification of Scleroderris canker in young red pine plantations. M.S. Thesis, SUNY College of Environmental Science and Forestry, Syracuse, N.Y. 13210.

Merrill, W. 1967. Analyses of some epidemics of forest tree diseases. Phytopathology. 57:822.

Van der Plank. 1963. Plant diseases: Epidemics and control. Academic Press, Inc. N.Y.

PRUNING AS A SILVICULTURAL TOOL IN THE MANAGEMENT OF *Pinus resinosa* INFECTED WITH *Gremmeniella abietina*

D.R. Bergdahl and T.M. Ward[1]

ABSTRACT

Pruning all branches from the lower one-third of red pine (*Pinus resinosa*) crowns infected with *Gremmeniella abietina* (Lagerb.) Morelet resulted in reduced infection, lower inoculum densities and shorter periods of inoculum production as compared to unpruned controls. Red pine seedlings (3-years-old) planted in the understory had fewer infections than those in unpruned areas. Pruned branches did not have to be removed from the plantation and pruning wounds did not serve as infection courts.

INTRODUCTION

Gremmeniella abietina (Lagerb.) Morelet was first found in young plantations of red pine (*Pinus resinosa* Ait.) in the Lake States (USA) by Ohman (1966). Later Setliff *et al.* (1975) and Skilling (1977) reported a more virulent form of *G. abietina* associated with widespread mortality of older growth red and Scots pines (*P. sylvestris* L.) in New York. Dorworth et al. (1977) established that isolates of *G. abietina* from this area of New York were serogically similar to isolates from Northern Europe and different from isolates from the Lake States region (including Canada) of North America. On this basis the current European and North American race (strain) differences for *G. abietina* were formed.

The European race of *G. abietina* appears more virulent on red and Scots pines than the North American race. The European race is capable of attacking all sizes of trees, but the North American race only attacks the lower branches to a height of about 2 m (Skilling et al. 1979). Because of this difference in virulence the European race is considered a serious threat to not only red and Scots pines, but also many other species of North American conifers (Setliff et al. 1975, Skilling 1977).

---

[1] Associate Professor and Research Technician, Department of Forestry, Aiken Center, University of Vermont, Burlington, Vermont 05405.

This reasearch was supported in part by funds provided by the USDA Forest Service North Central Forest Experiment Station, RMU FS-NC-2205 and by the McIntire-Stennis Cooperative Forestry Research Program (USDA/CSRS and The University of Vermont).

Gremmeniella abietina was first reported causing lower branch mortality on red and Scots pines near Belvidere, Vermont by Tattar et al. (1973). This find was not considered significant until the European race of G. abietina was isolated from Scots pine near Greensboro, Vermont in 1976 (H.B. Teillon, Personal communication). Since 1976 an intermediate serotype (race) has also been reported by Wendler et al. (1979). To date, G. abietina has been isolated from about 100 plantations of red and/or Scots pines all of which are located in the ten northern-most counties of Vermont. The average size of plantations infested with G. abietina is less than 8 ha (D.R. Bergdahl, unpublished data).

Skilling and O'Brien (1969) have concluded that the infestation of red pine plantations in Michigan was the direct result of planting diseased nursery stock. The distribution of G. abietina in Vermont suggests a similar relationship in that a small nursery in the north was adjacent to a 35-year-old stand of red pines most of which were infected (D.R. Bergdahl, unpublished data). It is postulated that this nursery, which was growing red and Scots pines, played some role in the distribution of this disease as the county in which the nursery is located has the highest incidence of G. abietina in plantations of red and Scots pines. Unfortunately nursery records were not available so infestations could not be traced.

Skilling et al. (1979) has briefly summerized the symptoms and damage caused by the North American and European races of G. abietina. The early symptoms and progression of G. abietina in some red and Scots pine plantations is simiilar to that reported for New York except that overall damage has not been nearly as extensive. This is probably due to the small scattered plantation sites in Vermont.

As summarized by Skilling 1972 and Skilling et al. (1979) both North American and European races of G. abietina primarily produce conidia (macroconidia) in pycnidia. These spores are primarily rainsplash disseminated downward over relatively short distances, but some are known to become airborne under certain weather conditions.

Zajchowski and Bergdahl (1983) have shown that many macroconidia can remain viable for extended periods of time (at least 20 da) after being placed on the foliage of red pines. In North America macroconidia have also been observed in cryptopycnidia (Cauchon and Lachance 1980, Zajchowski and Bergdahl, unpublished data) and microconidia are commonly found in both pycnidia and on mycelium in culture (VanLuijk 1927, Zajchowski and Bergdahl 1982). Both the function and dispersal mechanism for spores produced in this manner is not understood.

Cultural control of the North American race of G. abietina can best be accomplished by planting disease-free nursery stock in uninfested areas (Skilling and Cordell 1966, Skilling and O'Brien 1969, Dorworth 1972, Gremmen 1972). Prescribed fire, the use of resistant species (or varieties), thinning and pruning are other cultural practices that have been tried as summerized by Gremmen (1972) and Skilling et al. (1979).

Pruning has been shown to be an effective method of eliminating or reducing the incidence of certain forest pests. Acccording to Weber (1964), Lehrer (1982) and Hungerford et al. (1982) the pruning of lower branches of white pines (especially P. strobus L. and P. monticola Dougl.) has been effective in the management of white pine blister rust caused by Cronartium ribicola J.C.

Fischer ex Rabenh. The removal of all branches from the lower one-third of these white pines eliminated all known and a high proportion of the latent infections. This pruning also improved the quality of the final product. Wilson and Rudolph (1970) also suggested that G. abietina can be controlled by pruning the lower most branches, but no data or literature were given to support their statement. It is also assumed they were referring to the use of pruning in the management of the North American race of G. abietina.

In the early stage of disease development G. abietina usually infects the lower branches and then progresses up the tree. Considering this progression and the fact that the rainsplashed conidia are believed to be primarily disseminated downward it would appear that the suggestion by Wilson and Rudolph (1970) should be considered in an attempt to control the European race of G. abietina. The pruning procedure should reduce inoculum levels and reduced inoculum should result in reduced incidence of infection. Dorworth (1979) and Hudler et al. (1981) have shown that inoculum levels are related to levels of infection. According to Dorworth (1979a) the "relation of inoculum density to infection is important because apparent resistance among certain Pinus spp. breaks down under extremely high inoculum loads (C.E. Dorworth, unpublished)."

The objectives of this study were to evaluate the effectiveness of pruning as a means of reducing infection and inoculum levels in plantations of red pine in Vermont and to learn more about the epidemiology of the European race of G. abietina.

METHODS

Two northern Vermont red pine plantations, infested with G. abietina were selected for pruning. One plantation, near Milton (4 ha), was planted in 1966 and another plantation, near Wolcott (2 ha), was planted in 1960. Both plantations are located on excessively well drained sandy soils. The Milton plantation is located in a topographical depression and Wolcott is in a high elevation valley with relatively good air drainage as compared to Milton. A survey of the Milton red pine plantation in October 1977 established that there were two distinct infection centers. Infected branches of red pine were cultured and serologically tested confirming the European serotype of G. abietina (Vermont isolates 6,7).

On May 25, 1978 both infection centers in Milton were mapped and the total number of infected branches and trees determined. One infection center was divided into four areas (subplots) and the trees were pruned to a uniform height of 1.4 m using a maylan hand saw (Figures 1 & 2). There were no known infections above 1.4 m. After pruning, these four subplots were treated as follows:

Subplot	Treatment
A	Pruned branches removed from A and chipped on B.
B	Pruned branches removed from B and chipped on B.
C	Pruned branches removed from C and piled near C.
D	Pruned branches left on ground on D.

The remaining infection center was left unpruned as the control (designated E).

Figure 1. Pinus resinosa (12 yrs) in Milton, Vermont infected with Gremmeniella abietina. Arrow indicates height of infection.

Figure 2. Pinus resinosa (same trees as in Figure 1) in Milton, Vermont pruned to a height of about 1.4 m. This method of pruning removed all branches infected with Gremmeniella abietina. All branches were removed from this subplot (C).

Vaseline-coated microscope slides were placed 0.3 m above ground to trap conidia. There were 24 slides per subplot and these were changed and evaluated weekly, May-November 1978-1980. This trapping procedure is like that recently described by Ostry and Nicholls (1982). Bioindicator seedlings (3-year-old red pine) were also planted in the understory in all areas (A-E) following pruning (Figure 3). A total of 114 and 720 seedlings were planted in the control (E) and pruned subplots (A-D), respectively. These trees and the overstory trees were evaluated for infections from 1978-1982. Temperature, relative humidity and precipitation were recorded during the periods of spore trapping previously described.

The Wolcott plantation was divided into three plots (A-C) each separated by a 30 m buffer strip of eastern white pine (P. strobus). The European serotype was also confirmed in this plantation. In June 1978 two one-tenth acre (0.04 ha) subplots were established in each plot to follow patterns of conidia dispersal and infection. Conidia were trapped as previously described from June, 1978 - November, 1980. Potted 3-year-old red pine seedling (10 per subplot) were exposed on a weekly basis (June, 1978 - November, 1979) to study seedling infection during periods of conidia dispersal. After each period of exposure the seedlings were moved to an uninfested area to await symptom development. A set of seedlings was also exposed in the uninfested area as a control. Percent infection was determined for all seedlings one year after exposure.

Figure 3. Bioindicator seedling (Pinus resinosa 3-year-old) and a spore trap (silicone coated microscope slide) in Wolcott, Vermont (plot B).

Figure 4. Pinus resinosa (8 yrs) infected with Gremmeniella abietina pruned to a height of about 2.1 m at Wolcott, Vermont. All Pruned branches in Plot B were left on the ground (Arrow).

All red pine in plots A and B were pruned to a height of about 2.0 m in October 1978. If pruning eliminated more than one-half the live crown, the tree was cut. After pruning, plots A & B were treated as follows:

Plot	Treatment
A	Pruned branches removed from A and chipped near A.
B	Pruned branches left on ground in B (Figure 4).

Plot C was left as the unpruned control.

RESULTS

The red pine at Milton were pruned during the major conidia dispersal period for *G. abietina* in 1978. Conidia were trapped from May 5-September 28, 1978 (weeks 21-39), but spore number were very low after July 20 (week 29). Conidia were trapped in the pruned area for 11, 5 and 11 weeks during 1978, '79 and '80, respectively, but for 25, 17 and 18 weeks in the unpruned area for those years (Table 1). Conidia numbers were highest for the unpruned area and precipitation appeared necessary for conidia to be trapped, even though peak numbers of conidia did not coincide with peak periods of precipitation (Figure 5).

Table 1. Percent infection, peak conidia production and number of weeks conidia were trapped in the Milton pruning study from 1978-1982.

		Tree[a] Infection	Branch[b] Infection	Seedling[c] Infection	Weeks Spores Trapped	Peak Spore Production
		%	%	%	#/mm^2	#/mm^2
1978	Unpruned	43	--	--	24	5.80
	Pruned	44	--	--	18	0.08
1979	Unpruned	51	35	32	17	4.50
	Pruned	<1	<1	<1	7	.08
1980	Unpruned	54	58	45	21	2.88
	Pruned	<1	<1	<1	11	.06
1981	Unpruned	60	80	--	--	--
	Pruned	<1	<1	<1	--	--
1982	Unpruned	77	150	54	--	--
	Pruned	<1	<1	<2	--	--

[a] Percent infection based on 248 and 570 trees in the unpruned and pruned plots, respectively.

[b] Percent infection based on 325 and 1038 branches infected in the unpruned and pruned plots, respectively (percent increase based on branches infected in 1978).

[c] Percent infection based on 114 and 720 (3-year-old) red pine seedlings planted in unpruned and pruned plots, respectively.

The first infections on bioindicator seedlings appeared in June 1979. In the unpruned area 32 percent of the bioindicator seedlings became infected as compared to <1 percent for the unpruned area (Table 1). In 1982 the infection of bioindicator seedlings in the unpruned area had risen to 54 percent as compared to still <2 percent for the pruned area.

The increase in branch infections in the pruned area remained low (<1.0 percent) as compared to a 150 percent increase after five years in the unpruned area (Table 1). The percentage of trees infected increased to 77 percent in the unpruned areas between 1978 and 1982 (Table 1). Pruning wounds did not serve as infection courts.

Prior to pruning at Wolcott all plots (A-C) were evaluated to determine baseline levels of infection, height of infection, inoculum production and bioindicator seedling infection during the summer of 1978. These data indicate that Wolcott had a higher initial level of infection and higher conidia loads than did Milton (Table 2). Bioindicator seedlings also showed high levels of infection for all plots prior to pruning (Figure 6).

Inoculum levels after pruning were greatly reduced as compared to the unpruned plots. Infection levels in the bioindicators, however, remained high for both pruned and unpruned plots (Figure 7) as did infections on pruned trees (Table 2). Height of infection also increased for all plots from 1978-1982. There were no differences in infection levels between plots where branches were either left on the site or removed and chipped. Precipitation also appeared to be important in the dispersal of conidia, but high amounts of precipitation again did not result in high levels of conidia trapped or high levels of infection in the bioindicator seedlings (Figures 6 & 7).

Table 2. Percent and average height of infection for the Wolcott pruning study (Plots A-C) from 1978-1982.

Plot	1978		1979		1980		1981		1982	
	%	HT[a]	%	HT	%	HT	%	HT	%	HT
A[b]	78	1.3	0	0	26	2.2	35	2.2	39	2.4
B[c]	59	1.4	0	0	19	2.9	33	2.6	40	2.9
C[d]	100	1.7	100	1.9	100	2.2	100	2.5	100	3.0

[a] Average height of infection based on maximum height of infection per infected tree.

[b] Plot pruned (October 1978) and branches removed.

[c] Plot pruned (October 1978) and branches left on site.

[d] Unpruned control.

DISCUSSION

Previous reports have established that pruning is an effective means of managing certain canker disease of conifers (Weber 1964, Lehrer 1982, Hungerford et al. 1982, Wilson and Rudolph 1970).

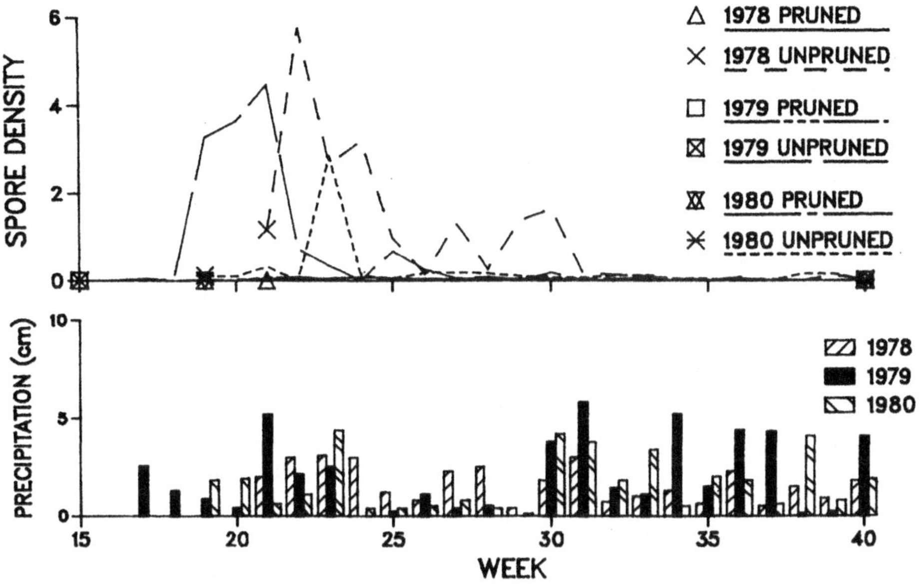

Figure 5. Density of conidia and precipitation records for the Milton, Vermont pruning study from 1978-80. Density = spores/mm^2.

These data show that inoculum of G. abietina (European race) and levels of infection in red pine can be managed by pruning. Pruned branches did not have to be removed from the site and pruning wounds did not serve as infection courts. These data (Wolcott) also suggest that pruning should be done in the spring after symptom development. Pruning at this time will help insure that all symptomatic branches are removed and therefore fewer latent infections will remain in the residual stand. The high infection levels for Wolcott in 1980 were most likely the result of latent infections being missed during fall pruning in 1978. These latent infections at Wolcott remained high in the tree so average height of infection remained similar to the control. Also, a follow up pruning should be considered (3-5 years after first) to remove any new or missed latent infections.

Dorworth (1979) and Hudler et al. (1981) have both suggested that inoculum densities influence the amount of infection on inoculated seedlings. The Wolcott seedling exposures during 1978-79 show that inoculum densities in the field may be variable and no correlation could be made with the previously cited studies. However, it must be pointed out that these previously cited studies were done with known inoculum concentrations, where as this study reports density of inoculum in terms of conidia trapped from the natural system.

It appears that until more is known about the epidemiology of the European race the best cultural control strategy to follow is to establish plantations in disease free areas using healthy nursery stock. For plantations that have had a history of G. abietina it would be advisable to consider planting less susceptible species.

For the management of plantations of red pine infected with the European race of G. abietina the overall objective should be to limit losses such as those described by Dorworth (1979b) and attempt to bring the current crop to a reasonable rotation age. This study has shown that pruning can be used as a silvicultural tool to lower inoculum densities and to reduce infection in plantations of red pine infected with the European race of G. abietina.

LITERATURE CITED

Cauchon, R. and D. Lachance. 1980. Research on cryptopycnidia for rapid diagnosis of Gremmeniella abietina. Can. J. Plant Pathol. 2:232-234.

Dorworth, C.E. 1972. Epidemiology of Scleroderris lagerbergii in Central Ontario. Can. J. Bot. 50:751-765.

Dorworth, C.E. 1979a. Influence of inoculum concentration on infection of red pine seedlings by Gremmeniella abietina. Phytopathology 69:298-300.

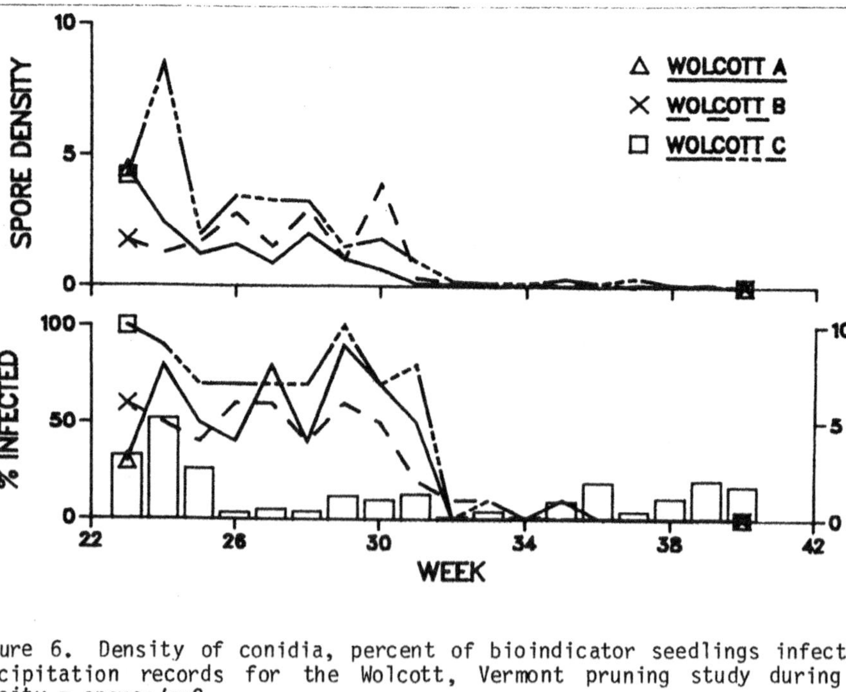

Figure 6. Density of conidia, percent of bioindicator seedlings infected and precipitation records for the Wolcott, Vermont pruning study during 1978. Density = spores/mm2.

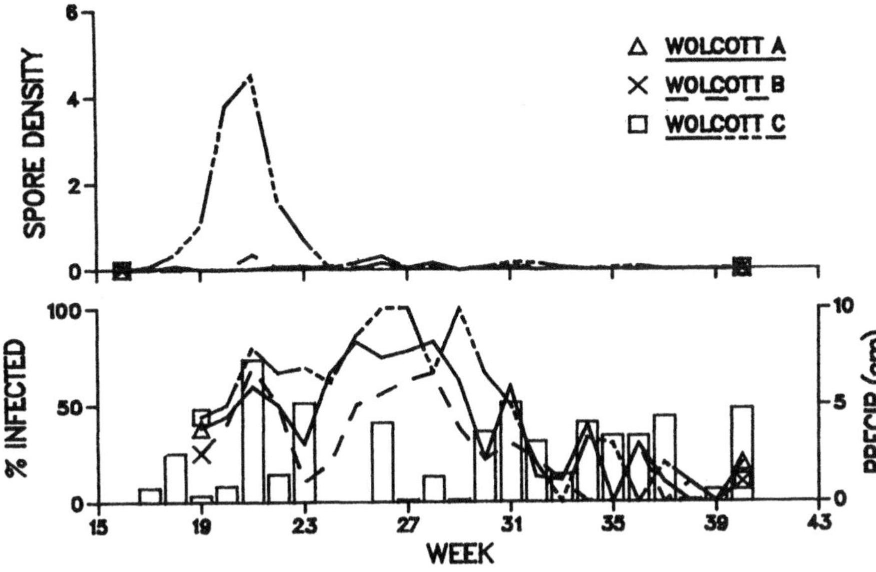

Figure 7. Density of conidia, percent of bioindicator seedlings infected and precipitation record for the Wolcott, Vermont pruning study during 1979. Density = spores/mm^2.

Dorworth, C.E. 1979b. Stand reduction of red pine by Gremmeniella abietina. Can. J. For. Res. 9:316-322.

Dorworth, C.E., E.J. Krywienczyk, and D.D. Skilling. 1977. New York isolates of Gremmeniella abietina (Scleroderris lagerbergii identical in immunogenic reactions to European isolates. Plant Dis. Rep. 61:887-890.

Gremmen, J. 1972. Our present-day knowledge of Scleroderris canker control. Eur. J. For. Pathol. 2: 40-43.

Hudler, G.W., G.R. Knudsen, and M.A. Beale. 1981. Dose-response relationship of pines and spruces to Gremmeniella abietina. Phytopathology 71:561.

Hungerford, R.D., R.E. Williams, and M.A. Marsden. 1982. Thinning and pruning western white pine: A potential for reducing mortality due to blister rust. U.S. For. Serv. Int. Mount. For. and Range Exp. Stn. Res. Note INT-322. 7 pp.

Lehrer, G.F. 1982. Pathological pruning: A useful tool in white pine blister rust control. Plant Dis. 66:1138-1139.

Ohman, J.H. 1966. Scleroderris lagerbergii Gremmen: The cause of dieback and mortality of red and jack pines in upper Michigan plantations. Plant Dis. Rep. 50:402-405.

Ostry, M.E. and T.H. Nicolls. 1982. A technique for trapping fungal spores. U.S. For. Serv. Res. Note NC-283.

Setliff, E.C., J.A. Sullivan, and J.H. Thompson. 1975. Scleroderris lagerbergii in large red and Scots pine trees in New York. Plant Dis. Rep. 59:380-381.

Skilling, D.D. 1972. Epidemiology of Scleroderris lagerbergii. Eur. J. For. Pathol. 2: 16-21.

Skilling, D.D. 1977. The development of a more virulent strain of Scleroderris lagerbergii in New York State. Eur. J. For. Pathol. 7:297-302.

Skilling, D.D. and C.T. Cordell. 1966. Scleroderris canker on National Forests in upper Michigan and Northern Wisconsin. U.S. For. Serv. Res. Pap. NC-3. 10 pp.

Skilling, D.D. and J.T. O'Brien. 1969. Scleroderris canker in the Lake States. A situation report. U.S. For. Serv. Res. Note NC-73.

Skilling, D.D., J.T. O'Brien and J.A. Beil. 1979. Scleroderris canker of northern conifers. USDA Forest Insect and Disease Leaflet 130. 7 pp.

Tattar, T.A., H.B. Teillon, and E.B. Walker. 1973. Scleroderris canker on red and Scots pine in Vermont. Plant Dis. Rep. 57:338.

Van Luijk, A. 1927. Brunchorstia destruens Erikss. auf Pinus Laricio var. Corsicana und ihve reinkultur. Med. Phytopath. Lab. Willie Commelin Scholten 11:52-57.

Weber, R. 1964. Early pruning reduces blister rust mortality in white pine plantations. U.S. For. Serv. Lake States For. Exp. Sta. Res. Note LS-38. 2 pp.

Wendler, P.B., A.R. Gotlieb, and D.R. Bergdahl. 1980. A newly discovered serotype of Gremmeniella abietina. Can. J. Bot. 58:1923-1928.

Wilson, L.F. and V.J. Rudolph. 1970. Early basal pruning to control forest pests. J. For. 68:632-634.

Zajchowski, S.J. and D.R. Bergdahl. 1982. Observations of a micro-conidiospore stage associated with Gremmeniella abietina. Phytopathology 72: 267.

Zajchowski, S.J. and D.R. Bergdahl. 1983. Viability of conidia of Gremmeniella abietina sprayed on new shoots of Pinus resinosa. Phytopathology 73: (In press).

THE GROWTH OF TREES AFFECTED BY *Gremmeniella abietina*[1]

Timo Kurkela[2]

ABSTRACT

The height and diameter increment of the pines classified according to disease and severity were compared. The growth patterns of the most affected trees showed that they had been suffering for several years. The causes for the growth differentation among the classes were adverse climate factors and/or the first phase of the disease, during which, some cankers appeared and replacement of leader shoot occurred. Areas with slowly growing trees were most often affected by the disease. Tree growth was most affected by topography or soil fertility. Thus, suppression of the trees seemed to be of little importance if any, in the development of susceptibility.

INTRODUCTION

The fungus *Gremmeniella abietina* causes growth losses in pines by killing those young shoots with the most actively assimilating needles, and also by blocking the xylem and phloem through canker formation. In healthy pine stands the fungus promotes natural pruning (Siepmann 1975), although it may be difficult to make a clear distinction between normal and pathological pruning. According to Read (1968) *G. abietina* occurs on the weakest trees and on branches with low rates of assimilation. During cold and rainy growing seasons, when poor resistance development in pines coincides with the optimum spore dispersal of a local *G. abietina* population, serious epidemics may arise. In susceptible pines, the shoots in the upper crown can also be infected.

The value of lost transplants, reforestation costs and quality losses in diseased trees have been discussed (Teich 1972, Dorworth 1979). Read (1968b) compared healthy and diseased Corsican pines and observed a clear dependence of poor growth on disease severity. Butin and Hackelberg (1978) classified sample trees in a diseased stand of *Pinus nigra* according to the annual rate of radial increment. By comparing the growth rate of healthy and diseased trees, they determined the year of infection. In their study, the intensity of the attack was characterized according to the number of dead or living branch whorles and internodal lengths.

---

1 A more detailed description of this study work by Tuovi Kallonen and Timo Kurkela will be published in "Communicationes Instituti Forestalis Fenniae."

2 The Finnish Forest Research Institute Unioninkatu 40 A, SF-00170 Helsinki 17, Finland.

In the present work, the growth losses of Scots pines classified according to the disease severity were calculated. The value of the stands for further growing are discussed on the basis of the density of the remaining trees.

MATERIAL AND METHODS

Six diseased stands were surveyed in 1979. The age of the stands varied from 18 years to about 40 years. The trees for the measurements were obtained by a line circular plot survey or by choosing trees systematically from the planted rows. The line direction was mainly perpendicular to the contour lines. The elevation of the sample plots was also recorded. Site classification of the sample plots was made according to the floristic forest type system (Cajander 1949). The height, breast height diameter, height increment, and radial increment were measured from the sample trees. The disease severity classification was based on the condition of the crown of the trees. The following scale was used:

Class 1: Healthy tree

Class 2: Occluded cankers only in the stem, the top, branches; and needles in good condition without symptoms.

Class 3: Stem cankers, the last top shoot healthy, dead shoots in lateral branches with brown needles.

Class 4: Stem cankers, numerous dead shoots in the crown, top shoot dead, lateral branches still mainly living.

Class 5: Several internodes from the top dead, only a few living branches left.

Class 6: Dead tree.

The effect of the disease on the growth of pines was determined by comparing the diseased pines to the group of the healthiest pines, or to the growth rate before the outbreak of the epidemics.

RESULTS AND DISCUSSION

During the warm period until 1973 or 1974, the height growth increased in tree Classes 1-5. In some areas, Class 6 separated from the other Classes a few years earlier. The reasons for this could be a continuous, minimal influence of the disease on the most susceptible trees. A remarkable number of cankers were already established in the trees in 1970 or 1972. The height growth in all trees decreased strongly in 1975, apparently caused by the direct effect of unfavorable weather. In 1975 the influence of the disease also increased. A possible indication of this increase was the significant increase in the number of replacement leaders.

There was a statistically significant separation in the height growth of trees according to the disease intensity classes. In some of the locations, the most affected trees had the lowest growth rate even before the outbreak of the epidemic. This difference in growth was mainly explained by the different site fertility which interacted with elevation. The disease incidence was usually highest at the bottom of depressions.

If it is assumed that without the influence of the disease, the trees should have grown continuously in relation to each other, as they did before the epidemic in 1967-1969. The calculated average annual decrease in height growth during 1974-1978 ranged from a negligible 0.1% to about 30%, and in 1978 from 30% to 49%, of the expected growth (Table 1). Because of competition among trees after closing of the stands, the suppressed trees were not significantly more infected than the dominant ones. However, the effect of infection seemed to be stronger in small trees.

Table 1. Losses in height increment as percentages of the expected growth in the studied stands of Scots pine.

Studied stand and Location	Period of epidemic 1974-1978	1978
	Percent	
1. Kuru	14.9	30.9
2. Hyytiälä	0.1	30.5
3. Multia I	18.5	45.5
4. Multia II	30.8	49.0
5. Palogjärvi	28.0	41.0

The disease seemed to affect radial growth earlier than height growth. That was to be expected, since the amount of radial increment is mainly affected by factors prevailing in the current season while height growth is greatly determined during bud formation in the previous season. Generally the same correlation was found between radial growth and disease incidence as between height growth and disease incidence.

One remarkable thing is that often the killed trees (Class 6) were growing better than less affected ones in 1968-1972, or they were able to increase their growth relatively more. The growth of the trees in Class 6 declined very rapidly after the first infection (the onset of epidemic). Later they seemed to be the main source of inoculum when more resistant trees or neighboring stands were infected.

For practical forestry, it is important to know when a diseased forest is no longer profitable for growing. During the field work of this study, we mapped the diseased areas and then gave recommendations for clear cuttings or for sanitary thinnings in less affected areas. It was allowed that trees in disease incidence Classes 1-4 would be able to survive, although the growth rate might remain low for a long time. When the number of these trees was more than 700/ha, we recommended that the stand be allowed to continue growing. According to Vuokila (1981) about 1500 pines per hectare gives a maximal volume growth after the first commercial thinning. When the number of stems/ha is 1000 or 600, the average loss in volume increment will be 11% and 26% respectively during the next seven years. These numbers were calculated for healthy trees. In a diseased stand the growth in surviving trees is also decreased. They may not produce timber of good quality. For that reason, the lower limit for profitable stocking should be higher than the usual 600-700 stems/ha which is accepted for healthy stands.

LITERATURE CITED

Butin, H., L. Hackelberg. 1978. Über den Verlauf einer Scleroderris lagerbergii - Epidemie in einem Schwarzkiefernbestand. Eur. J. For. Pathol. 8:369-379.

Cajander, A.K. 1949. Forest types and their significance. Acta For. Fenn. 56 (6):1-71.

Dorworth, C.E. 1979. Stand reduction of red pine by Gremmeniella abietina. Can. J. For. Res. 9: 316-322.

Read, D.J. 1968a. Some aspects of the relationship between shade and fungal pathogenicity in an epidemic disease of pines. New Phytol. 67:39-48.

Read, D.J. 1968b. Dieback disease of pines with special reference to Corsican pine Pinus nigra var. calabrica Schn. III. Mycological factor. Forestry 41:72-82.

Siepmann, R. 1975. Scleroderris lagerbergii Gr. als Schwächeparasit in gesunden Schwärzkiefernbestanden (Pinus nigra Arnold). Eur. J. For. Pathol. 5: 137-142.

Teich, A.H. 1972. Damage caused by the fungus Scleroderris in plantations and nurseries in Europe and North America. Eur. J. For. Pathol. 2:11-15.

Vuokila, Y. 1981. Nuoren männikön kasvureaktio ensiharvennuksen jälkeen (The growth reaction of young pine stands to the first commercial thinning). Folia For. 468:1-13.

PHYSIOLOGICAL AND BIOCHEMICAL FACTORS IN AUSTRIAN PINE CLONES WITH DIFFERENT SUSCEPTIBILITY TO *Gremmeniella abietina*

B.R. Stephan, F. Scholz and U.P. Singh[1]

ABSTRACT

In a seed orchard with 14 clones of Austrian pine (*Pinus nigra* ssp. *nigra*) significant clonal differences in susceptibility to *Gremmeniella abietina* were observed during the last 15 years. As a physiological trait, the buffering of pH in homogenized bark tissue of one-year shoots was studied. There were significant clonal differences and a good correlation with the mean susceptibility. Clones with higher resistance to *G. abietina* possess a higher buffering capacity. As a biochemical trait, the monoterpenes of the bark tissue of one-year shoots were investigated by gas chromatography. The clones differed significantly in the amount of the 5 monoterpenes. Myrcene- and limonene-content showed significant negative correlations with the attack by *G. abietina*. The possibility of indirect selection of less susceptible Austrian pines on the basis of these results is discussed.

INTRODUCTION

Austrian pine (*Pinus nigra* Arnold) could be a promising tree species in mixed coniferous forests on special sites of the coastal area of northern Germany. This pine species is less susceptible to wind and possesses higher resistance to needle cast disease (*Lophodermium seditiosum* Minter, Staley and Millar) than Scots pine (*P. sylvestris* L.). Successful cultivation is affected very strongly by *Gremmeniella abietina* (Lagerb.) Morelet, causing a dieback disease.

The first serious, and at some places total, damage caused by this parasite were recorded in northern Germany between 1960 and 1965 (Jaeger 1967, Gremmen 1967). Since that time, many Austrian pine plantations have been destroyed. Afforestations with this species have decreased to an insignificant level. During the last 20 years, several epidemics of *G. abietina* have occurred. A spreading on new host species has been recorded (Stephan 1970a, 1979).

---

[1] Federal Research Centre for Forestry and Forest Products, Institute of Forest Genetics and Forest Tree Breeding, Sieker Landstrasse 2, D-2070 Grosshansdorf, Fed. Rep. of Germany and Dept. of Genetics and Plant Breeding, Institute of Agricultural Sciences, Banaras Hindu University, Varanasi - 221005, India.

The award of a scholarship to one of us (U.P.S.) by the German Academic Exchange Service (DAAD) is gratefully acknowledged.

Despite the great variation of the fungus, with respect to morphological and cultural characters no occurrence of races has been observed in northern Germany until now. The typical form of G. abietina infects Austrian pine, especially causing an annual repeated dieback of current year shoots in the lower part of the crown of 10 to 25 year old trees. In northern Germany, G. abietina occurs mainly in its pycnidial form. The apothecial form can rarely be observed and is obviously less important for spreading.

In connection with the occurrence of Gremmeniella damage on Austrian pine studies have been started on the intraspecific variation in resistance to this parasitic fungus. A seed orchard with P. nigra ssp. nigra in Schleswig-Holstein gave a good opportunity for detailed investigations. First records showed significant clonal differences in the degree of attack (Stephan 1970b).

Of special interest was the question whether physiological and biochemical differences existed between susceptible and less susceptible Austrian pine clones. In the following paper, results of studies on buffering capacity and monoterpenes of bark tissue are given.

MATERIALS AND METHODS

Plant Material

In the forest district Rendsburg (Schleswig-Holstein), a seed orchard with 14 clones of Austrian pine was established in 1952. Grafted trees of the same clones were also planted in the clone collection of the institute at Schmalenbeck. From here, 2 to 3 trees per clone were used for the physiological and biochemical studies. The attack of G. abietina was scored in the seed orchard in 1969 and 1972, using a scale from 0 (no attack) to 10 (dead tree) (Stephan 1970b).

Because of the severe damage, only a few trees and clones started flowering. It has not been possible to make any controlled crosses, but in a few cases, open pollinated seed were harvested. A small, random sample of the progeny of clone Re 11 was included in these studies. The trees were 10 years old.

Evaluation of Buffering Capacity

From 3 trees of each of the 14 Austrian pine clones, samples of uninfected one year shoots were collected in October. Preparation of the homogenized bark tissue, and procedures for assessing buffering capacity are described in detail by Stephan and Scholz (1979).

Analysis of Monoterpenes

Oleoresin samples were taken from 2 trees of each of 13 clones in July. One clone (Re 12) could not be included in these investigations. After cutting the shoots, the oleoresin exuding from cortical tissues was collected with capillary tubes, sealed in glass tubes, and stored at -18C until analysis. Monoterpene analysis was carried out on a gas chromatograph Perkin-Elmer, type 990, with the following specifications: capillary column 50 m x 0.25 mm, 95C, N_2 3 ml/10 min. For the head space analysis, the glass tubes containing the oleoresin samples were pretreated for 30 min in a water bath at 60C. Seventy ul were injected for each analysis (temperature of injection tube 65C). The monoterpene content was expressed quantitatively in integration units as well as in the percentage of the total monoterpene amount. The monoterpenes were identified by comparing the peaks with the relative retention times of standard substances.

RESULTS

Differences in the Infection of Austrian Pine Clones

The 14 clones of Austrian pine show significant differences in the degree of infection by *G. abietina*. The mean scores of 1969 and 1972 are given in Table 1. The damage increased from 1969 to 1972. The damage values of the two years are correlated very strongly (r = 0.80). None of the clones was absolutely free of *Gremmeniella* infections, but e.g. clone Re 11 was considerably less attacked than clones Re 13 or Re 20, of which many individual trees were already killed.

Buffering of pH in Homogenized Bark Tissue

On the basis of the phenotypic observations of differing susceptibility of the seed orchard clones to *G. abietina*, the buffering capacity of homogenized tissue was investigated. As the results have already been published (Stephan and Scholz 1979), they will only be summarized here. The buffering capacity in homogenized bark tissue showed great variation between the pine clones (Table 1). Clone Re 18 with the highest buffering capacity differed from clone Re 13 with the lowest buffering capacity by the factor 1.6 (Figure 1). The clonal differences were statistically significant (p = .01%). It was shown that the buffering capacity and the resistance score were negatively correlated with r_S = -0.5 (p = .05%) (Table 2). Austrian pine clones with the highest susceptibility to *G. abietina* (e.g. Re 13 and Re 20) had low values of buffering capacity (Table 1).

Monoterpenes of Bark Tissue

Five monoterpenes were detected in the cortical oleoresin of 13 Austrian pine clones. They were identified, in the order of their retention time, as α -pinene, camphene, β -pinene, myrcene, and limonene (Table 1).

The predominant monoterpene was α -pinene ranging from 49.6% to 98.4% of the total monoterpene content. The other monoterpenes were present in the following concentrations: camphene 0% to 1.9%, β -pinene 0% to 11.9%, myrcene 0% to 1.6%, and limonene 0.4% to 49.6%. Variation between clones and trees of clones was high, regarding the qualitative and quantitative monoterpene content. On the basis of the relative monoterpene concentrations, the analyses of variance resulted in significant clonal differences for α -pinene and limonene, where 82% and 88% resp. of the total variance could be explained by differencs between clones. On the basis of the absolute monoterpene content, expressed in integration units, only limonene showed significant clonal differences with a variance component of 62% at the clonal level and 38% at the level of ramets.

Table 1. Attack scores, buffering capacity, and monoterpene content of 13 Austrian pine clones

clone no.	attack 1969 score	attack 1972 score	buff. cap. integr. units		α-pinene abs.	α-pinene %	camphene abs.	camphene %	β-pinene abs.	β-pinene %	myrcene abs.	myrcene %	limonene abs.	limonene %
Re 11	1.8	2.2	266.56		302	49.6	0	0	1	0.2	10	1.6	308	49.6
					360	67.0	0	0	0	0	5	0.9	172	32.0
				$\bar{x}$	331.0	58.3	0	0	0.5	0.1	7.5	1.3	240.0	40.8
Re 10	2.9	4.3	298.13		1896	91.2	20	0.9	90	4.3	10	0.5	62	3.0
					380	91.0	0	0	30	7.1	0	0	10	2.4
				$\bar{x}$	1138.0	91.1	10.0	0.5	60.0	5.7	5.0	0.3	36.0	2.7
Re 18	4.4	5.4	319.66		1067	84.3	16	1.3	120	9.5	10	0.8	53	4.2
					1030	86.6	10	0.8	76	6.4	8	0.7	66	5.5
				$\bar{x}$	1048.5	85.5	13.0	1.1	98.0	8.0	9.0	0.7	59.5	4.9
Re 22	2.4	5.4	277.81		910	96.4	6	0.6	8	0.8	0	0	20	2.1
					1290	92.9	10	0.7	12	0.9	7	0.5	69	5.0
				$\bar{x}$	1100.0	94.7	8.0	0.6	10.0	0.8	3.5	0.3	44.5	3.6
Re 16	4.3	5.6	261.22		1176	94.4	9	0.7	27	2.2	8	0.6	26	2.1
					1170	96.2	10	0.8	20	1.6	0	0	16	1.3
				$\bar{x}$	1173.0	95.4	9.5	0.7	23.5	1.9	4.0	0.3	21.0	1.7
Re 23	3.2	5.9	300.17		1574	95.6	12	0.7	28	1.7	8	0.5	25	1.5
					1780	96.6	8	0.4	24	1.3	7	0.4	24	1.3
				$\bar{x}$	1677.0	96.1	10.0	0.6	26.0	1.5	7.5	0.4	24.5	1.4
Re 15	3.6	6.0	234.13		444	92.5	0	0	2	0.4	0	0	34	7.1
					765	89.5	6	0.7	8	0.9	0	0	76	8.9
				$\bar{x}$	604.5	91.0	3.0	0.4	5.0	0.7	0	0	55.0	8.0
Re 14	2.9	6.5	278.85		1124	98.4	6	0.5	8	0.7	0	0	4	0.4
					456	96.6	0	0	0	0	0	0	16	3.4
				$\bar{x}$	790.0	97.5	3.0	0.3	4.0	0.4	0	0	10.0	1.9
Re 19	4.4	6.5	234.45		628	95.7	8	1.2	8	1.2	0	0	12	1.8
					1149	90.8	24	1.9	42	3.3	13	1.0	38	3.0
				$\bar{x}$	888.5	93.3	16.0	1.6	25.0	2.3	6.5	0.5	25.0	2.4
Re 21	3.9	6.6	293.13		334	97.1	0	0	0	0	0	0	10	2.9
					1460	88.1	18	1.1	38	2.3	6	0.4	136	8.2
				$\bar{x}$	897.0	92.6	9.0	0.6	19.0	1.2	3.0	0.2	73.0	5.6
Re 17	4.3	6.6	259.86		1046	91.4	8	0.7	10	0.9	0	0	80	7.0
					1544	91.5	12	0.7	23	1.4	7	0.4	102	6.0
				$\bar{x}$	1295.0	91.4	10.0	0.7	16.5	1.2	3.5	0.2	91.0	6.5
Re 13	6.0	8.0	203.89		281	83.6	0	0	40	11.9	2	0.6	13	3.9
					1608	92.3	18	1.0	70	4.0	3	0.2	44	2.5
				$\bar{x}$	944.5	88.0	9.0	0.5	55.0	8.0	2.5	0.4	28.5	3.2
Re 20	7.3	8.7	238.23		246	91.8	0	0	9	3.4	0	0	13	4.9
					645	89.6	4	0.6	36	5.0	0	0	35	4.9
				$\bar{x}$	445.5	90.7	2.0	0.3	22.5	4.2	0	0	24.0	4.9

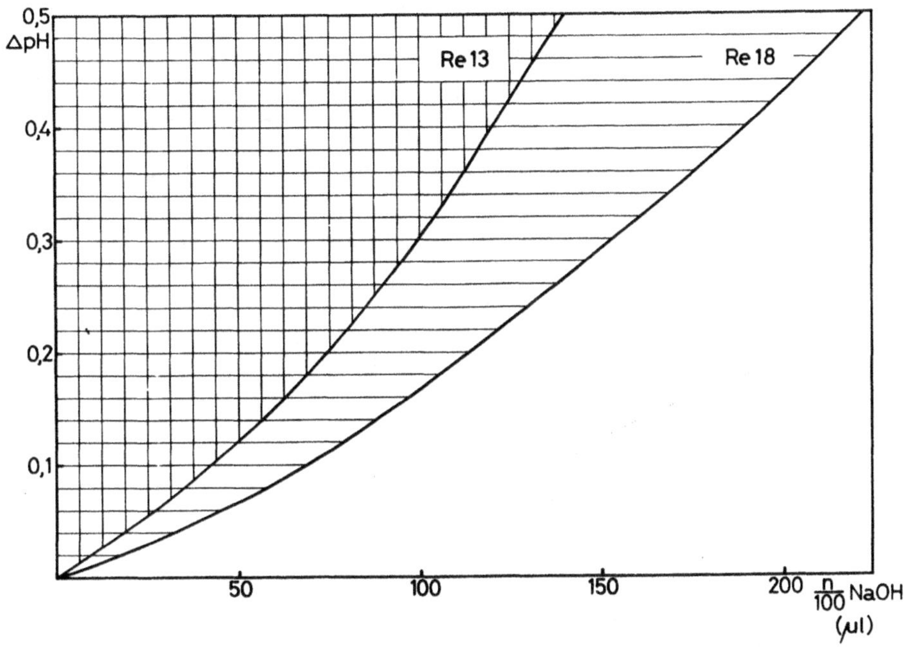

Figure 1. Buffering capacity of homogenized bark tissue of Austrian pine clones with different susceptiblity to Gremmeniella abietina (from Stephan and Scholz 1979).

Table 2. Correlation coefficients between attack scores, buffering capacity, and monoterpene content (percentage values: upper part, absolute integration values: bottom part of the table).

	1	2	3	4	5	6	7	8
1 attack 1969		.83***	-.51[1]*	.27	.19	.54	-.38	-.40
2 attack 1972	.83***		-.52[1]*	.63*	.20	.27	-.72**	-.67*
3 buff. capacity	-.51[1]*	-.52[1]*		.04	.05	-.03	.18	-.03
4 α-pinene	-.13	.06	.41		.39	.02	-.84***	-.96***
5 camphene	.12	.14	.19	.67*		.26	.01	-.48
6 β-pinene	.31	.06	.31	.30	.57*		.08	-.29
7 myrcene	-.34	-.58*	.50	.37	.52	.47		.76**
8 limonene	-.44	-.69**	.08	-.44	-.41	-.25	.36	

[1] Spearman Rank Correlation; one-sided test.

significance: * = 5 %, ** = 1 %, *** = .1 %

It was further investigated whether these clonal differences of monoterpene content showed any relation to the differing susceptibility of the 13 clones. The most conspicuous difference could be found in Austrian clone Re 11, which was actually the least susceptible one. This clone had a low α-pinene content, no camphene, and only traces of β-pinene, but the highest myrcene and limonene concentrations (Table 1). These differences in monoterpene content observed between the Austrian pine clones, lead to the question of a possible relation between monoterpenes and resistance to *G. abietina*. The correlations between the average values of resistance scores and the individual monoterpenes are given in Table 2. On the basis of the relative monoterpene content, significant correlations were obtained between the resistance score 1972 and α-pinene (0.63), myrcene (-0.72) and limonene (-0.67). Since the absolute amount of a monoterpene presumably plays a more important role in connection with disease resistance, that correlation was also tested (Table 2). Significant negative correlation coefficients were found only between the resistance score 1972 and myrcene (-0.58) and limonene (-0.69). The resistance and absolute -pinene amount were not significantly correlated. Thus, the less attacked Austrian pine clones tend to have higher contents of myrcene and limonene. There were also some relationships between individual monoterpenes. Thus, for example, clones with higher limonene content also had a higher amount of myrcene, while clones with high limonene and myrcene content had less α-pinene.

Progenies of Austrian Pine Clones

Ten trees from an open-pollinated off-spring of the relative resistant Austrian pine clone Re 11 were included in these studies on monoterpene content. They too showed a wide variation, but two distinct groups of trees could be distiguished one group (5 trees), which resembled the mother tree, had relative limonene contents between 23.0% and 41.2% with an average of 33.7%. This group also had the highest myrcene content (average of 2.0%) and a low α-pinene content (average of 62.9%). The second group (5 trees) was distinguished by a low limonene content between 3.4% and 13.6% (average of 6.7%), a low myrcene content (average of 0.7%), and a high α-pinene content (average of 88.9%).

Analysis of inheritance of monoterpene content and their relation to resistance will be investigated.

DISCUSSION

Infection of Austrian pines by *Gremmeniella abietina* is influenced by various factors, e.g. by site and environmental conditions, by silvicultural practice and others. Striking differences in susceptibility in an Austrian pine clonal seed orchard show that intraspecitic genetic factors are involved in resistance. These differences seem to be related to physiological and biochemical traits.

As in other host-parasite systems, here too, a relationship between the physiological trait buffering capacity and the resistance to *G. abietina* was demonstrated. The present knowledge of these relationships has been discussed in detail by Scholz and Stephan (1982). The pH regulation seems to be an important factor in the interactions between host and parasite. Several points, e.g. its inheritance, are not yet well known for *Pinus nigra*. In *Picea* abies, however, a remarkable heritability of buffering capacity was found (Scholz and Reck 1977). At the moment, however, buffering capacity cannot be recommended generally as a trait for use in indirect selection.

The monoterpene chromatograms are in good agreement with results of Arbez et al. (1974) who investigated Pinus nigra provenances of different subspecies. Even within a provenance they found great differences between individuals in the monoterpene composition, e.g. in the α-pinene and limonene concentrations.

Monoterpenes are useful biochemical markers, because they are easy to study, are hardly influenced by environmental factors, and are under strong genetic control (Arbez et al. 1974, Squillace 1976, Bridgen and Hanover 1982). Monoterpenes are widely distributed, especially in conifers, and show a great inter- and intraspecific variation. In some cases, a causal relation between monoterpenes, single or in combination, and resistance to fungal diseases and insect pests was demonstrated (Schuck 1982, Bridgen and Hanover 1982). In host-parasite systems studied, the individual monoterpenes can be stimulating or inhibiting, depending on the host and on the parasite. The results presented here suggest that, under the given conditions, limonene and myrcene play a role in the resistance of the Austrian pine clones to G. abietina. However, a causal relationship between resistance and these monoterpenes is not yet proven. But even without causal relations to resistance, monoterpenes can be useful markers for indirect selections, when they are genetically linked to resistance (v. Weissenberg 1976, Bridgen and Hanover 1982). Investigations on the effect of monoterpenes on the germination and growth behaviour of G. abietina, and on their possible use as markers for indirect selections will be continued with Austrian pine clones and progenies.

The results of these studies show that several different characters, e.g. buffering capacity and monoterpene content, can be involved in the resistance. This corresponds with the fact that in forest trees resistance is mostly polygenic.

LITERATURE CITED

Arbez, M., C. Bernard-Dagan, C. Fillon. 1974. Variabilité intraspécifique des monoterpènes de Pinus nigra Arn. Bilan des premiers resultats. Ann. Sci. Forest. 31:57-70.

Bridgen, M.R., J.W. Hanover. 1982. Indirect selection for pest resistance using terpenoid compounds. In: Heybroek, H.M., Stephan, B.R., Weissenberg, K.v. (Eds.): Resistance to diseases and pests in forest trees. Pudoc, Wageningen. 161-168.

Gremmen, J. 1967. Das Triebsterben der Schwarzkiefer, insbesondere in den Niederlanden und in Nordwestdeutschland. Forst Holzwirt 22:136-142.

Jaeger, K. 1967. Pilzliches Triebsterben an der Schwarskiefer im Küstengebiet. Forst Holzwirt 22:31-32.

Scholz, F., S. Reck. 1977. Effects of acids on forest trees as measured by titration in vitro, inheritance of buffering capacity in Picea abies. Water, Air, Soil Pollut. 8:41-45.

Scholz, F., B.R. Stephan. 1982. Buffering of pH in plant organs and resistance against fungi. In: Heybroek, H.M., Stephan, B.R., Weissenberg, K.v. (Eds.): Resistance to diseases and pests in forest trees. Pudoc, Wageningen. 176-186.

Schuck, H.J. 1982. Monoterpenes and resistance of conifers to fungi. In: Heybroek, H.M., Stephan, B.R., Weissenberg, K.v. (Eds.): Resistance to disease and pests in forest trees. Pudoc, Wageningen. 169-175.

Squillace, A.E. 1976. Analyses of monoterpenes of conifers by gas-liquid chromatography. In: Miksche, J.P. (Ed.): Modern methods in forest genetics. Springer, Berlin. 120-157.

Stephan, B.R. 1970a. Beitrag zu der durch _Scleroderris lagerbergii_ verursachten Triebspitzenkrankheit an verschiedenen Kiefernarten. Pflanzenkra. 77:417-424.

Stephan, B.R. 1970b. Klonabhängiges Verhalten bei _Pinus nigra_ Arnold gegenuber Scleroderris lagerbergii Gremmen. Allg. Forst Jagdztg. 141:60-63.

Stephan, B.R. 1979. Beobachtungen uber das Vorkommen und die Variation von _Scleroderris lagerbergii_ Gremmen in Norddeutschland. Z. Mykol. 45:45-53.

Stephan, B.R., F. Scholz. 1979. Weitere Untersuchungen zur unterschiedlichen Anfalligkeit von _Pinus nigra_-Klonen gegenuber _Scleroderris lagerbergii_. Eur. J. For. Pathol. 9:46-51.

Weissenberg. K. von. 1976. Indirect selection for improvement of desired traits. In: Miksche, J.P. (Ed.): Modern methods in forest genetics. Springer, Berlin. 217-228.

SCREENING CONIFERS FOR RESISTANCE TO *Gremmeniella abietina*

Darroll D. Skilling and Donald E. Riemenschneider[1]

ABSTRACT

Seed from 125 seed sources were grown as greenhouse container stock. Seedlings were inoculated with either the North American or European strain of *G. abietina*. Inoculum was supplied as infected branches on a screen placed over trees. Inoculation techniques resulted in infection levels up to 100%. Results have produced susceptibility ratings for most conifers grown in North America. Variation in resistance by seed sources was found in Scots pine.

INTRODUCTION

When the European strain of *Gremmeniella abietina* (Lagerb.) Morelet was identified in New York in 1977 (Dorworth et al. 1977), foresters were uncertain as to the susceptiblity of North American conifers to this fungus. Field observations in northern New York indicated that red pine, jack pine, Scots pine, and ponderosa pine were highly susceptible.

Pole-size eastern white pine showed some resistance and white spruce, showed only minor branch tip infection. The primary objective of this study was to determine the relative susceptibility of North American conifers to the European strain of *G. abietina* when inoculated under controlled conditions. A secondary objective was to determine the range of resistance to *G. abietina*

---

1 Principal plant pathologist, North Central Forest Experiment Station. USDA Forest Service, St. Paul, MN 55108; and plant geneticist, North Central Forest Experiment Station, USDA Forest Service, Rhinelander, WI 54501, respectively.

The field work for this study was done by Bruce Schneider, James Sullivan (decease), Don Fasking, Terry Podolski, Mike Hubschmidt, and Steve Bazan of the New York Department of Environmental Conservation. Container stock production was by Mariann Kienzler and Eileen Haynes of the North Central Forest Experiment Station and the New York Department of Environmental Conservation's Saratoga Nursery. Dr. Owe Martinsson from the Swedish Royal College of Forestry and Ralph Read formerly at the Rocky Mountain Forest and Range Experiment Station supplied Scots pine seed for this study.

Dr. Carl Mohn, College of Forestry, University of Minnesota and R. Schantz -Hansen, Potlatch Corporation provided the stock for the red pine seed source test. L. Oscar Selin of Georgia-Pacific Corporation supplied the larch seed sources.

observed within each conifer species and to determine if this population resistance could be used to develop resistant seed for future planting programs. Dietrichson (1968), in an international test of Scots pine provenances found that populations from western and southern Europe were most susceptible. He considered winter injury a primary cause of differences between sources in resistance to *G. abietina*. Stephan (1970), working with *Pinus nigra* var. *austriaca* found that of 14 clones grafted onto *P. nigra* stock, three appeared to be fairly resistant and two highly susceptible. Teich and Smerlis (1969), tested 92 provenances of jack pine for resistance to *G. abietina* in Canada. Three provenances appeared to be resistant. These three studies are a strong indication that resistance to *G. abietina* exists within some pine species.

MATERIAL AND METHODS

Screening for Species Resistance

Two sites in northern New York State were selected for the screening tests. The Boonville area was located 8 km NE of Boonville, New York. The Lake Clear site was located 14 km NW of Saranac Lake, New York. The European strain of *G. abietina* was present in both areas. When possible, 100 seedlings of each test species were planted in prepared transplant beds at each site. Each test plot was approximately 0.5 x 0.75 m. In most cases 2-0 seedlings were planted. In all tests the trees were inoculated in May or June by placing infected red pine branches with pycnidia of *G. abietina* over them on a vinyl-coated wire screen. Inoculum remained over test seedlings throughout the entire year. A total of 125 seed sources of 43 species were tested. For most species trees were planted for two different years to determine an average infection rate at each site. One year after planting, the trees were rated for resistance on the basis of the number of infected seedlings. Seedlings that were infected at time of rating, but later outgrow the infection were counted as infected. If the seedlings showed major differences in infection during the two years of testing both infection ratings were given in the tables to show the range of infection.

Testing Within Species Resistance

Scots pine. Seed from 15-full-sib families of Scots pine from the Swedish seed orchard Y 403 Nedansjo as well as seed from seven northern Europe and Asian sources were grown as container stock. The seedlings were outplanted and inoculated in the spring of 1979 at the Boonville and Lake Clear test sites.

In the fall of 1980 Scots pine seed from 21 sources was collected at the Natural Bridge seed orchard in northern New York. Container stock was produced during the winter of 1980-81 at St. Paul, Minnesota. In the spring of 1981 approximately 150 seedlings of each source were outplanted at the Boonville and Lake Clear test sites.

Red pine. In 1981 approximately 50 seedlings of each of 39 red pine seed sources were outplanted at Boonville, New York using the same inoculation procedure used for the Scots pine tests. Seedlings from these sources were also tested against the North American strain of *G. abietina* at a site northeast of Loretta, Wisconsin.

RESULTS

Screening for Species Resistance

Infection rates varied widely for a given species between test sites and between years (Tables 1 and 2). For example, in 1979 at Lake Clear only eight percent of the mugo pine seedlings were infected by G. abietina. In 1980 at the same site, however, the infection rate was 100 percent. Thus the potential for severe infection was present in 1979 but environmental conditions were not as favorable as in 1980.

G. abietina was able to infect and produce conidia on all 16 pine species tested. Every species in at least one test plot had over 50 percent infection. Species such as ponderosa, Monterey, red, and western white pine consistently had over 90 percent seedling infection. Most of these infected seedlings died shortly after symptom expression. Both Scots and jack pine showed some recovery of infected trees although many of the recovered trees became infected and died during the 2nd test year.

Within the genus Picea, Sitka spruce was the most susceptible. In 1979 this species was 100 percent infected at both sites. Mortality was 74 percent one year later. Englemann and Norway spruce were also severely infected in at least one test plot. The infected branch tips on these seedlings dropped off however, and the trees recovered. White, red, and black spruce showed only minor branch symptoms, these seedlings also later recovered. Conidia of G. abietina were found on the dead branch tips of all three species.

Infected branch tips with conidia were found in all Abies species tested. Infection was very light on all species except one plot of silver fir where 65 percent of the test seedlings had infected tips. One year later all fir seedlings had overgrown the infection and appeared to be healthy.

Douglas-fir seedlings showed moderate to heavy infection as measured by dead branch tips but these seedlings also recovered from the infection.

Both eastern and western hemlock were heavily infected (90+ percent) in these tests but recovered within one year. Natural eastern hemlock seedlings show considerable resistance to G. abietina in northern New York.

Northern white cedar and western red cedar appear to be immune to G. abietina in our tests.

Tamarack and Japanese larch appears quite resistant with only a few minor branch tips infected. The other Larix species have only been tested for one year, but they appear to be quite susceptible.

Testing Within Species Resistance

Scots pine. Results from the 15-full-sib families of Scots pine from Sweden are shown in Table 3. No single family showed high resistance at both test sites. Trees resistant in Sweden were not necessarily resistant under our test conditions. Our field inoculation procedure may be so severe that some resistance mechanisms may be overpowered. The infection levels were consistently higher at the Boonville test site. These plots are more shaded than the Lake Clear site and this may be more conducive to higher infection rates.

Table 1. Susceptibility of seedlings of genus Pinus and picea to European strain of G. abietina.

Species	Susceptibility rating[1]
Pinus spp. (pine)	
P. virginiana (Virginia)	3
P. echinata (shortleaf)	3
P. thunbergii (Japanese black)	3
P. taeda (loblolly)	3
P. resinosa (red)	3
P. strobus (eastern white)	2-3
P. nigra (Austria)	3
P. mugo (mugo)	2-3
P. monticola (western white)	3
P. ponderosa (ponderosa)	3
P. lambertiana (sugar)	3
P. contora (lodgepole)	3
P. banksiana (jack)	2-3
P. sylvestris (Scots)	2-3
P. rigida (pitch)	3
P. radiata (Monterey)	3
Picea spp. (spruce)	
P. sitchensis (Sitka)	3
P. engelmannii (Engelmann)	3
P. abies (Norway)	2-3
P. glauca (white)	1-2
P. mariana (black)	1
P. pungens (blue)	2-3
P. rubens (red)	1

[1] Susceptibility rating is based on the number of seedlings infected one year after inoculation. Seedlings that outgrew the infection later were still counted as being infected. A dual rating such as 2-3 for a species means that in one year the average infection was 2 but the other year the infection was 3.

0 = immune; 1 = 1-9% of seedlings infected; 2 = 10-49% of seedlings infected; 3 = 50+% of seedlings infected.

These ratings are based on 100 tree plots at two sites. The tests were repeated a second time so the total sample size was 400 trees for each species.

Results of the seven northern Scots pine seed sources are shown in Table 4. As in the previous test no evidence of resistance was shown by any seedlings tested.

Because of the unusually warm weather at the time of planting, seedling survival from the 21 Scots pine seed sources at Lake Clear was too low to be evaluated. Results from the Boonville test are shown in Table 5. The two best seed sources were 3-SUP-1 and Ohio with 69 and 72 percent infection respectively. Seedlings from these sources were replanted again in 1982 and the results will be available in June of 1983.

Table 2. Susceptibility of seedlings of genus Abies, Thuja, Tsuga, Larix, and Pseudotsuga to European strain of *G. abietina*.

Species	Susceptibility rating[1]
Abies spp. (fir)	
A. balsamea (balsam	0-1
A. fraseri (Fraser)	1-2
A. grandis (grand)	1
A. magnifica (California red)	1
A. procera (noble)	1
A. concolor (white)	1-2
Thuga spp. (cedar)	
T. plicata (western red)	0
T. occidentalis (northern white)	0
Tsuga spp. (hemlock)	
T. heterophylla (western)	1-3
T. canadensis (eastern)	3
Pseudotsuga menziesii (Douglas-fir)	1-3
Larix spp. (larch)	
L. leptolepis (Japanese)	1
L. laricina (tamarack)	1
L. occidentalis (western)	3
L. gmelini dahurica (Mongolian)	3
L. sibirica (Siberian)	3
L. x eurolepis (hybrid)	3
L. sukaczewii (Siberian)	3
L. sudetica (Sudeten)	3

[1] Susceptibility rating is based on the number of seedlings infected one year after inoculation. Seedlings that outgrew the infection later were still counted as being infected. A dual rating such as 2-3 for a species means that in one year the average infection was 2 but the other year the infection was 3.

0 = immune; 1 = 1-9% of seedlings infected; 2= 10-49% of seedlings infected; 3= 50+% of seedlings infected.

These ratings are based on 100 tree plots at two sites. The tests were repeated a second time so the total sample size was 400 trees for each species.

Red pine. At the end of one year in the test bed most of the red pine seedlings from the 39 seed sources were infected by *G. abietina* (Table 6). Seedlings from the best seed source, no. 20201, had 72 percent infection at Boonville. Seedlings from the same seed source, when tested against the North American strain had 98 percent infection at the Wisconsin site. These seedlings will be maintained for one more year but so far very little evidence of resistance is apparent in this material. All infected trees were dead at the end of the 1982 growing season.

Table 3. G. abietina inoculation results of seedlings from 15-full-sib families of Scotch pine at Lake Clear and Boonville, N.Y.

Seed Source No.	Lake Clear No. trees planted	Lake Clear Infection percent	Boonville No. trees planted	Boonville Infection percent	Resistance Rating in Sweden by Martinsson
403-77	44	100	50	100	Inter.
403-126	42	69	49	100	High
403-142	34	99	39	100	High
403-148-50	45	44	40	100	Low
403-88	48	29	50	96	Inter.
403-148	29	27	29	100	Low
403-166	42	28	49	88	High
403-136	38	10	39	100	High
403-67	51	37	50	100	High
403-52	57	16	49	100	Low
403-63	50	28	48	85	Inter.
403-180	32	9	30	100	Low
403-16	45	0	37	100	Inter.
403-103	43	53	50	100	Inter.
403-137	42	52	50	98	High

Table 4. G. abietina results of seedlings from seven Scots pine sources from northern Europe and Asia at Lake Clear and Boonville, New York.

Seed Lot No.	Seed Origin	Lake Clear No. trees planted	Lake Clear Infection percent	Boonville No. trees planted	Boonville Infection percent
250	Urals	86	100	100	100
257	Urals	62	100	98	100
522	Sweden	95	100	40	100
265	Scotland	57	98	60	100
230	Finland	173	92	200	99
256	Siberia	113	85	100	100
233	USSR	87	75	100	100

DISCUSSION AND CONCLUSION

Based on the results of this work we know that the European strain of G. abietina is able to infect most of our North American conifers. The most serious damage will be to the hard pine species. Most of the infected hard pine seedlings eventually died. Sitka spruce also appears to be highly susceptible to the European strain with resulting severe mortality. If the European strain of G. abietina spreads west into the natural range of Sitka spruce and ponderosa and lodgepole pine, it appears that these three species could be severely damaged especially at the higher elevations. Our results indicate that the European strain will also be very damaging in the Lake States, especially to red pine. No evidence of resistance was found in 39 red pine seed sources tested. With the exception of Sitka and perhaps Engelmann spruce, this genus appears to have good resistance to G. abietina. Infection was light and seedlings that were infected usually recovered. Forest managers

Table 5. *G. abietina* inoculation results of seedlings from 21 Scots pine seed sources at Boonville, New York.

Seed Source	No. trees planted 6-81	No. trees living 9-81	Infection 6-82[1]
Riga	94	94	100
11-SE-3	150	134	84
FEP-2	150	146	92
Czech	100	91	98
SCH-3	150	147	98
12-SE-2	150	138	100
SL-30	100	98	90
FEP-3	150	150	99
3-SVD-1	150	144	69
Er. Park	150	148	94
Dallarmi	150	118	100
1-FP-1	150	0	-
30-T-1-66	150	104	91
PV-1	150	150	99
SCH-2	144	143	94
Ohio	62	62	72
SUP-2	150	150	94
Low. Wat.	150	144	96
SCH-1	150	143	99
14-SC	150	150	99
SCH-4	150	136	97

1 Infected seedlings from all seed sources had conidia present 6/82.

should be able to plant the Genus *Picea* with little fear that Scleroderris canker will be a major problem. All the firs in this test were very resistant and no problems should be experienced with this group. This is also true for Douglas-fir, eastern and western hemlock, eastern larch, northern white cedar and western red cedar. Although Japanese larch appears to be quite resistant the European larch varieties were badly infected. At this point land managers should use caution with this group until more data are available.

LITERATURE CITED

Dorworth C.E., J. Krywienczyk and D.D. Skilling. 1977. New York isolates *Gremmeniella abietina* (*Scleroderris lagerbergii*) identical in immunogenic reaction to European isolates. Plant Dis. Rep. 61:888-890.

Dietrichson, J. 1968. Provenance and resistance to *Scleroderris lagerbergii* Fremmen (*Gremmeniella abietina* Lagerb.). The international Scots pine provenance experiment of 1938 at Matrand.

Stephan, B.R. 1970. Clonal variation in susceptibility of *Pinus nigra* var. *austriaca* to attack by *Scleroderris lagerbergii* Allg. Forst-Und Jagdzeitung 141:60-63.

Teich, A.H. and E. Smerlis. 1969. Jack pine resistance to *Scleroderris lagerbergii*. Bi-mon. Res. Notes. Ottawa 25:47.

Table 6. Inoculation results of seedlings from 39 red pine seed sources. European strain of G. abietina used at Boonville, New York and North American strain used at Loretta, Wisconsin.

Seed Sources	No. Trees planted 6-81	Infection 6-82 %	No. Trees planted 6-81	Infection 6-82 %
43401	37	84	40	95
30003	48	100	40	100
32702	40	88	40	97
33001	46	78	40	95
20301	49	78	40	95
44802	48	85	40	95
43301	47	94	40	84
20402	48	88	40	95
44602	46	83	40	97
43302	48	81	40	90
20201	50	72	40	98
20602	48	97	40	92
32901	48	94	40	98
44503	47	94	40	90
32704	48	98	40	85
44601	49	90	40	95
44502	48	85	40	97
20304	48	96	40	90
20401	48	96	40	100
20701	47	96	40	98
30101	43	86	40	100
30302	47	100	40	97
20501	45	100	40	100
44402	46	98	38	92
20102	48	98	40	100
20403	45	100	40	80
20302	49	100	40	85
30903	45	100	40	98
45401	45	100	40	100
44604	48	96	40	85
31804	50	98	40	95
43402	48	96	40	88
30501	47	91	40	88
20103	48	98	40	100
30305	45	98	40	100
32804	47	98	40	88
44504	44	100	40	85
20101	46	100	40	100
30901	48	96	40	100

RESPONSE IN CANADA OF JACK PINE PROVENANCES TO *Gremmeniella abietina*

Christopher W. Yeatman[1]

ABSTRACT

In a series of range-wide jack pine provenance tests planted in boreal and eastern Canada, resistance to Scleroderris canker was consistently found among sources from western, northern, and eastern Quebec. Low infection levels were also evident in particular provenances from New Hampshire, New Brunswick, and Cape Breton. Generally, provenances from the milder southern limits of the jack pine range were highly susceptible to the disease, as well as being subject to winter dessication, when planted in severe boreal or high altitude environments. Winter-hardy provenances from western Canada, Ontario and the maritime provinces were intermediate in susceptibility to Scleroderris canker. No Scleroderris canker was found in test plantations located within the milder parts of the jack pine range in central and eastern Canada. In addition to seed origin, level of infection and mortality were related to climate, exposure, inoculum level, and micro-environment within a site. Operational trials of selected provenances of jack pine are warranted in areas and environmental conditions conducive to development of Scleroderris canker.

INTRODUCTION

Range-wide jack pine (*Pinus banksiana* Lamb.) provenance trials were established during the 1960s at widely dispersed locations in Canada and the adjacent United States with seed and seedlings distributed from the Petawawa Forest Experiment Station (now Petawawa National Forestry Institute) by Mark J. Holst (1960, 1962). The first evidence of pronounced provenance differences in susceptibility to Scleroderris canker (*Gremmeniella abietina* (Lagerb.) Morelet) was found in five-year-old trees growing in a nursery trial located within the Boreal Forest Region (Rowe 1972) at Long Lac, northern Ontario (Teich 1967). Subsequently, infection was reported for selected provenances growing in range-wide trials planted in Quebec (Yeatman 1976a) and in Ontario (Yeatman 1976b). The objective of this report is to summarize data on infection at these and at additional trials planted in Ontario and New Brunswick, and to suggest ways in which losses to Scleroderris canker of jack pine in Canada might be avoided or reduced.

[1] Project Leader, Tree Genetics and Breeding, Petawawa National Forestry Institute, Canadian Forestry Service, Chalk River, Ontario, Canada, K0J 1J0.

METHODS

Seed of jack pine was obtained from some 100 stands located throughout the range of the species which extends from the Atlantic coast in the east to the Mackenzie River Valley in the Northwest Territories. Replicated experiments were established in cooperation with provincial forest services and forest industries. These trials included two nursery tests and twelve plantation trials established in a range of contrasting climates from northwestern Ontario to New Brunswick (Yeatman 1974). Experiments referred to here include the nursery trial at Long Lac, northern Ontario, sown in 1962 (Teich 1967), the 1962 nursery trial at Petawawa used for inoculation tests (Teich and Smerlis 1969) and four field trials planted in 1966, two in northern Ontario, one in northwestern Quebec and one in New Brunswick (Figure 1, Table 1). An additional test of four provenances was planted in 1972 at Long Lac on the basis of earlier results in the nursery trial.

In the Long Lac nursery, 92 provenances were sown to establish 100 trees per provenance in 10-tree plots within 10 replications. Infection by Scleroderris canker was recorded in 1966 when the trees were six years of age (Teich 1967).

The four plantation trials listed in Table 1 included different numbers and identities of provenances, but 39 common provenances are identified by location in Figure 1. These trials were established with 2-0 bareroot stock in 5- or 6-tree plots replicated 8 or 10 times, i.e. either 48 or 50 trees were planted per provenance per site. Plant spacing was 1.8 x 1.8 m (6x6 ft). Scleroderris infection, mortality, winter injury, and height of living trees were recorded in the fall of 1973 when the trees were 10 years old, or eight years from planting.

Table 1. Scleroderris infection at 10-years-of-age of 39 range-wide jack pine provenances common to 4 test-sites in Canada.

Site Location		Trees Infected[1] Provenance Min. (%)	Max. (%)	Range (%)	Site mean (%)
Caramat, Ont. For. Sect. B.8	Lat. 69° 22'N Long. 86° 00'W	0	19	19	5
Fraserdale, Ont. For. Sect. B.4	Lat. 49° 46'N Long. 81° 30'W	4	68	64	33
Sugarloaf, Que. For. Sect. B.7	Lat. 47° 54'N Long. 75° 38'W	2	46	44	20
Little Bald Mtn., N.B. For. Sect. A.1	Lat. 47° 15'N Long. 66° 20'W Alt. 460 m	3	48	45	22
Four sites combined		6	32	26	21

[1] Percent based on trees (living and dead) present 8 years after planting.

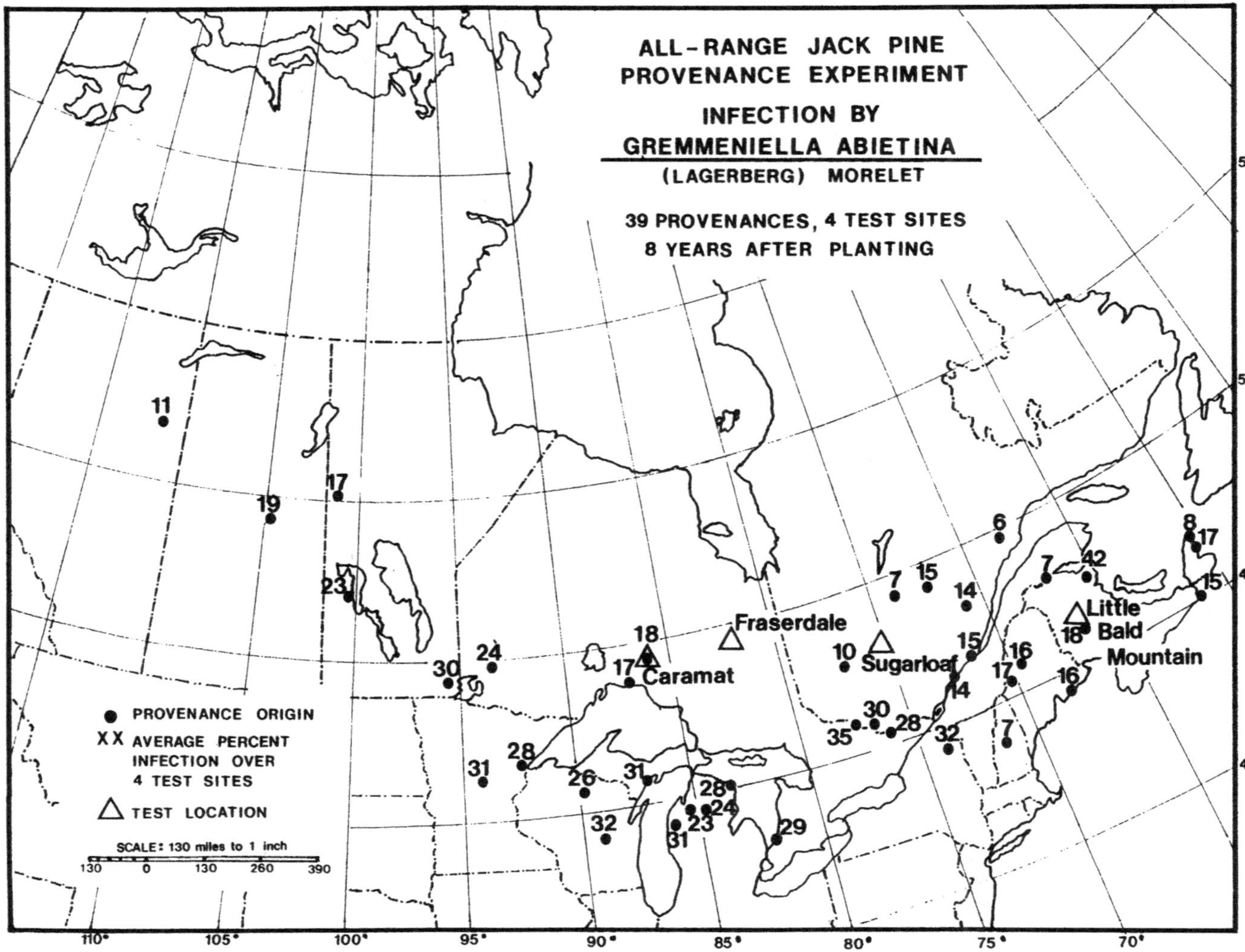
ALL-RANGE JACK PINE
PROVENANCE EXPERIMENT
INFECTION BY
GREMMENIELLA ABIETINA
(LAGERBERG) MORELET
39 PROVENANCES, 4 TEST SITES
8 YEARS AFTER PLANTING
PROVENANCE ORIGIN
XX AVERAGE PERCENT INFECTION OVER 4 TEST SITES
TEST LOCATION
SCALE: 130 miles to 1 inch
130 0 130 260 390
Fraserdale
Sugarloaf
Caramat
Little Bald Mountain
55°
50°
45°
40°
110°
105°
100°
95°
90°
85°
80°
75°
70°
11
17
19
23
30
24
31
28
26
32
31
18
17
28
24
23
31
29
35
30
28
10
7
15
14
15
14
32
17
16
7
6
7
42
18
16
8
17
15

Two provenances, Trinity Bay and Lac Sault au Cochon from eastern Quebec identified in the Long Lac nursery test for superior tolerance to Scleroderris infection, were sown in the nursery at Petawawa in 1969 together with seed of Caramat and Petawawa origin (Rudolph and Yeatman 1982). The four seedlots were planted at Long Lac in 1972 as 2-1 stock in 25-tree plots randomized within 10 replications. The Caramat seed originated some 25 miles southeast of Long Lac and is taken to be representative of the local Long Lac populations. Petawawa is used as a common provenance in all field trials of this material. The site chosen was a flat field surrounded on three sides by mature jack pine. The large, open area was cleared originally for a horse paddock and subsequently planted in the early 1960s to jack pine and Scots pine (*Pinus sylvestris* L.). This plantation failed due to heavy infection and mortality from Scleroderris canker. Scattered survivors were cleared for the new planting but many remained in the vicinity and bore clear evidence of chronic infection at the time of planting, and also in 1978 when the trial was assessed. The presence of these older trees undoubtedly assured a high inoculum level of *Gremmeniella* spores surrounding the new plantations.

Survival was recorded at the end of the first growing season to measure success of establishment. Survival was next recorded six years later together with infection by Scleroderris canker, mortality, and height of living trees. Plot trees remaining at this time were classified into one of four categories with respect to infection by Scleroderris canker, viz.:

0 - all needles green, no apparent infection
1 - one or two branch tips killed
2 - one or two basal whorls killed
3 - most branches killed, some green needles remain
4 - tree killed

In all instances Scleroderris infection was identified as the cause of the needle and/or plant necrosis on the basis of yellow-green discoloration of the wood beneath the dead bark.

RESULTS

Of 92 provenances sown in the Long Lac nursery test, only seven had four or more trees free of disease symptoms at age five. The three best were from northern and eastern Quebec. The number of disease-free trees in the nursery test decreased with increasing distance from a shelterbelt of red pine oriented north-south on the east side of the experiment in front of one replication (Teich 1967). Heavily infected Scots pine shelterbelt trees elsewhere in the nursery were considered the original source of inoculum (Punter 1967). Subsequent inoculation of provenances growing in the nursery test at Petawawa confirmed the relative resistance of the three Quebec provenances. Resistance correlated well with dry matter content of the needles (Teich and Smerlis 1969).

Levels of infection observed eight years after planting differed among test sites of the all-range plantation trials where Scleroderris canker was found (Table 1). Infection was least at Caramat (5 percent) and most frequent at Fraserdale (33 percent). However, low stocking (less than 50 percent, data on file) at Little Bald Mountain was indicative of heavy post-planting infection and low survival due to early and severe Scleroderris infection of the planted trees at this high altitude location. Fertility of the podzolized shale soils characteristic of this area of New Brunswick is low and exposure is extreme over large expanses on which the jack pine, spruce, and fir were clear cut in the early 1960s. In 1973, provenance survival here ranged from a high of 65 percent to a low of six percent.

Scleroderris infection was not observed at the one remaining provenance trial in the Boreal Forest Region located near Red Lake, northwestern Ontario, perhaps due to lack of a local source of inoculum. No infection was found at any of six sites in the more moderate climate of the Great Lakes-St. Lawrence Forest Region or at the lowland site in New Brunswick (Acadian Forest Region).

Effects of microtopography on level of infection were evident at the three Boreal sites in Ontario and Quebec. Because replications did not coincide with topographic patterns, the effects of provenance, microsite, and replication were somewhat confounded but differences among provenances in susceptibility were clearly evident. Trees of susceptible provenances were heavily infected when located in hollows and flat areas, but not as heavily or frequently infected when planted on humps and gentle slopes. On the other hand, trees of resistant provenances showed only minor infection in low areas and none on high ground.

Provenance minima, maxima and ranges of infection are listed in Table 1 for each of the plantation trials. Some inconsistency in provenance ranking (provenance x site interaction) is evidenced by the reduced range, increased minimum, and reduced maximum derived when provenances were averaged over the four sites (Table 1). Provenances most susceptible to infection and mortality orginated in the south of the range of jack pine, from southeastern Manitoba through the Lake States and southern Ontario to upper New York state (Figure 1). Provenances of intermediate and variable susceptibility ranged from Alberta to the Atlantic coast. Provenances having consistently low levels of infection over all four sites come from western, northern, and eastern Quebec. Provenances, from lower Gaspe, Quebec, northern Cape Breton, Nova Scotia, and Welch Mountain, New Hampshire, also exhibited low average levels of infection. (Figure 1) The results are consistent with those from the earlier nursery trial at Long Lac reported by Teich (1967). However, in the Long Lac nursery infection was so severe that the best provenance did not exceed 14 healthy trees out of the 100 established initially (Table 2).

The later plantation trial at Long Lac of four selected provenances was prompted by the results of the nursery test. Table 3 provides a summary of the evaluation made seven years after planting when trees were ten years of age. There were no significant differences among sources in first-season (post planting) survival, which was consistently high. Six years later the Petawawa trees had failed completely; only 50 percent of the Caramat trees were still living, of which 35 percent were infected; both the eastern Quebec provenances had excellent survival (87 and 92 percent). Only 16 percent of trees from Trinity Bay on the north coast of the gulf of St. Lawrence were infected, none of them severely.

The three provenances with 50 percent survival or more did not differ significantly in average height of surviving trees (Table 3).

Table 2. Location and survival following Scleroderris infection in a nursery test (Teich 1967) of four provenances subsequently planted in a field trial at Long Lac, Ontario.

Provenance	Latitude	Longitude	Healthy trees[1] 1966
	(N)	(W)	(No)
Trinity Bay Que.	49° 40'	67° 15'	14
'ac Sault au Cochon, Que.	49° 18'	69° 53'	6
Caramat, Ont.	49° 35'	85° 58'	0
Petawawa, Ont.	45° 50'	77° 25'	0

1 Number of healthy trees remaining out of 100 per provenance.

Table 3. Survival, mortality, resistance to Scleroderris canker and average height seven years after planting four jack pine provenances at Long Lac, Ontario (Lat. 69· 47'N, Long. 86· 30'W)

Provenance	1972 Fall survival[1]	Summer 1978				
		Living	Dead	Missing[2]	Disease-free[3]	Tree height[4]
	(%)	(%)	(%)	(%)	(%)	(cm)
Trinity Bay, Que.	96	92	3	5	84	153
Lac Sault au Cochon, Que.	97	87	8	5	67	149
Caramat, Ont.	99	50	31	19	35	152
Petawawa, Ont.	94	1	5	94	0	-

1 Percent based on 250 trees planted per provenance.
2 Percent based on trees accounted for (living and dead) fall 1972.
3 Percent based on trees living summer 1978.
4 Average height of living trees.

DISCUSSION

Observations reported here of Scleroderris canker of jack pine growing in provenance trials are consistent with other reports of infection by the North American strain in the boreal forest, with the additional, and potentially critical, genetic dimension of seed origin. Scleroderris canker is not considered a serious threat to natural regeneration of jack pine in Ontario. The disease may cause serious mortality in planted trees, especially when infection becomes widespread within 1-3 years after planting (Dorworth and Davis 1982). Jack pine is intermediate between red and Scots pine in susceptibility to Scleroderris canker when tested to the south of the Boreal forest (Cordell et al. 1968, Dorworth 1983). However, in the northern forests red pine is beyond its natural range except for occasional outliers on lake shores and, generally, Scots pine is not as well adapted to the Canadian forest environment as jack pine on typical pine sites.

European experience confirms that species or provenances planted outside certain environmental limits may be highly susceptible to Scleroderris canker (Dietrichson 1968, Donaubauer 1972, Roll-Hansen 1972, Teich 1972, Pelkonen et al. 1982, Donaubauer this volume) and this is certainly the case with jack pine, especially on exposed and low fertility sites. Conditions that weaken the host or subject it to stress favor infection (Skilling and O'Brien 1969, Dorworth 1971, 1974, Dorworth and Buchan 1972).

Scleroderris is recognized for its ability to develop within the host at low temperatures (Skilling 1969a, Dorworth 1972b) and characteristically it was most prevalant in jack pine provenance trials located at northern and high altitude sites in Canada. The significance of epidemic conditions, "wave years" (Kohh 1964, Skilling and O'Brien 1969), was evident at the Long Lac nursery test in the mid sixties. Both here and in the later test at Long Lac, the heavy spore load generated by infected trees nearby ensured a high level of infection and mortality among susceptible planting stock. This is consistent with the observations of Dorworth and Buchan (1972) and tests of concentration of inoculum reported by Hudler et al. (1981).

At all test locations, provenance response was conditioned by microclimate in relation to shelter or microtopography. Higher levels of infection are commonly associated with depressions and troughs subject to radiation fog which prolongs the incubation period afforded the fungus (Dorworth 1972a, 1974, Doworth and Buchan 1972).

Well defined differences in susceptibility among jack pine provenances were evident. Provenances from western, northern, and eastern Quebec show outstanding promise in terms of suitability for planting in the Boreal forest Region of Ontario and Quebec and in cold upland areas in the maritime provinces. This is particularly evident where exposure within large clearcuts creates severe environmental stresses for planted stock. Under these circumstances, the use of Scleroderris resistant seed sources will help ensure prompt and successful regeneration, and minimize build up of inoculum that might otherwise jeopardize other healthy regeneration nearby, whether natural or introduced.

There is evidence of tree-to-tree variation in susceptibility within the more disease resitant jack pine provenances. In Europe, well defined differences in infection were observed among 14 clones of Austrian pine growing near Rendsberg, Germany (Stephan 1970). Correlations of dry matter content of needles in fall with disease resistance have been found for Scots pine (Deitrichson 1968) and for jack pine (Teich 1968). The possibility exists for further improvement of provenances selected for use within specific regions by early selection.

CONCLUSIONS

1. Environments, land types, and forest conditions conducive to development of Scleroderris canker can be recognized as areas of high hazard for artificial regeneration.

2. Losses to Scleroderris canker can be minimized in these situations by using sources of jack pine known to be winter hardy, fast-growing, and resistant to disease.

3. The use of Scleroderris-prone planting stock will result in escalation of inoculum levels, failure to regenerate, and increased infection of other jack pine seedlings in the vicinity.

4. "Wave" years are unpredictable, but occurrence of Scleroderris infection can be reduced and severity minimized by insistence on use of climatically adapted planting stock.

5. Resistance to Scleroderris canker of jack pine should be considered seriously in creating improved breeding populations of jack pine for seed production and regeneration in the Boreal Forest Regions.

Genetic management is the first step in biological control. Known Scleroderris-resistant populations of jack pine need to be developed and used for seed production and to be tested widely on an operational scale within the Boreal Forests of Canada.

LITERATURE CITED

Cordell, C.E., D.D. Skilling and J.W. Benzie. 1968. Susceptibility of three pine species to _Scleroderris lagerbergii_ in Upper Michigan. Plant Dis. Rep. 52:37-39.

Deitrichson, J. 1968. Provenance and resistance to _Scleroderris lagerbergii_ Gremmen (_Gremmeniella abietina_ Lagerb.). The _international Scots pine provenance experiment of 1938_ at Matrand. Meddr Norske Skogfors Ves. 25:395-410.

Donaubauer, E. 1972. Environmental factors influencing outbreak of _Scleroderris lagerbergii_ Gremmen. Eur. J. For. Pathol. 2:21-25.

Donaubauer, E. Experiences with Scleroderris canker on _Pinus cembra_ L. in afforestation of high altitude. International Symposium on Scleroderris Canker of Conifers. Syracuse, New York, June 21-24, 1983.

Dorworth, C.E. 1971. Diseases of conifers incited by _Scleroderris lagerbergii_ Gremmen: a review and analysis. Can. For. Serv. Publ. No. 1289. 42 p.

Dorworth, C.E. 1972a. Epidemiology of _Scleroderris lagerbergii_ in central Ontario, Can. J. Bot. 50:751-765.

Dorworth, C.E. 1972b. Longevity of _Scleroderris lagerbergii_ Gremmen in pine slash. Bi-Month. Res. Notes 28:5.

Dorworth, C.E. 1974. Disease problems in intensively managed forests - _Scleroderris lagerbergii_. Possibilities for selection and breeding for resistance. Eur. J. For. Pathol. 4:228-232.

Dorworth, C.E. 1983. Impact of _Gremmeniella abietina_ in a jack pine plantation. Tree Planters Notes 34:21-24.

Dorworth, C.E. and P.E. Buchan. 1972. _Scleroderris lagerbergii_ Gremmen in the boreal forest of Ontario. Can. For. Serv., Great Lakes For. Res. Cent. Info. Rep. O-X-156. 7 pp.

Dorworth, C.E. and C.N. Davis. 1982. Current and predicted future impact of the North American race of Gremmeniella abietina on jack pine in Ontario. Can. For. Serv., Great Lakes For. Research Centre. Info. Rep. O-X-342. 18p.

Holst, M.J. 1960. Forest tree breeding and genetics at the Petawawa Forest Experiment Station. Biennial report, April 1, 1958, to March 31, 1960. Proc. 7th Meetg. Comm. For. Tree Breed. Can., Part II: K-1 - K-27.

Holst, M.J. 1962. Forest tree breeding and genetics at the Petawawa Forest Experiment Station. Biennial report, April 1, 1960, to March 31, 1962. Proc. 8th Meetg. Comm. For. Tree Breed. Can., Part II: M-1 - M-25.

Hudler, G.W., G.R. Knudson and M.A. Beale. 1981. Dose-response relationships of pines and spruces to Gremmeniella abietina. Phytopathology. 71:561.

Kohh, E. 1964. Om tallens gren- och granens topptorka och dess be kampning. Skogen 51:200-203 (from Punter 1967).

Pelkonen, H., P. Tuomi and J. Valtanen. 1982. Survival of pine on reforested sites in northern Finland Folia For. 511. 23 p. (Finish with Engl. summary).

Punter, D. 1967. Scleroderris lagerbergii Gremmen, a new threat to nurseries in Northern Ontario. For. Chron. 43:161-164.

Rowe, J.S. 1972. Forest Regions of Canada. Dept. of Environ., Can. For Serv., Ottawa. 172 p + map.

Rudolph, T.D. and C.W. Yeatman. 1982. Genetics of jack pine. U.S.D.A., For. Serv. Res. Pap. WO-38. 60 p.

Roll-Hansen, F. 1972. Scleroderris lagerbergii: resistance and difference in attack between pine species and provenances. Eur. J. For. Pathol. 2:26-39.

Skilling, D.D. 1969. The effect of temperature on ascospore release by Scleroderris lagerbergii. Plant Dis. Rep. 53:289-290.

Skilling, D.D. and J.T. O'Brien. 1969. Scleroderris canker in the Lake States - a situation report, 1968. U.S.D.A. For. Serv. Res. Note NC-73. 4 p.

Stephan, B.R. 1970. Klonabhangiges Verhalten bei Pinus nigra Arnold gegenuber Scleroderris lagerbergii Gremmen. Allgem. Forst. Jagdzeit. 141:60-63.

Teich, A.H. 1967. Resistance of jack pine to Scleroderris lagerbergii Gremmen. Bi-month. Res. Notes, Can. Dept. For. 23:5.

Teich, A.H. 1968. Foliar moisture content as a criterion for resistance to frost and Scleroderris canker in jack pine. Can. Dep. Forest. Rur. Develop., Bi-mon. Res. Notes 24:3.

Teich, A.H. 1972. Damage caused by the fungus Scleroderris in plantations and nurseries in Europe and North America. Eur. J. For. Pathol. 2:11-15.

Teich, A.H. And E. Smerlis. 1969. Jack pine resistance to Scleroderris lagerbergii. Can. Dep. Fish. Forest., Bi-mon. Res. Notes 25:47.

Yeatman, C.W. 1974. The jack pine genetics program at Petawawa Forest Experiment Station 1950-1970. Dept. Environ., Can. For. Serv. Publ. No. 1331. 30 p.

Yeatman, C.W. 1976a. A Canadian example of Government-industry collaboration in tree improvement. For. Chron. 52:283-288.

Yeatman, C.W. 1976b. Seed origin-first, last and always. Can. For. Serv., Petawawa For. Exp. Sta., Infor. Rep. PS-X-64, 12 p.

RESISTANCE OF LODGEPOLE PINE TO SCLERODERRIS CANKER IN NORTHERN SWEDEN

Owe Martinsson[1]

ABSTRACT

Seedlings of eleven lodgepole pine progenies from western Canada and one local progeny of Scots pine growing on two sites with a great difference in site conditions were artificially inoculated with Gremmeniella abietina.

The interaction between site and the genetic origin or progenies was highly significant on the survival of the seedlings. On both sites the survival of Scots pine was lower than lodgepole pine.

INTRODUCTION

The introduction of lodgepole pine into the forestry of Scandinavia has become of great economic importance during the last 10-15 years, especially in Sweden where 50 million seedlings of lodgepole pine have been planted annually during the last 5 years.

Upon introduction of a foreign tree species special attention must be payed to problems of diseases, native as well as the risk of introducing new ones. Scleroderris canker, caused by Gremmeniella abietina, is native to Scandinavia but epidemic outbreaks of this disease have been rare and have never taken any proportion comparable to the outbreak in northeastern United States in the 1970's. However, in addition to snow blight (Phacidium infestans) and pine twist rust (Melampsora pinitorqua), Scleroderris canker is considered the most common disease in 4-9 year old plantations of Scots pine in northern Sweden and can locally attack 25-30% of the seedlings (Martinsson et al, 1983). This fungus is also involved in a disease-complex on spruce in southern Scandinavia called sub-top-dying (Barklund & Rowe 1981). So far, plantations of lodgepole pine have usually been healthier than the Scots pine with respect to any diseases.

The objective of the experiment described here was to test the field resistance to Scleroderris of different populations of lodgepole pine and compare the resistance of Scots pine.

[1] Research leader at the Swedish University of Agricultural Sciences, Faculty of Forestry, Dep. of Silviculture, S-901 83 Umeå, Sweden.

MATERIAL AND METHODS

Seedlings from 5 provenances and 6 open pollinated progenies of lodgepole pine and 1 full-sib-family of Scots pine were cultivated in a nursery and outplanted as 1-year old transplants on two experimental sites in northern and central Sweden (Figure 1). The origin of the experimental material is listed in Table 1. The Scots pine progeny were selected from an earlier test of 193 progenies as having a medium resistance to Scleroderris (Martinsson 1975). One hundred seedlings of each population were outplanted on each site. The northern site, Svartberget, was a poor lichen pine site with frost regularly during the growing season. Several attempts had been made to regenerate Scots pine on this site without success. The southern site, Langmor, was abandoned farm land with good frost drainage. Because of a heavy grass vegetation, this site was properly site prepared before planting.

Table 1. Origin of the experimental material.

Population no.	Reg.no.	Seed collection site	Latitude N	Longitude W	Altitude m, a.s.l.
1	Is 737	Rocky Mt. House	52°38'	115°23'	1150
2	Is 705	Chulty Lake	53°50'	123°30'	730
3	Is 601	-	54°32'	128°30'	300
4	Is 708	Chetwynd, North	55°45'	121°40'	670
5	Is 668	Hudson Hope	56°00'	122°00'	700
6	SCA tree 21	-	58°43'	123°39'	615
7	" " 13		62°06'	135°41'	630
8	" " 14		62°03'	135°47'	590
9	" " 2		63°19'	136°24'	990
10	" " 6		63°28'	136°26'	660
11	" " 8:a		63°23'	136°49'	500
12	Y 411-50	Y 3025 x Z 4033 Domsjöänget, seed orchard			

The seedlings were outplanted in the spring in complete randomized 10-seedling plots in 10 replications on each site. In October the same year the material was inoculated with conidia of G. abietina suspended in water. The inoculum had been growing in a growth chamber on artificial medium. The suspension which was prepared contained 120000 germinable conidia per ml. With a medicine syringe, 4 drops of the suspension (= 0.1 - 0.2 ml) were put on the top bud of each seedling.

Symptoms of the disease started developing in the spring of the following year. The first complete assessment was not done until the spring of the following year, i.e. 18 months after the inoculation. Killed, damaged and undamaged seedlings were assessed. A second assessment was done in the fall, four years after the inoculation, when seedlings were assessed as either living or killed and the top height of living seedlings were measured.

RESULTS

The mean survival on the harsh-climate site, Svartberget, was 42% and the range from 7% (the Scots pine family) to 72% (lodgepole pine pop. no. 6). On the mild site, Långmor, the mean survival was 58% and the range from 29% (1.p. pop. no.1) to 76% (1.p. pop. no.10) (Figure 1 and Table 2). The best surviving populations on the harsh-climate site were lodgepole pine originating from northern British Columbia and the Yukon. On the mild-climate site some of the populations from central British Columbia performed better than those coming from the Yukon and northern British Columbia. On the harsh-climate site the Scots pine was more susceptible to the disease than any of the lodgepole pine populations and on the mild-climate site only one of the lodgepole pine populations was more susceptible than the Scots pine.

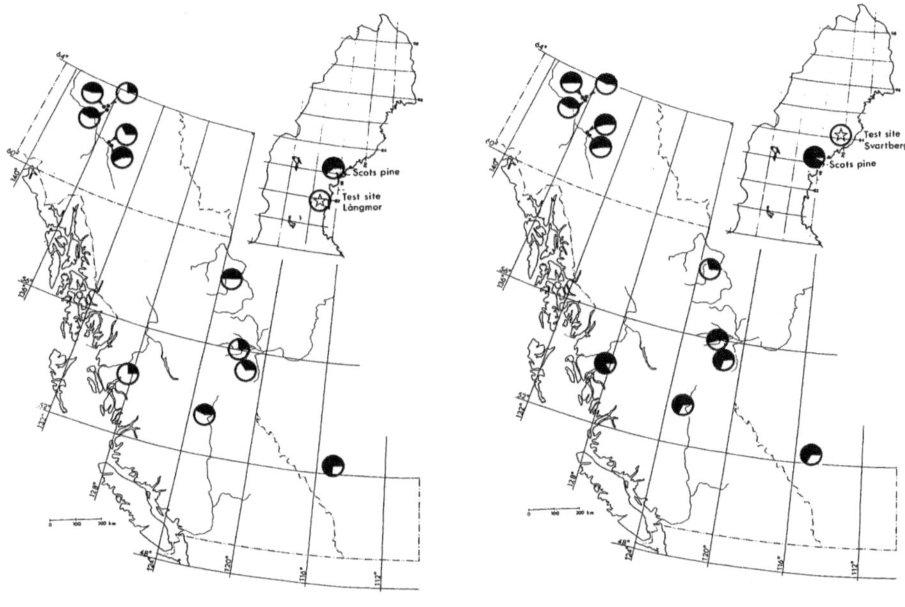

a b

Figure 1. Mortality of one Swedish Scots pine and eleven Canadian lodgepole pine populations tested for resistance to <u>Gremmeniella abietina</u> in Langmor (Figure 1a) and Svartberget (Figure 1b). The black sector indicates % mortality.

DISCUSSION

There is a big difference in survival of the same populations on the two experimental sites, or in other words there is a strong population site interaction (Table 3). The reason for this difference in survival of the same populations on the two experimental sites, or in other words there is a strong population site interaction (Table 3). The reason for this difference between the sites is probably the large difference in climatic conditions, which is

Table 2. Survival and mean height of 5 year old seedlings growing on two experimental sites four years after inoculation of Gremmeniella abietina.

Population no.	Survival, %		Means heights of undamaged 5 year old seedlings, cm		Mean height of damaged but living 5 year old seedlings in percent of the height of undamaged	
	Svartberget	Långmor	Svartberget	Långmor	Svartberget	Långmor
1	34	29	80	133	51	66
2	32	60	71	146	55	72
3	16	75	57	123	55	64
4	31	66	90	144	52	55
5	45	74	48	141	92	62
6	72	52	70	111	72	77
7	47	67	84	138	46	57
8	48	43	90	121	57	73
9	58	61	80	127	63	62
10	59	76	96	150	51	65
11	50	56	67	136	68	50
12	7	38	-	120	-	91

also reflected in the height growth of undamaged seedlings (Table 2). In Svartberget, the experimental site was localized in an area where severe frost occurred regularly through the summer and the survival of the seedlings would probably have been rather low even without inoculation of the disease. In Långmor, the disease was the primary reason for mortality. On both sites, however, the attack of the disease increased from one year to the next. Consequently, a part of the material was killed by secondary inoculation spread from neighboring seedlings damaged or killed by the artificial inoculation.

Table 3. Analysis of variance of 240 plot means.

Source of variation	D.f.	Mean Square
Between populations (a)	11	0.33
Between test sites (b)	1	1.62
Interaction test sites x populations (c)	11	0.25
Error (d)	216	0.04
	239	

$F_{a/d} = 9.49^{***}$ $F_{b/d} = 46.29^{***}$ $F_{c/d} = 7.03^{***}$

Those seedlings which were damaged at the first assessment and still living at the second had a more or less reduced growth rate compared to seedlings which were undamaged on both occasions (Table 2).

One would have expected that on the northern site the best survivor should be found among the Yukon populations. The reason this was not so could be that all populations 6-11 were open pollinated single tree progenies. The variation of a trait between trees within a stand could be large and the progeny of a single tree cannot represent the provenance. The populations 1-5 were bulk seed collected from several trees of the stands.

CONCLUSIONS

The field resistance to Scleroderris canker of lodgepole pine seems to be dependent on the genetic origin of the seed only to some degree. The interaction between the environment and the genetic traits of the population strongly influences resistance. A population which is weakened by the unfavorable environment is more sensitive to the disease than a population which is well adapted to the environment.

LITERATURE CITED

Barklund, P. and J. Rowe. 1981. Gremmeniella abietina (Scleroderris lagerbergii) a primary parasite in a Norway spruce die-back. Eur. J. For. Pathol. 11:97-108.

Martinsson, O., M. Karlman and J-E. Lundh. 1983. Mortality and damage in semi-practical trials of Scots pine and lodgepole pine 4-9 years after plantation (in Swedish with English summary). - SLU, Inst. Skogskötsel. Rapp. 10, 33 p.

Martinsson, O. 1976. Resistensprovning mot parasitsvampar. 1975. Skogshögskolan. Inst. För. Skogsbotanik Stencil, 54 p.

TESTING FUNGICIDES FOR CONTROL OF *Gremmeniella abietina* IN NEW YORK

Darroll D. Skilling, Bruce S. Schneider, and Terence B. Podolski[1]

ABSTRACT

Six fungicides were tested for controlling the European strain of *G. abietina* on Scots pine stock. Chlorothalonil and a combination of chlorothalonil + cycloheximide were most effective. Four sprays of chlorothalonil at 1% ai were as effective as 12 sprays at 0.3% ai. Severe infection developed in all test plots regardless of treatment during February 1981 when temperatures were above 0C and free water was present on exposed seedlings. These infected seedlings showed no symptoms until mid-June 1981.

INTRODUCTION

Scleroderris canker, caused by the fungus *Gremmeniella abietina* (Lagerb.) Morelet, was first identified in North America in Ontario in 1962 and in the United States in 1964 (Ohman 1966). The fungus was responsible for major losses in both conifer plantations and forest nurseries in Northern United States and Canada during the 1960's. Damage from this fungus was limited to pine nursery stock and to young pines less than 2 meters tall. Red pine (*Pinus resinosa* Ait.) and jack pine (*P. banksiana* Lamb.) were primarily hosts, but Scots pine (*P. sylvestris* L.) and black spruce (*Picea mariana* (Mill.) B.S.P.) were also moderately attacked.

During the 1960's, research showed that most damage in plantations resulted from planting infected nursery stock. Fungicide controls were developed for nursery stock and by 1970 the losses from this disease had been greatly reduced. Although losses continued where new pine plantations were established near infected stands, plantation losses dropped from an average of 40% before nursery control to less than 10%.

In 1975, the European strain of *G. abietina* was identified in northern New York. It appeared to be more pathogenic than the North American strain because it was able to infect a greater number of conifer species and to attack conifers of all ages.

---

[1] Principal plant pathologist, North Central Forest Experiment Station, USDA Forest Service, 1992 Folwell Avenue, St. Paul, MN 55108; principal forestry technician, New York State Department of Environmental Conservation, Lowville, NY 13367 and forestry technician, New York State Department of Environmental Conservation, Lowville, N.Y. 13367, respectively.

Field studies in northern Wisconsin by the North Central Forest Experiment Station had shown that the fungicide chlorothalonil alone or in combination with cycloheximide gave excellent control of the American strain of G. abietina when aplied to red pine seedlings seven times during the growing season at a concentration of 1.0% ai (Skilling and Waddell 1974). The purpose of this study was to evaluate these and other candidate fungicides for controlling the European strain of G. abietina.

MATERIALS AND METHODS

A forested area 8 km NE of Boonville, New York, was used as a test site for all fungicide trials. The European strain of G. abietina was present in the immediate area.

1979 Tests

During the first week of May 1979, 42 plots (0.5 x .75 m) were each planted with 100 2-0 Scots pine seedlings of the Danish variety. Infected red pine branches with fresh pycnidia of G. abietina were positioned 0.4 m from the soil over the test seedlings. These branches were removed prior to each fungicide application and replaced after each spray. Two spray schedules were used: plots were sprayed eight times -- the 1st and 3rd weeks in May and June and the 1st week in July, August, September, and October; and plots were sprayed only four times -- 1st week in May, June, July, and August. We varied both spray schedules some-what depending on frequency and intensity of rain.

Plots were sprayed with portable 7.5 liter sprayers equipped with pressure gauges. The spray nozzles were calibrated to deliver 100 ml of spray in 15 seconds at 275 KPa (40 psi). Each block of 100 trees received 100 ml of fungicide at each spray application. Separate sprayers were used for each fungicide tested to avoid contamination.

Four fungicide mixtures were tested in 1979: (1) tetrachloroisophthalonitrile (chlorothalonil) (Bravo 6F) applied at 0.3% and 1.0% concentration of active ingredients (ai); (2) a mixture of chlorothalonil (Bravo W75) at 0.3% ai + 0.1% or .2% ai of 3 (2-(3, 5-Dimethyl-2-oxocyclohexyl) -2-hydroxyethyl) -glutarimide (cycloheximide);(3) 2, 3-dichloro-1. 4-naphthoquinone (dichlone) (Phygon 50W) at 0.3% or 1.0% ai; and (4) N-(Trichloromethylthio) phthalimide (folpet) (Phalton W50) at 0.3 or 1.0% ai.

Three unsprayed check plots were also inoculated with G. abietina. Each treatment was replicated three times and treatments were randomly applied. In June 1980, seedlings were harvested and examined for evidence of Scleroderris canker. If disease symptoms were not obvious, cultural isolations were made to determine whether the trees were infected.

1980 Tests

The 1980 tests consisted of 51 plots (0.5 x .7 m) each with 100 2-0 Scots pine seedlings. Inoculation procedures were the same as in 1979. Two spray schedules were used: (1) plots were sprayed four times -- on May 6, June 2, July 7, and August 4; and (2) plots were sprayed 12 times -- on May 6, 19, June 2, July 7, 21, August 4, 18, Sept. 3, 15, 29, Oct. 14, 29. More applications were used in 1980 because spore trapping studies adjacent to the fungicide test in 1979 indicated that viable spores were produced in late October and November after the spray program was completed. Spray equipment and procedure were the same as in the 1979 test. Each plot again received 100 ml of spray solution.

The following five fungicides were tested in 1980: (1) chlorothalonil applied at 0.3% and 1.0% ai; (2) a mixture of chlorothalonil at 0.3% ai + 0.1% or 0.2% ai cycloheximide; (3) dichlone at 0.3% or 1.0% ai; (4) folpet at 0.3% or 1.0% ai, and (5) cycloheximide at 0.1% or 0.2% ai. Three unsprayed inoculated check plots were included. Each treatment was replicated three times and applied randomly.

1981 Tests

In 1981 23 plots each with 100 2-0 Scots pine seedlings were used. Inoculation procedures and spray equipment were the same as in previous tests. Three spray schedules were used: (1) plots were sprayed four times -- on May 4, June 1, June 29, and August 3; (2) plots were sprayed eight times -- on May 4, 18, June 1, 15, 29, August 3, 31 and October 5; and (3) plots were sprayed 12 times on May 4, 18, June 1, 15, 29, July 13, August 3, 31, Sept. 21, October 5, 19, and November 2.

Three fungicides were test: (1) chlorothalonil at 0.3 and 1.0% ai; (2) 1-[[2-(2,4 dichlorophenyl)-4-ethyl-1, 3-dioxolan-2yl] methyl] -1H-1,2,4-trizole (etaconazole) (Vangard 10W) at 0.025 and 0.050% ai; and (3) cycloheximide at 0.2 and 0.4% ai. Two inoculated check plots were included. Each fungicide treatment was replicated three times.

RESULTS

1979 Tests

No evidence of Scleroderris canker was observed on the test seedlings until mid-April 1980. At that time we observed orange discoloration at the base of infected needles, needle drooping, loose needles, and stem lesions.

We saw no evidence of phytotoxicity in any of the test plots regardless of fungicide or concentration. The eight sprays of chlorothalonil at 0.3% ai gave the best control of *G. abietina* (Table 1). The eight-spray applications of chlorothalonil + cycloximide gave some control but variation in plot infection in this test was high. Dichlone and folpet gave only minimal control of *G. abietina*. In all tests eight applications were more effective than four applications.

1980 Tests

Although the spray schedule was extended in 1980, it was not sufficient to eliminate winter infection. During February 1981 three weeks of unusually warm weather in northern New York melted all the snow off the test beds. For several days the temperature remained above freezing with fog and rain at frequent intervals. This unusual weather resulted in a heavy infection by *G. abietina* during February. The first symptoms did not show on the test seedlings until late June 1981.

Table 1. Effectiveness of fungicides for control of the European strain of Scleroderris canker on 2-0 Scots pine seedlings.

Fungicide formulation	% Active ingredient	No. sprays	No. trees	% Trees infected
1979 Results				
Chlorothalonil	0.3	4	288	43
(Bravo 6F)	0.3	8	289	7
	1.0	4	284	24
Chlorothalonil +	0.3	4	273	62
0.1% cycloheximide	0.3	8	263	14
Chlorothalonil +	0.3	4	269	55
0.2% cycloheximide	0.3	8	257	29
Dichlone	0.3	4	242	84
(Phygon 50W)	0.3	8	253	50
	1.0	4	274	69
Folpet	0.3	4	261	82
(Phalton W50)	0.3	8	259	34
	1.0	4	263	54
Check inoculated	--	0	264	96
1980 Results				
Chlorothalonil	0.3	4	287	64
(Bravo W75)	0.3	12	290	26
	1.0	3	289	25
Chlorothalonil +	0.3	4	288	53
0.1% cycloheximide	0.3	12	284	26
Chlorothalonil +	0.3	4	289	59
0.2% cycloheximide	0.3	12	294	12
Dichlone	0.3	4	286	96
(Phygon 50W)	0.3	12	291	55
	1.0	4	292	70
Folpet	0.3	4	290	92
(Phalton W50)	0.3	12	285	52
	1.0	4	287	69
Cycloheximide	0.1	4	295	100
(Acti-dione TGF)	0.1	12	293	98
	0.2	4	288	199
Check inouclated	--	0	294	100
1981 Results				
Chlorothalonil	0.3	12	286	31
(Bravo S75)	1.0	4	291	38
Etaconazole	0.025	8	289	100
(Vangard 10W)	0.025	12	278	100
	0.050	4	276	100
Cycloheximide	0.2	12	272	99
(Acti-dione TGF)	0.4	4	275	99
Check inoculated	--	0	188	100

The best results again were produced by chlorothalonil and chlorothalonil + cycloheximide. Dichlone and folpet gave only minimal control. Cycloheximide alone showed no control activity.

1981 Tests

Chlorothalonil was again the most effective fungicide in the 1981 test. Under the heavy inoculation loads used in this test, however, the best treatment still resulted in 31 percent infection. Etaconazole showed no effectiveness in field tests although this material was highly active in laboratory screening tests. As in 1980, cycloheximide alone showed no control activity.

CONCLUSIONS

Based on these fungicide studies, chlorothalonil is the best fungicide tested to date for controlling *G. abietina*. The results, however, are disappointing compared with previous tests using this same fungicide against the North American strain of *G. abietina*. At this point it appears that further tests are needed using chlorothalonil at higher concentrations and more applications. The current study was artificial because very heavy inoculum loads were placed directly over the test seedlings. In a nursery situation the inoculum levels would be much lower. In previous tests, however, using the same heavy inoculum loads only one percent of the red pine test seedlings were infected when chlorothalonil was applied seven times during the growing season at 1.0% ai (Skilling and Waddel 1974).

In the current tests dichlone, folpet, and etaconazole did not control *G. abietina* under heavy inoculum loads. The results of this study indicate that because of the long infection period for the European strain, it will be difficult for nursery managers to protect seedlings throughout the entire year.

LITERATURE CITED

Ohman J.H. 1966. *Scleroderris lagerbergii* Gremmen: the cause of dieback and mortality of red and jack pines in upper Michigan plantations. Plant Dis. Rep. 50:402-405.

Skilling, D.D. and C.W. Waddell. 1974. Fungicides for control of Scleroderris canker. Plant Dis. Rep. 58:1097-1100.

INTERACTIONS BETWEEN EPIPHYTIC BACTERIA AND CONIDIA OF *Gremmeniella abietina*

G.R. Knudsen and G.W. Hudler[1]

ABSTRACT

Washings from red pine needles treated with a strain of *Pseudomonas fluorescens* and a nutrient source remained inhibitory to conidial germination of *Gremmeniella abietina* for 72 hr. after application of bacteria. In field studies, strains of *P. fluorescens* were applied to red pine seedlings that were subsequently inoculated with conidia of *G. abietina*. Disease incidence was lowest when bacteria were applied in a nutrient solution immediately prior to inoculation with conidia. Results suggest that temperature, availability of free water and nutrients affect survival of applied antagonists and subsequent control of *G. abietina*. Environmental effects on antagonist and pathogen survival and interaction are being modeled.

INTRODUCTION

Scleroderris canker, caused by the ascomycete *Gremmeniella abietina* (Lagerb.) Morelet, causes branch dieback on species of *Pinus*, *Picea*, *Abies*, *Larix* and *Pseudotsuga* in New York State. Most infections occur on current year's shoots following deposition of conidia by splashing rain. High humidity or moisture on the plant surface is presumed necessary for conidial germination.

Previous investigations showed that bacteria, including strains of *Pseudomonas fluorescens* were common on red pine shoots, and they inhibited germination of *G. abietina* conidia on agar and on needles of red pine seedlings in a growth chamber (Knudsen 1981; Knudsen and Hudler 1983). One mechanism of inhibition was determined to be production of an antibiotic substance. Germination on needles was reduced from 54% in the absence of *P. fluorecens* to 2% when 10^8 colony-forming units (CFU)/ml of *P. fluorescens* in dilute nutrient broth were added.

Although *P. fluorescens* inhibited germination of *G. abietina* conidia on red pine needles when the two organisms were applied together, it was not known whether the inhibitory effect would persist on the plant surface for any significant length of time. Persistence of either bacteria or the inhibitory compounds they produce is critical if applied bacteria are to protect seedlings from an influx of natural inoculum over time. In this report, an experiment is

1 Graduate Research Assistant, and Assistant Professor, Department of Plant Pathology, Cornell University, Ithaca, N.Y. 14853

described to test for persistence of an inhibitory effect on seedling surfaces for 72 hr. following addition of P. fluorescens. Preliminary field trials to test the ability of applied P. fluorescens to prevent infection of red pine seedlings by G. abietina are also described.

Numerous other discoveries of epiphytic bacteria antagonistic to pathogenic fungi have been reported (Blakeman and Brodie 1976). However, there has been little research on the effects of environment on antagonist survival and interactions between antagonist and pathogen on plant surfaces. Knowledge of these effects may help to determine when and in what concentration bacteria should be applied. It may also help to predict the effects on germination of conidia and subsequent infection of the plant host. Data obtained from axenic systems may be inappropriate for these purposes; the use of whole plants or excised plant parts in a controlled environment provides a compromise between "unnatural" axenic systems and less easily controlled field experiments. In this report, initial efforts to determine the effects of temperature, leaf wetness and added nutrients on growth and survival of P. fluorescens, and temperature effects on germination of conidia on pine needles are described. Use of these results in a systems analysis approach to biological control of G. abietina is discussed.

MATERIALS AND METHODS

Strain of P. fluorescens and G. abietina Used

Three strains of P. fluorescens were used for experiments described herein. Strains R10a and S5a, collected from red pine in New York State, have been previously described (Knudsen 1981). Both strains inhibited germination of G. abietina conidia on agar. Strain R10a-r was selected for resistance to rifampin (Calbiochem-Behring Corp., La Jolla, CA 92037) by spreading 0.1 ml of a suspension of strain R10a on plates of Difco nutrient agar containing 20 g/ml rifampin. Resistant colonies were streaked again on the same medium, and a single resultant colony was subsequently maintained on the rifampin-supplemented medium.

Conidia of G. abietina used in the following experiments were obtained from 3-wk-old cultures of our isolates Ga39 or Ga6 growing on a defined medium (Knudsen 1981). Both isolates were obtained from infected red pine trees in New York State and were serologically identical with European isolates of the fungus (D.D. Skilling, personal communication).

Persistence of Inhibitory Effect on Seedling Surfaces

Three-yr-old red pine seedlings growing in pots were sprayed to runoff with either sterile water (SW), 10^8 CFU/ml of R10a in SW, 10^8 CFU/ml of R10a in 0.08% Difco nutrient broth (NB), or 10^8 CFU/ml of R10a plus 10^5 G. abietina conidia/ml in SW. Conidia were included in the latter treatment as a potential nutrient source for bacteria. Seedlings were sealed in large plastic bags to maintain leaf wetness and held at 20C. At 1,12,48 and 72 hr, seedlings were misted with sterile water until drops formed on the needles. A capillary pipette was used to collect 4 0.01-ml samples of these drops per treatment. Each 0.01 ml sample was placed on a 0.5 cm^3 block of Difco Bacto agar (WA) and covered with a 0.2 m-pore size filter (Gelman Sciences, Inc. Ann Arbor, MI 48106). Then, 0.1 ml of conidial suspension (10^5 conidia/ml) was placed onto each filter. After 36 hr at 20 C, conidia were stained with cotton blue in 70%

ethanol. At least 100 conidia were counted along one radius of each filter, and percent germination was determined. A conidium was considered germinated if it had a distinct germ tube at least as long as the width of the spore.

At 1,12 and 48 hrs. 3 0.1-ml sample drops were removed from needle surfaces on each seedling and placed in test tubes containing 9.9 ml SW. Ten-fold dilutions were made, and 0.1 ml of each dilution was placed onto Difco *Pseudomonas* isolation agar (PSA). After four days at 20C, colonies were counted and average numbers of CFU/ml in drops were determined.

Field Trials to Assess Control of G. abietina by Antagonistic Bacteria

Field experiments were conducted to determine if *P. fluorescens* applied to red pine seedlings could reduce the incidence of infecton by *G. abietina*. In May 1981, 1000 3-yr-old seedlings were planted in 50 plots of 20 seedlings each. Three weeks later, seedlings were sprayed to runoff with suspensions of *G. abietina* conidia (10^3 conidia/ml). Prior to inoculation with conidia, some plots were sprayed with 10^8 CFU/ml of *P. fluorescens* (strain S5a or R10a) in SW or NB or with sterile NB according to the schedule in Table 1, Part A. Treatments were replicated three times. Strains of *P. fluorescens* had been grown for 24 hr on nutrient agar prior to application. Inoculations were performed after sunset to minimize initial evaporation. Disease incidence, expressed as the proportion of seedlings exhibiting symptoms of infection by *G. abietina* (basal discoloration of needles and necrotic cambial tissue), was assessed in May 1982.

Two test plots were sprayed with the rifampin-resistant strain R10a-r in SW or NB respectively. Seedlings from these plots were sampled as follows to determine persistence of applied bacteria: At 0,1,12,16,24 and 36 hrs. after inoculation, the distal 5 cm of shoots from two seedlings in each plot were washed for 1 hr. in 100 ml SW, and serial dilutions were plated onto PSA amended with 20 g/ml rifampin. Colonies were counted after four days at 20C. Shoot samples were dried and weighed, and average numbers of CFU/g of shoot were determined.

Survival and Growth of Strain R10a on Red Pine Needles

Population changes of strain R10a applied to continuously wet red pine needles were determined under different temperature and nutrient regimes. Needles (1-2-yr-old) were obtained from 3-yr old red pine seedlings transplanted from a nursery and kept 5-7 months in pots in a greenhouse. Each experimental unit consisted of a pair of needles sharing a basal sheath and trimmed to 5 cm in length. Needles were immersed for 1 min in an aqueous suspension of R10a (3 x 10^6 CFU/ml) or in a suspension of R10a (3 x 10^6 CFU/ml) in NB, and then placed on plastic mesh screens over moist filter paper inside glass petri dishes. Petri dishes were sealed with Parafilm M (American Can Co., Greenwich, CT 06830) and incubated at 5,10,18 or 28C. Five needle pairs per treatment were removed at the following time intervals for sampling: 0,12, 24,48, 72,96,120 and 144 hr. Needles were visibly wet when sampled. Each sample was placed in a test tube containing 10 ml sterile water, and test tubes were then placed in a water bath (30 C) in the tank of an ultrasonic cleaner and exposed to ultrasound for 1 min to dislodge bacteria. Ten-fold serial dilutions were made from each tube, and 0.1 ml. samples were plated on PSA and incubated for four days at 22C. Colonies were counted to determine numbers of bacteria per needle pair recovered in each treatment.

Population changes of strain R10a on red pine needles at different temperatures in the absence of free water were determined by immersing needle pairs (as described above) for 1 min in an aqueous suspension of R10a (10^9 CFU/ml), and then placing them on plastic mesh screens over dry filter paper in petri dishes. Petri dishes were left open and incubated at 5,10,18,21 or 28C. Five needle pairs per treatment were sampled at 0,1,4,12, 24,28,72,96, 120 and 144 hr.

Effect of Temperature on Germination of Conidia on Red Pine Needles

Red pine needle pairs were immersed in an aqueous conidial suspension (10^5 conidia/ml) and placed on plastic mesh screens over moist filter paper in glass petri dishes. Petri dishes were sealed and incubated at 5,10,18,21 or 28C. Needle pairs (5-10) were sampled at 0,12,24,36,48,72,96,120 and 144 hr. Needles were visibly wet when sampled. Conidia were removed by pressing adaxial surfaces of needles onto double-stick transparent tape affixed to microscope slides. Conidia were observed by bright-field microscopy, and percent germination of 400 conidia per sample was determined. In order to estimate maximum viability, a sample of conidia from the original suspension was incubated on WA for 48 hr at 18C.

Table 1. 1981 field trials: application of *Pseudomonas fluorescens*[1] to red pine seedlings prior to inoculation with conidia of *Gremmeniella abietina*.[2]

Part A: Treatments		Part B: Results	
		% Seedlings Diseased	
Treatment	Date of Application	Range	Mean
1. R10a in NB	6/17	35-41	37
2. R10a in SW	6/17	39-71	60
3. S5a in NB	6/17	29-100	55
4. S5a in SW	6/17	29-67	53
5. Sterile NB	6/17	8-64	28
6. R10a in NB	6/23	22-53	40
7. R10a in SW	6/23	22-64	44
8. S5a in NB	6/23	19-72	44
9. S5a in SW	6/23	43-89	69
10. Sterile NB	6/23	19-59	45
11. R10a in NB	6/24	20-33	27
12. R10a in SW	6/24	17-93	58
13. S5a in NB	6/24	7-22	13
14. S5a in SW	6/24	50-73	63
15. Sterile NB	6/24	15-65	35
16. No prior treatment		47-64	61

[1] *P. fluorescens* strain R10a or S5a in sterile water (SW) or 0.8% nutrient broth (NB), 10^8 CFU/ml sprayed to runoff.

[2] 10^3 conidia/ml in aqueous suspension, sprayed to runoff on 6/24.

RESULTS

Persistence of Inhibitory Effect on Seedling Surfaces

Germination of conidia exposed to filtered water drops from needles are shown in Figure 1. Recovery of fluorescent bacteria from water drops collected from the same seedlings are shown in Figure 2. Germination was consistently above 50% when conidia were exposed to drops from needles treated with SW only, and relatively few fluorescent bacteria (<1000 CFU/ml in drops) were recovered from these seedlings (Figure 2). Since R10a was not added in this treatment, any bacteria recovered presumably represented normal microflora and/or possibly contaminants from other treatments. Where needles received drops of SW+R10a, germination of conidia increased with longer incubation time of drops on needles, and recovery of fluorescent bacteria decreased. Where needles received either NB+R10a or SW=R10a+Ga39, germination was low (<10%) regardless of incubation time of drops. Populations of fluorescent bacteria in these two treatments fluctuated but were relatively high (2-7 x 10^5 CFU/ml in drops) at 48 hr.

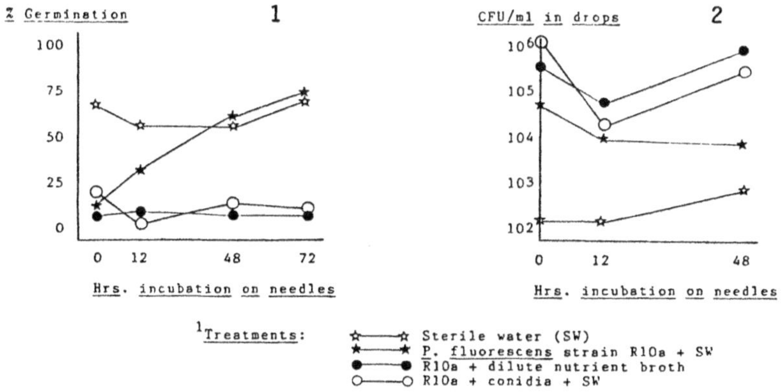

Figure 1. Germination of conidia exposed to drops[1] incubated on red pine needles.

Figure 2. CFJ/ml of fluorescent bacteria recovered from drops[1] incubated on red pine needles.

Field Trials

Table 1, Part B shows the levels of disease incidence observed in each of the treatments. Considerable variability was observed within treatments; when all treatments were considered independently, analysis of variance detected no significant differences among treatments ($P > 0.05$). Similarly, partitioning the treatment sums of squares to compare all treatments with S5a or R10a to all treatments without bacteria detected no significant differences between the two groups. However, the two treatments in which bacteria were applied in NB concurrently with conidial inoculation had significantly lower disease incidence ($P > 0.01$) when compared to all other treatments as a group.

Mean values for recovery of strain R10a-r are shown in Figure 3. During the course of the experiment we observed that needles remained visibly wet approximately 12 hrs. following inoculation (i.e., overnight), but appeared dry during the following day. Populations of bacteria declined most sharply after about 12 hrs., and the rate of decline was faster for bacteria applied in SW than for those applied in NB.

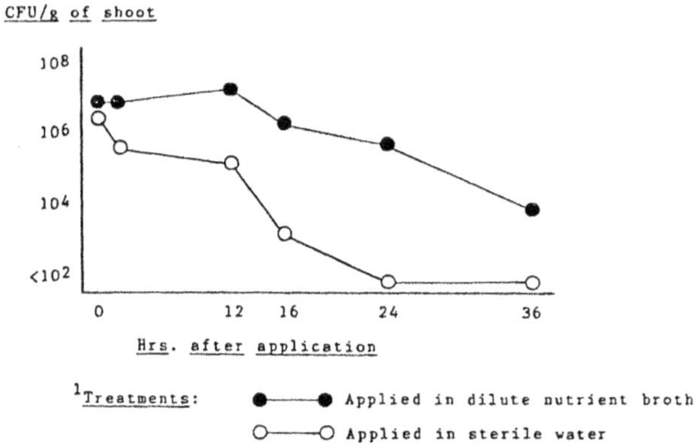

Figure 3. Recovery of rifampin-resistant bacteria[1] from seedlings.

Survival and Growth of Strain R10a on Pine Needles

Changes in numbers of fluorescent bacteria recovered from red pine needles under the various nutrient, temperature and leaf wetness regimes are shown in Figures 4 and 5. Populations of bacteria increased whenever leaf wetness was maintained; the rate of increase was greater in nutrient solution than in water, and varied with temperature. With constant leaf wetness, numbers of bacteria continued to increase throughout the sampling period. By 144 hrs. however, saprophytic fungi (e.g., Alternaria spp.) were observed on needle surfaces, and bacteria may have derived nutrients from decomposition of the excised needles. In the absence of free water, bacterial numbers declined rapidly at first, then increased for several hours and declined again. Rates of decline were faster at higher temperatures. Surfaces between needles in each pair were observed to remain wet for several hours after outer needle surfaces appeared dry.

Effect of Temperature on Germination of Conidia

Cumulative germination of conidia over time at different temperatures is shown in Figure 6. Maximum observed germination on needles was 61%; germination of conidia on WA was 85%. Conidia did not germinate at 28C. Many lysed conidial cells were seen in samples from needles, and these cells were often surrounded by unidentified bacteria or yeasts.

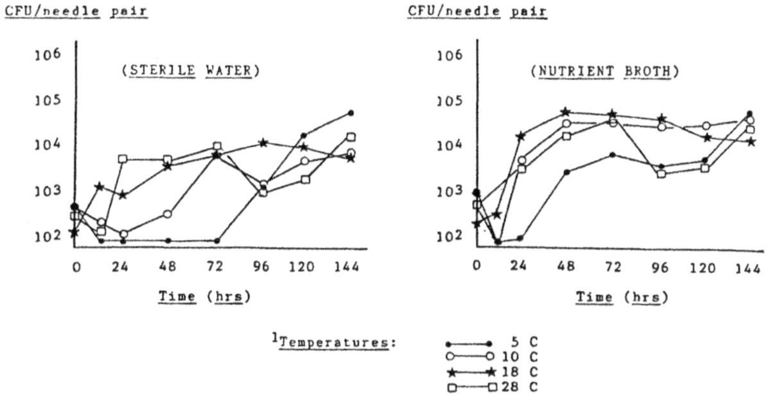

Figure 4. Population dynamics of P. fluorescens on red pine needles at various temperatures[1]; leaf wetness maintained.

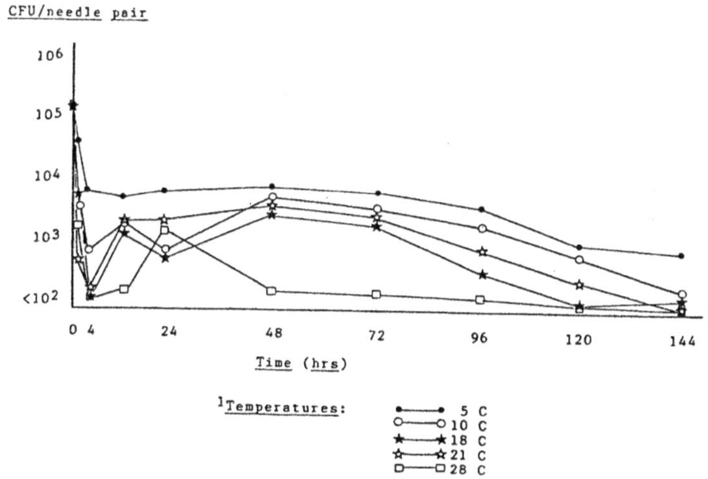

Figure 5. Population dynamics of P. fluorescens on red pine needles at various temperatures[1]; leaf wetness not maintained.

DISCUSSION

An inhibitory effect persisted on pine needles when P. fluorscens strain R10a was applied in nutrient broth or with conidia, and inhibition of conidial germination corresponded with persistence of applied bacteria. Persistence of any inhibitory effect is important if applied bacteria are to be used for field control of foliar pathogens, as conidia may be deposited on seedlings throughout the growing season.

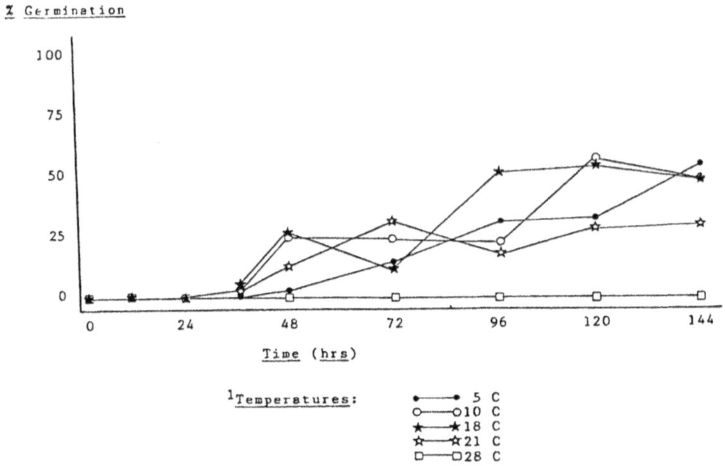

Figure 6. Effect of temperature[1] on germination of conidia.

Field trial results indicated that disease incidence was lowest when P. fluorescens was applied concurrently with conidia and in a nutrient solution. However, disease incidence was still higher than expected based on previously observed inhibition of conidial germination on seedlings in a growth chamber. We suspect that environmental conditions in field experiments were less conducive to bacterial survival than those in growth chamber experiments. For example, some bacteria may have dried out and died on seedlings in the field, whereas constant free water was available to those in growth chambers (Knudsen and Hudler 1983). Failure of bacteria to reduce disease incidence if applied one week or one day prior to inoculation with conidia may also have been due to death of bacterial cells, as intervening weather included warm, sunny periods.

These conclusions are supported by results of preliminary experiments to estimate survival and growth of R10a applied to excised pine needles. Bacterial populations on excised needles increased either with or without added nutrients if free water was present. Without free water present, bacterial populations declined rapidly at first, then rose slightly for several hours before declining again. It is possible that some bacteria survived in the film of water initially preserved between needles under dry conditions, but as this film dried out, bacterial numbers again declined. Recent observations (unpublished) indicate that a film of water persists between needles for several hours, and more bacteria can be isolated from this area than from elsewhere on the needle surface.

Rates of conidial germination were slower than we have previously observed on agar (unpublished), and a smaller cumulative proportion of conidia germinated. Lysed conidial cells were surrounded by bacteria or yeasts, suggesting that biological control may occur even in the absence of applied antagonists.

Of course, the host may also produce compounds that inhibit germination of conidia. These results suggest that extrapolation from germination data obtained on agar should be avoided.

Zajchowski and Bergdahl (1983) reported that conidia of G. abietina may remain viable for up to 20 days after dissemination. If survival ability of conidia is superior to that of P. fluorescens, it may limit the ability of applied P. fluorescens to inhibit germination under alternating moist and dry periods. Further progress in the development of biological control agents for foliar pathogens will require a more exact knowlege of the environmental factors affecting survival and efficacy of antagonists. We are currently using results from experiments such as those described to develop a computer simulation model of population dynamics of applied bacteria, and their influence on conidial germination and subsequent incidence of Scleroderris canker under varying environmental conditions. This model will be useful in determining the potential of P. fluorescens as a biological control agent for Scleroderris canker, and may eventually serve as a more general framework for further research on other plant surface interactions between fungal pathogens and antagonistic microbes.

LITERATURE CITED

Blakeman, J.P. and I.D.S. Brodie. 1976. Inhibition of pathogens by epiphytic bacteria on aerial plant surfaces. Pages 529-557 In: C.H. Dickinson and T.F. Preece, eds. Microbiology of Aerial Plant Surfaces. Academic Press, London. 669 pp.

Knudsen, G.R. 1981. Interactions between conidia of Gremmeniella abietina and epiphytic bacteria from Pinus spp. MS Thesis, Cornell University, Ithaca, N.Y. 42 pp.

Knudsen, G.R. and G.W. Hudler. 1983. Inhibition on Gremmeniella abietina by Pseudomonas fluorescens on red pine seedlings. Phytopathology. 73:369 (Abstr).

Spurr, H.W. 1982. Experiments on foliar disease control using bacterial antagonists. Pages 369-381 In: J.P. Blakeman, ed. Microbial Ecology of the Phylloplane. Academic Press, London. 502 pp.

Zajchowski, S.J. and D.R. Bergdahl. 1983. Viability of conidia of Gremmeniella abietina sprayed on shoots of Pinus resinosa. Phytopathology. 73:378 (Abstr).

SCLERODERRIS CANKER SURVEY USING AERIAL PHOTOGRAPHY

Paul F. Hopkins and Lawrence P. Abrahamson[1]

ABSTRACT

Study plots with known infection levels of Scleroderris canker were interpreted on aerial 70 mm normal color and color infrared photographic transparencies. In the interpretation procedure, five infection levels, the identifying characteristics of Scleroderris canker, and the relationships between these identifying characteristics and the infection levels of the stand were defined. The defined relationships were used to develop photointerpretation keys which were evaluated on aerial photographs of test stands by personnel likely to use the keys in practice.

The evaluation results showed Scleroderris canker was detectable and classifiable into five infection classes to an acceptable degree of accuracy. Detection accuracies in excess of 90% and classification accuracies greater than 80% were obtained with the inexperienced personnel. It was concluded that photointerpretation can improve the present survey techniques by helping to make more efficient use of field personnel and time.

INTRODUCTION

The presence of a virulent new strain of Scleroderris canker in North America has led to extensive surveys in the United States and Canada to determine the distribution, severity, and characteristics of the disease (Skilling, 1979). Incorporation of remote sensing techniques (particularly aerial photograph interpretation) into survey procedures has improved many other similar forest stress surveys. The Manual of Remote Sensing (American Society of Photogrammetry 1975) provides examples of many research projects which applied remote sensing techniques to forest stress situations.

---

1 Assistant Professor, Department of Forest Engineering, and Senior Research Associate, School of Forestry/School of Biology, Chemistry and Ecology, State University of New York, College of Environmental Science and Forestry, Syracuse, N.Y. 13210

Research reported herein was partially supported by a grant from the United States Department of Agriculture, Forest Service, Northeastern Area, State and Private Forestry, Forest Insect and Disease Management, under Cooperative Agreement No. 42-239.

Our basic objective in this research was to ascertain if remote sensing techniques using aerial photography could similarly improve the present survey methods being used for Scleroderris canker. This basic objective can be stated more precisely in three specific objectives. There are: (1) to ascertain whether Scleroderris canker disease is or is not detectable on aerial photography; (2) if the disease is detectable, to determine if Scleroderris canker is classifiable by severity of infection, and (3) if the disease is detectable and classifiable, to develop an accurate and objective photointerpretation system for use in Scleroderris.

The photointerpretation system selected and developed for the detection and classification of this disease was a set of dichotomous photointerpretation keys. These keys provided the most convenient and logical method for describing the appearance of the various levels of Scleroderris canker infection on aerial photographs.

METHODOLOGY

Data Collection

Ground data. Eight study plots consisting of four red pine (*Pinus resinosa* Ait.) plantations and four Scots pine (*P. sylvestris* L.) plantations of approximately ten acres each were chosen from within the quarantine area of northern New York State. Each of the four study plots within each species of pine was chosen to coincide with a different level of Scleroderris canker infection ranging from no infection to heavy infection. The eight study plots provided the "ground truth" necessary for developing the photointerpretation keys for training the interpreters. In addition to the study plots, thirteen test plots were chosen from the air for use in evaluating the developed photointerpretation keys. The "true" level of infection of the thirteen test plots was unknown when they were chosen but was determined later by visual ground inspection.

Aerial data. The aerial data used in this study was 70 mm normal color (NC) and color infrared (CIR) photography acquired using Hasselblad 500 El/M cameras with 80 mm focal length lenses. The 70 mm transparencies were acquired at a large scale (1:6000) and a medium scale (1:12,000). The normal color film was exposed through a haze filter and the color infrared film was exposed through a Wratten 15 filter.

Development of the Photointerpretation Keys

The development of the photointerpretation keys consisted of three activities in which the infection levels, the identifying characteristics of the disease, and the relationships between these identifying characteristics and the level of infection were defined.

Definition of infection levels. The photointerpretation keys are based on five infection classes: severe, heavy, medium, light, and very light to no infection. The severe classification was added to the study because the original heavy infection category could be easily and accurately subdivided into two separate classes (the difference between the severe and heavy classes is based entirely on the amount of mortality in the stand). The four study plots for each species were chosen by ground evaluation to coincide with each of the four original infection classes. Figure 1 is a graphical illustration of four typical trees which are representative of the four original infection classes.

Identifying characteristics of Scleroderris canker. A number of characteristics which were indicative of the presence and intensity of Scleroderris canker were identified from aerial photographs of the study plots (with known levels of infection). These identifying characteristics are outlined below in the approximate order of importance:

1. Tree mortality. Dead trees (defoliated or not) are easily distinguished on NC and CIR aerial photography as long as they are not overtopped by other trees.

2. Color variation. Two types of color variation, between different but neighboring crowns and between different parts of the same crown, are evident on the aerial photography.

3. Edge tree appearance. Edge trees are located along the border of the stand being studied and their exposed foliage should extend nearly to the ground. The condition of this exposed foliage often helps indicate the infection level of the stand.

4. Nearby small pine trees. The condition of any small pine trees situated near the stand being investigated can be indicative of the infection level of the stand.

5. Small and thin crowns. The occurrence of small and thin crowns in more than natural amounts is one of the more subtle characteristics of Scleroderris canker.

6. Stand density and the degree of crown closure. These characteristics may be related to the level of infection by diseases, insects, or forest management practices (such as thinning).

7. Texture. Texture is the perception of coarseness in a photograph and can be related to Scleroderris canker infection.

8. Color of the stand. The general color of the stand is the overall or underlying color of the whole stand. This is the least important characteristic because it can be related to many factors besides the level of Scleroderris canker infection.

Relationships of identifying characteristics with infection level. Relationships between these eight characteristics and the intensity of infection were identified and then used to develop the photointerpretation keys. These relationships are summarized below and in Figure 2 for each of the eight identifying charactertics:

1. The number of dead trees visible on the photographs increases with an increasing amount of infection.

2. Both the between crown and within crown color variations in a stand increase as the amount of infection increases.

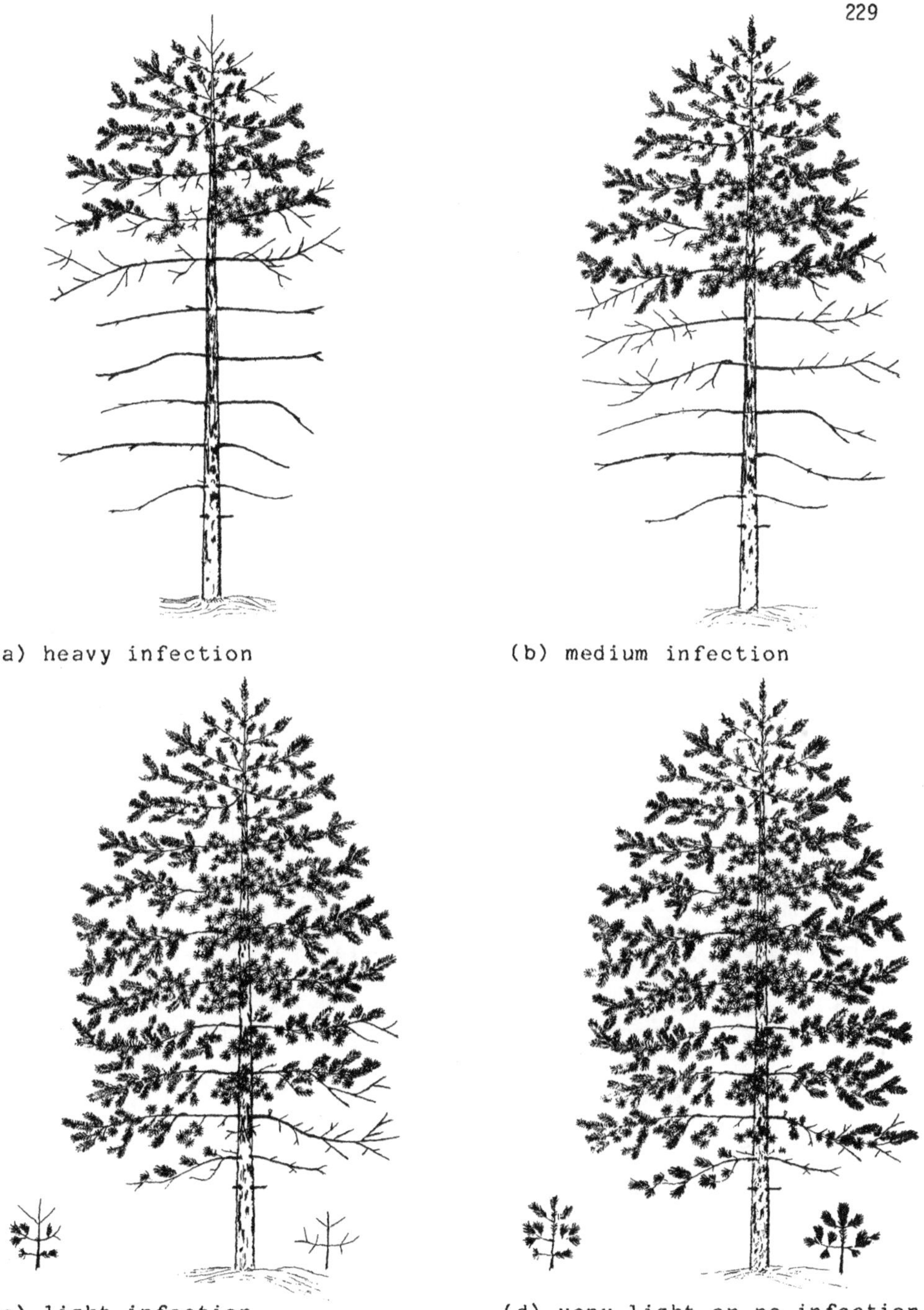

Figure 1. Graphical illustrations of "typical" trees from stands with Scleroderris canker infection.

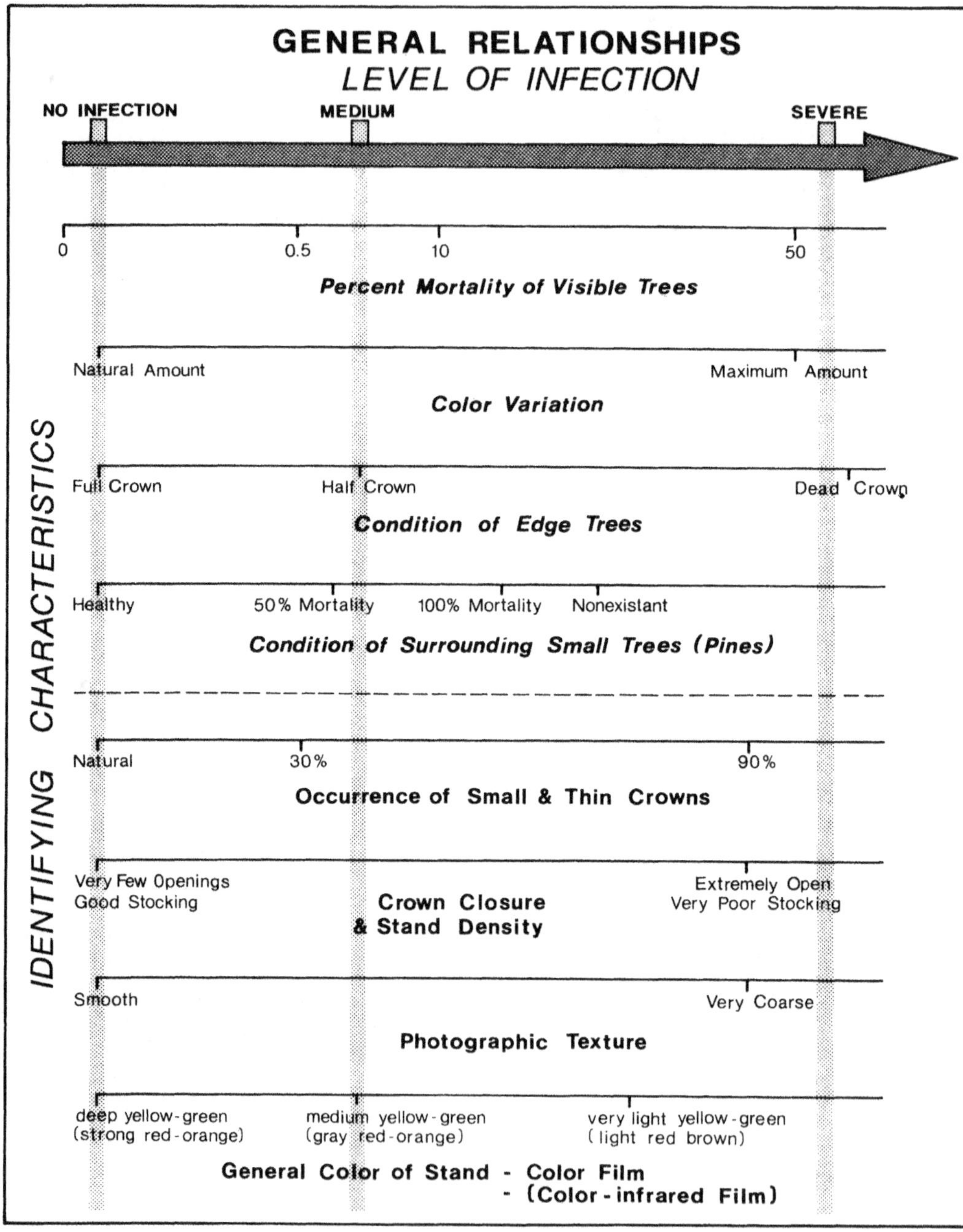

Figure 2. Illustration of the relationship between the identifying characteristics of Scleroderris canker and the level of infection.

3. The limit to which the exposed foliage of an edge tree extends to the ground decreases as the level of infection increases.

4. The condition of small pine trees near the stand quickly deteriorates with higher levels of infection.

5. The number of small and thin crowns increases with increasing severity of the disease.

6. The degree of crown closure and stand density decreases with increasing presence of the disease.

7. The photographic texture of a stand becomes more coarse with increasing amounts of infection.

8. The general color of the stand becomes progressively lighter for higher levels of infection.

In the photointerpretation keys, specific descriptions of each level of infection were developed using these eight relationships.

Evaluation of Photointerpretation Keys

Evaluation of the photointerpretation keys was accomplished by using the keys to classify each of the thirteen test plots into one of the five infection classes. The interpreters used in the evaluation were personnel who were likely to use the keys in actual practice. The evaluation was performed in three stages using a different group of interpreters and a progressively better training session in each stage. The first stage involved four research technicians from the College of Environmental Science and Forestry (CESF) who had varying amounts of ground experience with Scleroderris canker. The second stage involved three field technicians from the New York State Department of Environmental Conservation (DEC) who were familiar with the disease on the ground. The third stage involved three graduate students at CESF who had varying degrees of ground experience with Scleroderris canker. All of the people used as interpreters in the study were totally inexperienced in photointerpretation methods and only one of the graduate students had any knowledge of remote sensing.

The evaluation procedure consisted of two units: a training session and the actual photointerpretation of the test plots. The training sessions served to introduce the interpreters to some simple photointerpretation principles, Scleroderris canker, the photointerpretation keys, and the photointerpretation equipment. Progressive improvements were made to the training sessions for the three stages. The actual photointerpretation trials involved using the photointerpretation keys to classify each test plot into one of the five infection classes. During these trials, the "true" infection classes of the test plots were unknown to the interpreters and the test plot photographs were masked to prevent the ground experienced interpreters from recognizing the plots.

RESULTS

Explanations of two terms are necessary before proceeding with the results. "Overclassifying" a plot means the plot was interpreted to have a higher infection level than it actually had. In the same way, "underclassifying" a plot means the plot was interpreted to have a lower infection level than it actually had. The results of the photointerpretation key evaluation are reported in terms of detection and then classification of the disease.

The accuracies of detecting the infected stands by separating them from the stands with no infection varied with the interpreter. Table 1 summaries the detection accuracies for each of the ten interpreters. The detection accuracies ranged from a low of 61% to a high of 94% for an average of 84%. The average accuracy for group one was 76% and the average accuracies for groups two and three were 89%. It was hardest to detect the presence of the disease in stands verified on the ground to have a light infection, accounting for 98% of the detection mistakes.

The accuracy of classifying the stands into the correct level of infection category (severe, heavy, medium, light, or no infection) also varied with the interpreter. Table 2 summarizes the classification accuracies for each of the ten interpreters. The classification accuracies ranged from a low of 55% to a high of 87% for an average of 68%. The average classification accuracies for each of the three groups were 58% for group one, 71% for group two, and 76% for group three. The accuracy of classification within one category of the ground verified class ("true" class + one class) ranged from 86% to 99.5%.

DISCUSSION

The ability to detect the presence of Scleroderris canker using aerial photography as exemplified in this study is very encouraging. The high levels of detection accuracy (94% for one interpreter) were obtained by people who had no previous experience in photointerpretation and with only a short training session (3-4 hours). High accuracy was obtained while using all combinations of the emulsions and scales even through some of these combinations may not have been as effective as others.

The classification accuracies (the number of times a stand was interpreted into the correct class of infection) of this study were also very encouraging. They show that an interpreter can not only detect the presence of the disease from aerial photographs, but can also make a consistent and accurate determination of the level of infection. Again, these results were obtained under conditions (such as experience) which were less than ideal. Using better trained and more experienced interpreters with the optimal combination of photointerpretation technique, scale, emulsion, and timing of photography could very well improve upon both the detection and classification accuracies.

Upgrading the training session appeared to be extremely important to the performance of the interpreters as the average accuracies in both detection and classification increased substantially for each stage (Tables 1 and 2). The improved training sessions may have also improved the consistency of the interpreters of each stage even though the interpreters were probably the largest source of variation in the study.

Species composition was another source of variation in the study. The Scots pine plots were generally overclassified and red pine plots with light infection had a tendency to be underclassified. The reason for the slight tendency to underclassify the red pine light plots is that in dense red pine plantations color variations do not seem to be as obvious. Also, lower parts of the crowns are usually not visible, thereby masking low levels of infection. The major reason for the definite tendency to overclassify Scots pine plots is the greater natural color variability of Scots pine over red pine on both color and color infrared film. Another possible reason for the overclassification of Scots pine is its generally poorer condition in this area of New York State. In addition to being susceptible to attack from Scleroderris canker, Scots pine suffers from snow damage, Woodgate rust (*Peridermium harnesii*), and poor growth due to the genetic variability of the plantings.

Another finding of the study was that the color infrared emulsion tends to overclassify the plots and the normal color emulsion tends to underclassify the plots. These tendencies to misclassify are due primarily to the important role that color variation plays in the detection and classification of the disease. The variations in color of a stand are evident on both emulsions, but are much more apparent on color infrared film. It is felt that this difference in apparent color variation causes the misclassification tendencies because of the inexperience of the interpreters and the inadequacies of the training session. It was concluded that the color infrared emulsion was better than the normal color emulsion for this application due to its better portrayal of color variations and other advantages.

Generally, significant and important differences exist between the two scales of photography with respect to their classification tendencies, but there are no important differences in the accuracies of the two scales. The 1:6000 scale seemed to overwhelm the inexperienced interpreters with detail causing them to overclassify the plots. Similarly, the lack of detail in the 1:12,000 scale caused a slight tendency to underclassify the plots.

The timing of the photography is an important consideration for Scleroderris canker photointerpretation. For best results, the photography should be taken when the disease symptoms are most evident and when shadow and differential shading problems are at a minimum. Shadows and differential shading result in a loss in detail (especially of the lower crowns) and exaggerate apparent color variations. Shadow and differential shading are at a minimum at the maximum solar elevation angles which occur in the month of June. Visual disease symptoms are most evident in the later parts of May and in June. Therefore, the month of June (at midday) would be the best time for the photography to be acquired. The timing of the photography used in this study (September) was, therefore, not optimal.

The prospect of using aerial photointerpretation (or other remote sensing techniques) in a time related study of Scleroderris canker is attractive. A temporal emphasis could reveal information on the direction and rate of disease spread and on factors which might be related to this spread. A time study using remote sensing imagery could also yield information on the rate of intensification of the disease once it has infected an area.

Existing programs for the survey of Scleroderris canker emphasize the detection of the disease in a stand by ground crews. Much of this work is done in connection with the quarantine being implemented in New York State by the New York State Department of Agriculture and Markets. Scleroderris canker survey and research programs are also being conducted by the DEC to locate forest stands and plantations with Scleroderris canker infection and determine the

Table 1. Detection accuracies of the interpreters.

Interpreter	Detection Accuracy (Percent)
1	61
2	79
3	76
4	88
5	85
6	89
7	93
8	94
9	87
10	86
Average	84
Group 1 average	76
Group 2 average	89
Group 3 average	89

Table 2. Classification accuracies (percent) of the interpreters.

Interpreter	Classified Correctly	Underclassified by one class	Overclassified by one class	Total
1	58.3	34.5	5.3	98.1
2	59.2	22.8	13.6	95.6
3	54.9	19.9	11.2	85.9
4	61.2	11.7	21.8	94.7
5	67.0	14.1	15.0	96.1
6	61.2	10.2	23.8	95.1
7	83.5	8.3	6.8	98.5
8	67.5	5.3	22.8	95.6
9	86.9	8.7	3.9	99.5
10	74.3	20.9	3.9	99.0
Average	67.4	15.6	12.8	95.8
Group 1 average	58.4	22.2	13.0	93.6
Group 2 average	70.6	10.9	15.2	96.7
Group 3 average	76.2	11.6	10.2	98.0

intensity of the infection. The DEC program is also trying to monitor the progress of the disease as it moves between stands and within infected stands. The results of this study show that the work involved in both types of ground surveys can be decreased and probably made more economical, efficient, and effective by using aerial photointerpretation. The improvement will be possible because the stands with heavy and severe infection classes will probably not need to be ground inspected. The stands with the other infection classes can be ground checked in accordance to a priority system based on the photointerpretation. Monitoring the progress of the disease to other stands or within a stand that has been determined to be infected will also be improved because once the presence of the disease has been verified, the need to go back to the stand on the ground will be minimal.

The economy of using this photointerpretation system will of course depend on the scale of the survey program and certain logistical considerations. If only ten or twenty readily accessible stands are to be surveyed then it would probably be more economical to survey them on the ground. On the other hand, if there are many stands to be surveyed, the stands are widely scattered, or the stands are somewhat inaccessible, photointerpretation methods will be over ground methods.

Based upon the results of this study, we have shown that Scleroderris canker is detectable and classifiable to an acceptable degree of accuracy using aerial photography and our photointerpretation keys. The detection and classification accuracies should improve when interpreters with more experience and a better remote sensing background are used and when variables such as photointerpretation technique and scale are optimized.

The research has shown that the photointerpretation of Scleroderris canker can best be accomplished through stereoscopic viewing of color infrared film. The scale of photography used would depend on interpreter preference and the magnification of the stereoscope used to view the photographs. We found the medium scale (1:12,000) was approximately as effective as the large scale (1:6000) as long as a 4X magnification or greater was used with the medium scale. The best time to obtain the aerial photography would be during early summer (June-July) and at midday.

Photointerpretation in general, and this photointerpretation system specifically, can improve Scleroderris canker survey efforts by reducing the amount of ground inspection needed. The efficiency and economy of present methods of ground survey would also be improved if the suspect stands are large in number and/or quite inaccessible on the ground. Aerial photography can also be used in evaluating the impact of the disease (especially in mortality estimation); in planning, monitoring, and evaluating salvage operations; and in monitoring quarantined Christmas tree stands.

LITERATURE CITED

American Society of Photogrammetry. 1975. Manual of Remote Sensing, Robert Reeves (ed.), 1st Edition, Falls Church, Virginia. 2144 pp.

Skilling, D.C. 1979. "Scleroderris Canker Study Progress Report No. 3," North Central Forest Experiment Station. U.S.D.A. Forest Service, St. Paul, Minn.

DISEASE MANAGEMENT OF SCLERODERRIS CANKER BY REGULATORY ACTION

Dorothy M. Laidlaw[1]

ABSTRACT

Regulatory actions are usually taken to prevent the introduction into or spread from an area of a pest designated as quarantine significant. There are international obligations to prevent the spread of pests into new areas. Legislative actions form an integral part of pest management strategies.

INTRODUCTION

There are two aspects of regulatory mangement of any disease. These include the legislative and the pathological aspects. The intertwining of these two may generate controversy, usually because there is a lack of concrete data falling into "black" and "white" categories.

DISCUSSION

Legistation

Canadian regulatory section is taken federally under the authority of the Plant Quarantine Act 1968-1969 (Revised Statutes 1970) and its regulations. The stated purpose of this legislation is "to prevent the introduction or spreading of pest injurious to plants." The prohibition which is the basis of the act is as follows:

> "3(1) Except as provided by this Act and the regulations no persons shall knowingly introduce or admit into Canada spread within Canada or convey within or from Canada any pest or any plant or other matter that is infested or likely to be infested with a pest or that constitutes a biological obstacle to the control of any pest."

The act permits the erection of regulations for preventing the spread of pests and can do so by regulating the movement of plants and all their parts, plant products, packing materials, conveyance (ships, rail cars, truck etc.), soil, or any other medium that may harbor or be suspected of harboring plant pests. The Plant Quarantine (PQ) Act is potentially quite broad in its scope.

[1] Clinical Pathologist, Plant Health Division, Agriculture Canada, K.W. Neatby Bldg., Ottawa, Ontario K1A 0C6

The provinces have the right and do enact plant health legislation. The Plant Protection Act of the province of Newfoundland is such an example.

Canada has international obligations regarding plant quarantine matters because she is a signatory to the International Plant Protection Convention (IPPC) 1951 FAO which has "the purpose of securing common and effective action to prevent the spread and introdution of pests of plants and plant products and to promote measures for their control." The convention defines the term "Quarantine Pest" as "a pest of potential national economic importance to the country endangered thereby and not yet present there, or present, but not widely distributed and being actively controlled." The European race of Gremmeniella abietina (Lagerb) Morelet would fall into the latter category.

As signatories to the IPPC, Canada is obliged to establish an official plant protection organization. This is the Plant Health Division, Agriculture Canada and has been in existence since 1910 when the first Destructive Insects and Pests Act was proclaimed by the Canadian parliament. We are obliged to distribute information regarding pests and means for the prevention and control, to carry out research in the field of plant protection, to provide for the issuance of phytosanitary certificates, to cooperate internationally in reporting pest outbreaks, and to take measures under our plant protection legislation to regulate the entry of plants and plant products unless this is made necessary by phytosanitary considerations alone. We cannot erect trade barriers under the guise of PQ legislation. The convention provides for the establishment of regional plant protection organizatons which are groups of cooperating nations. NAPPO, the North American Plant Protection Organization with Canada, the USA and Mexico as members, is one such organization and was established in 1976. EPPO, the European and Mediterranean Plant Protection Organization (Ebbels & King 1979) is another.

The idea of regulatory disease management is not a new phenomenon and has been entrenched in domestic and international law for a long time. McCubbin (1954) "The Quarantine Problem," the more recently published "Plant Health" (Ebbels & King 1979), or Dowling et al., (1982) "Plant Inspection and Certification" will provide some insight into quarantines and interesting reading.

Quarantine Significance

There has been controversy about whether G. abietina is or is not a "quarantine pest." Applying the IPPC definition automatically means that to some countries it could be considered as such, and to others not. For Canada it is regarded as quarantine significant. To assure that it is a pest of "potential national economic importance" to Canada, the following statistics are provided. Only 34 percent or 3.4 million square Km of the total land area of Canada estimated at 9.22 million square Km is forest land. Of this only 20 percent is productive forest land (Bonnar 1982). This 34 percent is more than twice as large as Europe's 1.4 million square Km of forest land (Environment Canada 1981). The province of Ontario in central Canada alone has a forest resource inventory equal to the area of forest land in Sweden, Norway and Finland combined (Bonnar 1982).

By contrast, only 13 percent of Canada's land mass can support agriculture of any kind, including rough grazing. This 13 percent incudes the 0.5 percent considered as Capability Class 1 land, that is with soils with no physical limitations to production and capable of supporting the widest range of crops, and the 5 percent in Capability Class 1, 2, and 3 including land free of severe physical limitations and capable of supporting field crops on a sustained basis. Eighty six percent of Canada's land area is Capability Class 7 and has no capacity for modern mechanized agriculture (Dummanski & Stewart 1981).

Canada's twenty-one most economically important tree species include hosts of all known races of G. abietina. Picea and Abies are first, Tsuga second Pseudotsuga menziesii fourth, Pinus contorta, P. banksiana, P. strobus, P. resinosa, P. ponderosa are fifth, sixth, seventh, eleventh and twelfth respectively and Larix spp. are twenty-first (Statistics Canada and Canadian Forestry Service).

It has recently been recognized that the Canadian forestry sector is undergoing a transition from exploitation of the forest resource to one of more systematic forest management and it has been recognized that the basis for strengthening the forest resource base is a combination of economic, social and environmental criteria (Environment Canada 1982). One component which will strengthen this base is better protection from disease. For example it has been revealed that losses due to insects and diseases exceed those due to forest fires (Bonnar 1982). Potential exotic pest introductions are a serious concern to Canada. European introductions into North America are about ten times as numerous as those carried in the opposite way, a process referred to as Europeanization (Rapport et al., 1976). Need we be reminded of the introduction of Endothia parasitica causing Chestnut blight, Ceratocystis ulmi causing Dutch elm disease or Cronartium ribicola causing white pine blister rust into North America. As the result of an analysis of exotic conifer pests, the commercial importantion of conifers for propagation purposes from sources off the North American continent will likely be curtailed in the near future. In practice this will impact only on the Netherlands as the importation of Picea, Pinus, Pseudotsuga, and Larix is already prohibited from other areas where the European race of G. abietina occurs.

Virulence

G. abietina is a serious conifer disease which has caused much confusion to mycologists and pathologists alike. It has been included in a number of fungal genera over the last one hundred years since one of its stages was described as Brunchorstia pinea (Karst) V. Hohn (anamorph stage). It has been known in western Europe and parts of western Asia (Georgia in USSR) for a long time and has been reported on Picea, Pinus, Larix and Pseudotsuga (CMI 1974, Dorworth 1971). It is reported to attack Abies in Japan (Yokata 1974)and Cedrus deodora in Russia (Shishkina and Tasavana 1970). Early references in the literature indicate it was present in western North America (Fink 1911, Weir 1921). In these references the problem was associated with or attributed to Cenangium abietis (Pers.) Rehn (= C. ferruginosum (Fr.) Fr.) the supposed perfect state at that time. While descriptions of the supposed syndrome and possibly of one of the fungal elements involved may seem to parallel that associated with G. abietina it seems prudent to treat this literature with due caution. Isolated locations in Alberta (Dorworth 1975) and British Columbia (Hiratsuka and Funk 1976) of the North American race are the only documentated finds in western North America. The presence of G. abietina in eastern North America is well documented and in numerous instances it was disseminated to new areas on infected forest nursery stock (Punter 1967, Skilling and O'Brien 1969).

During the late sixties and early seventies, studies were carried out and much information was published on the control of Scleroderris canker of nursery stock. For information consult works of C.E. Dorworth, Great Lakes Forest Research, Centre, Sault Ste Marie, Ontario and D.D. Skilling, North Central Forest Experimental Station St. Paul, MN. Apparently management practices, coupled with chemical fungicide treatments could overcome the nursery disease problem and there was hope that once plantation trees reached a critical size they were no longer susceptible and controls were no longer necessary. This proved to be the case until 1975. In 1975, G. abietina isolates from around the world, based on serological evidence were grouped, according to geographic origin. North American isolates formed the North American race, European isolates formed the European race and Japanese isolates which fitted neither of the above groups were constituted as a third group, the Asian race (Dorworth and Krywienczyk 1975).

In 1975 it was reported that large stands of 20-30 year old red and Scots pines in New York were severely attacked by G. abietina (Setliff et al., 1975). Eventually over a period of a couple of years the large trees were totally killed (Skilling 1977). This type of behavior by the pathogen was incongruous with its past behavior in North America. These trees were apparently being attacked by a more virulent population of the pathogen. Isolates of this virulent race which were termed tentatively New York Strain were serologically typed and found to fit into the European group (Dorworth, Drywienczyk and Skilling 1977). Because this serologic race had never before been observed in North America, and because of the obvious correlation with virulent behavior, its spread was considered to pose an extreme threat to North America's conifer forests, much of which is located in Canada.

As a result some US states invoked quarantine, the United States Department of Agriculture invoked an emergency quarantine and Canada adopted a policy aimed at preventing the introduction of the European race into Canada. For Canada this constituted the prohibition of imports of G. abietina hosts, Pinus, Picea, Larix, and Pseudotsuga from areas where the European race occurred (Straby 1981). The United States revoked its 1977 Emergency Regulation effective 22 November 1982.

The discovery of pockets of the European race in New Brunswick (1978), Quebec (1978) and in Newfoundland (1980) and later followed by descriptions of new races called intermediate (Skilling 1981, Wendler et al., 1980) have complicated the quarantine problem. The recovery of the European race in Canada required the imposition of domestic quarantine actions. In 1981 North America experienced another example of the virulence of the European race. This occurred in a plantation at Torbay in Newfoundland. The trees were destroyed by burning.

Serology

The intermediate race is defined with reference to an antiserum prepared against a typical Vermont isolate. Against this antiserum, the reference North American (N.A.) (C-1) the reference European (Eur.) (F-4) strains and the homologous Vermont isolate share a common antigen, hence the conclusion that the Vermont isolates are intermediate between the two (Wendler et al., 1980). The majority of Vermont isolates also exhibited the diagnostic European precipitin line in gel double diffusion tests but lacked the diagnostic N.A. precipitin line. This is where the uncertainties arise. Because these intermediate races have the diagnostic Eur. line, they are considered to be European race with relationships in common with the N.A. race. Because they lack the diagnostic N.A. line as established by the Dorworth and Krywienczyk (1975) race structure, they can not be considered as "hybrids" using that serological system. If they were they should then possess both the N.A. and the Eur. lines. It is for this reason they are considered European race by quarantine officials in Canada. The fact that the intermediate isolates lack the N.A. diagnostic line is more significant than the fact that they share other common factors. Indeed all races of G. abietina would be expected to show antigenic relationships and that is the reason why isolates have not been split into subspecies. The Vermont, or intermediate antiserum, is a new reference antiserum and before it can be accommodated satisfactorily within the already established race system it will have to be tested against all other previously race-typed isolates to determine if they also share the "intermediate" line. If they all do, there is little value gained by adding this antiserum to the present reference scheme for determining whether the race is North American or not.

The Serologically intermediate isolates of Skilling (1981) presents a more difficult problem because these represent individual isolates obtained from a single apothecium. Fortunately to date these have been found only in New York where the state takes quarantine action to prevent the further spread of all races of the pathogen. This points to the need for determining the virulence potential of isolates and it is recognized that this is an extremely difficult task; one that can not be determined in North America alone because we can not field test foreign isolates on mature trees. Therefore precaution must remain a major quarantine consideration.

The Canadian Situation

The Canadian situation regarding the European race is relatively static except for Newfoundland. In Quebec, affected pine plantations along the Quebec-New York border, a region not a great distance from known New York infected plantations, are all under premise quarantine. Inspections with disease control by pruning and burning is practiced. In the Canadian experience pruning and burning is proving an effective method for preventing disease spread. In New Brunswick, all known European sites were contained and sanitized. The most significant of these was in a forest tree nursery. All infected seedlings there were destroyed and intensive chemical control regimes applied to all remaining stock over a period of several years have left the nursery free of disease. As a precaution, the nursery is still closely monitored and no seedlings are permitted to leave the province. In Newfoundland, on the Avalon peninsula (around St. John's) the European race claimed, in 1981, a 4 acre plantation of red pine in a spectacular fashion, similar to the 1975-77 New York kills. This example serves to reinforce the observed correlation of virulence with the European race. Newfoundland instituted, in 1980, The Plant Quarantine Regulation 1980 (Newfoundland Regulation 115/80) under the authority of their provincial Plant Protection Act 1978. This establishes a quarantine area and prohibits the movement of Picea, Pinus, Pseudotsuga and Larix from it. All finds of G. abietina, with the exception of the plantation, consisted of small numbers of trees or of individuals. These were cut and burned or were pruned and subjected to further inspections.

CONCLUSIONS

Quarantine measures may not totally prevent the gradual spread of the European race of G. abietina in Canada. They should, if properly applied, reduce the losses and the rate of spread. Our forest is a valuable resource that requires protection until the plant breeders have selected and produced resistent planting stock. One final thought, lest all of this weigh heavily upon us; the importance of pine trees to a country are relative. While one tree in Canada's vast forest might not be missed this is not so for others, for not less than 15 people were killed and 12 injured in a dispute over the felling of a single pine tree in Pakistan (Anonymous 1982). We have an obligation to ourselves to solve this problem. Regulations may provide some breathing time.

LITERATURE CITED

Anonymous. 1982 Abst. 43:597.

Bonnar, G.M. 1982. Forest inventories in Canada. For. Abst. 43: CMI. 1974. Distribution Maps of Plant Diseases #423.

Dorworth, C.E. 1975. Gremmeniella abietina collected in Alberta Canada. Plant Dis. Rep. 59:272-73.

Dorworth, C.E. 1971. Disease of conifers incited by Scleroderris lagerbergii Gremmen: A review and analysis. Dep. Fish. For. Can. For. Serv. Publ. No. 1289.

Dorworth, C.E. and J. Krywienczyk. 1975. Comparison among isolates of Gremmeniella abietina by means of growth rate, conidia measurement and immunogenic reaction. Can. J. Bot. 53:2506-2525.

Dorworth, C.E., J. Krywienczyk and D.D. Skilling. 1977. New York isolates of Gremmeniella abietina (Scleroderris lagerbergii) idential in immunogenic reaction to European isolates. Plant Dis. Rep. 61:887-890.

Dowling, C.F., Jr., A.E. Graham and S.A. Alfieri, Jr. 1982. Plant inspection and certification. Plant Dis. 66:345-351.

Dumanski, J. and R.B. Stewart. 1981. Crop production potentials for land evaluation in Canada. Land Resources Research Institute, Research Branch, Agriculture Canada, Ottawa, Ontario.

Elbels, D.L. and J.E. King. Editors. 1979. Plant health, the scientific basis for administrative control of plant diseases and pests. Blackwell Scientific Publications Oxford, London.

Environment Canada. 1981. Canada forests. Dept., Can. For. Ser. Cat. No. F051/1981 E.

Environment Canada. 1982. Policy statement a framework for forest renewal, Can. For. Serv. Ottawa 2 Sept. 1982.

Fink, B. 1911. Injury to Pinus strobus caused by Cenangium abietis. Phytopathology 1:180-183.

Hiratsuka, Y. and A. Funk. 1976. Additional records of _Gremmeniella_ _abietina_ in Western Canada. Plant Dis. Rep. 60:631.

McCubbin, W.A. 1954. The quarantine problem. Ejnar Munksgaard, Copenhagen 255 p.

Punter, D.L. 1967. _Scleroderris lagerbergii_ in and near Ontario forest nurseries. Plant Dis. Rep. 51:357.

Rapoport E.H., E. Ezcurra and B. Drausal. 1976. The distribution of plant diseases: A look into the biogeography of the future. J. Biogeog. 3:365-372.

Revised Statues of Canada. 1970. Plant Quarantine Act. 1968-69, C.35, S.1, Chapter P-13, 6:58-63.

Setliff, E.C., J.A. Sullivan and J.H. Thompson. 1975. _Scleroderris_ _lagerbergii_ in large red and Scots pine trees in New York. Plant Dis. Rep. 59:380-381.

Shishkina, A.K. and N.I. Tsavana. 1970. O. novykh zabolenvaniyakl deboratívnykh rastenii v gruzinskoi ssr (on new diseases of ornamental plants in Georgian USSR.) Mikol. Fitopatol. 4:76-79. (English translation.)

Skilling, D.D. 1977. The development of a more virulent strain of _Scleroderris_ _lagerbergii_ in New York. Eur. J. For. Pathol. 7:297-307.

Skilling, D.D. 1981. Development of an intermediate strain of _Gremmeniella_ _abietina_ in New York. Phytopathology 71:255.

Skilling, D.D. and J.T. O'Brien. 1969. Scleroderris canker in the Lake States, a situation report 1968. U.S.D.A. Forest Service. Research Note NC-73.

Straby, A.E. 1981. Discussion, Scleroderris canker development of strains and potential damage in North America. Can. J. Plant Pathol. 3:265.

Weir, J.R. 1921. Notes on _Cenangium_ _abietis_ (Pers) Rehm. on _Pinus_ _ponderosa_ Laws. Phytopatology 11:166-169.

Wendler, P.B., A.R. Gotlieb and D.R. Bergdahl. 1980. A newly discovered serotype of _Gremmeniella_ _abietina_. Can. J. Bot. 58:1923-1928.

Yokota, S. 1974. Scleroderris canker of todo fir in Hokkaido, Northern Japan I. Present status of damage and features of infected plantations. Eur. J. For. Pathol. 4:65-74.

THE USE OF A QUARANTINE TO CONTROL SCLERODERRIS CANKER IN NEW YORK

Ronald O. Barrett[1]

ABSTRACT

The New York State Department of Agriculture and Markets quarantine, initiated in 1977, effectively monitors the movement of Scleroderris infectable commodities. Infected materials are not allowed out of the quarantine area. The area of the quarantine has not substantially increased since 1977.

INTRODUCTION

The Commissioner of the New York Department of Agriculture and Markets promulgated a Scleroderris canker quarantine on August 15, 1977. This document provided the Department with the authority to establish a program directed at preventing the artificial dissemination of Scleroderris canker pathogen, Gremmeniella abietina.

The quarantine had been developed after an extensive review of available research information, a study of potential host genera and a survey of the known infected area. We determined the threat of Scleroderris canker met the four basic fundamental prerequisites of quarantine in that: (1) the pest offered an actual or serious economic threat; (2) the quarantine was a necessary measure for which there was no other substitute involving less interference with normal activities; (3) the objective of preventing introduction or limiting spread was a reasonable expectation; and (4) the economic gains expected would outweigh the cost of administration.

The quarantine was effectuated with the following two major objectives: (1) to prevent the disease from spreading to uninfected forest area, nurseries and Christmas tree plantations, and (2) to permit the movement of uninfected and unexposed host commodities within New York and to other states and countries.

In structuring the quarantine we provided for: definitions, quarantined area, regulated articles, conditions governing the issuance of certificates and permits, inspection and disposition of shipments, marking requirements, shipments for experimental and scientific purposes and statement of origin of regulated articles.

---

1 Director, Division of Plant Industry, New York State Department of Agriculture and Markets, Albany, N.Y. 12235

Since it was not possible to quickly differentiate between the various strains of G. abietina, the quarantine was applied to all strains of the pathogen. The time consuming serological testing for strain differentiation, we believe, would reduce the effectiveness of the quarantine. Additionally, because no form of the disease was known to occur in New York outside of the Adirondack Mountains area we had no reservations about this provision.

The quarantined area originally consisted of portions of ten counties in our Adirondack Mountains region. In establishing the quarantined area, we included a buffer zone of at least one town around the perimeter of the known infection.

Regulated articles included nursery stock, Christmas trees, wreaths, branches, twigs or parts thereof except logs or pulpwood of any species of Larix, Pinus, Picea and/or Pseudotsuga.

Our inclusion of a provision for the issuance of certificates and permits gave us a means to develop an inspection program inside of the quarantined area. Regulated articles could be inspected and if found uninfected, unexposed or treated in an approved manner could be certified for movement from the quarantined area.

The provision for a statement of origin of regulated articles was necessary to assure compliance and facilitate enforcement of the quarantine. A certificate of origin must accompany each shipment of regulated articles moving within or outside of the quarantined area.

The remaining provisions are general features included in most plant quarantines.

METHODS

Methods for enforcing the quarantine and preventing disease dissemination include: (1) the inspection of nurseries and Christmas tree plantations within the quarantined area; (2) a delimiting survey around the perimeter of quarantined area; (3) a random survey throughout the state, and (4) the random inspection of Christmas tree sales lots throughout the state.

Annually, all uninfected nurseries and Christmas tree plantations growing susceptible plant material within the quarantined areas are inspected. Preferably, these inspections are completed during May and June when visual symptoms are vividly expressed. In addition to the actual growing site, the area within 1000 ft. of the site is inspected to determine disease presence or absence. Should infected material or suspected material be discovered, samples are obtained for microscopic examination. If the presence of infection is confirmed, the owner or custodian is informed and is served with a site quarantine order. After the site has been quarantined, regulated articles cannot be moved except under a limited permit within the quarantined area.

If the nursery or Christmas tree plantation is found to be free of disease and exposure to infection, the owner is advised and susceptible plant materials may be moved from the site. Such disease free material when moved is accompanied by certificates of inspection stating apparent freedom from Scleroderris canker disease.

A survey of conifers is completed annually in towns around the perimeter of the quarantined area to determine any possible spread of the disease. This survey is a cooperative project involving people from the Department of Agriculture and Markets, Department of Environmental Conservation, the College of Environmental Science and Forestry and the College of Agriculture and Life Sciences. If Scleroderris canker is found in a previously uninfected town, that town is added to the quarantined area and is subject to regulation.

All personnel of the Department of Agriculture and Markets have been trained to recognize and identify Scleroderris canker disease symptoms. Consequently, a continuous informal random survey is conducted throughout the state during each inspectors normal daily activities.

During New York's Christmas tree sales period, October through December, inspections are made at sales lots to determine the origin of the trees and articles being sold. These inspections are made to determine compliance with quarantine requirements and to prevent the movement of infected or exposed articles. Additionally we are interested in ascertaining the movement of regulated articles from other known infected areas in the United States and Canada.

If regulated articles are found in the marketplace without a certificate of origin or a comparable document, they are placed under quarantine. Disposition of the quarantine and action against the owner of the articles is dependent upon their origin and proof of that origin.

RESULTS

Since the quarantine's inception in 1977, the quarantined area has been expanded very little. This expansion we attribute more to the development and use of refined survey and inspection techniques than to a spread of the disease. The quarantined area now includes two complete counties and portions of ten other counties. With the exception of an isolated infection discovered in the City of Saratoga Springs in 1980, the disease has been contained to the Adirondack region. Figure 1 indicates the quarantine boundaries in 1977 and 1982.

The number and acreage of infected and quarantined nursery and Christmas tree sites has increased from 25 sites totaling 612 acres in 1978 to 56 sites and 2,110 acres in 1982. A 1978 growing site inspection summary is shown in Table 1 and a 1982 growing site inspection summary with a cumulative infection summary for 1978-82 is shown in Table 2.

The statewide inspection of Christmas tree sales lots has revealed a very low incidence of noncompliance with the quarantine regulations. A summary of the sales lot inspections during the years 1980 through 1982 appears in Table 3.

In 1978, department personnel discovered a shipment of 100 Scots pine Christmas trees moved illegally from the quarantined area. The trees were quarantined and eventually destroyed while the shipper was referred to our Legal Bureau for penalty action.

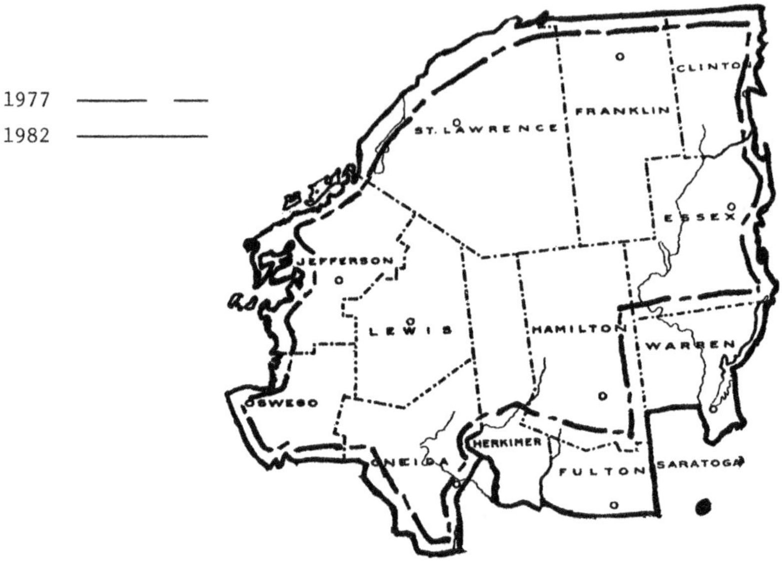

Figure 1. New York State Scleroderris canker quarantined area.

Table 1. Growing site inspection summary, 1978.

County	Sites inspected		Sites infected	
	Number	Acres	Number	Acres
Clinton	24	418	0	0
Essex	19	83	1	15
Franklin	37	982	4	11
Fulton	5	37	0	0
Hamilton	3	3	0	0
Herkimer	18	760	2	26
Jefferson	16	791	1	20
Lewis	44	859	13	244
Oneida	40	734	1	50
Oswego	13	170	1	25
St. Lawrence	32	1,062	2	221
Warren	4	27	0	0
Totals	255	5,926	25	612

Table 2. Growing site inspection, 1982, and cumulative infection summary for 1978-82.

County	Sites inspected 1982		Sites infected 1978-82	
	Number	Acres	Number	Acres
Clinton	27	252	0	0
Essex	28	94	3	17
Franklin	37	986	18	724
Fulton	9	32	0	0
Hamilton	5	19	0	0
Herkimer	18	700	2	190
Jefferson	15	987	2	70
Lewis	46	916	25	605
Oneida	53	697	3	160
Oswego	21	409	1	25
St. Lawrence	31	937	2	320
Saratoga	1	58	0	0
Warren	13	71	0	0
Totals	304	6,158	56	2,111

Table 3. Summary of Christmas tree sales lot inspections.

Year	Sales lots inspected	Trees inspected
1980	875	697,807
1981	904	574,236
1982	1,114	846,066

During 1982, two situations involving the illegal movement of approximately 700 assorted Christmas trees were encountered. These trees were returned to the quarantined area under Departmental supervision for disposition.

CONCLUSIONS

Surveys have not detected any dramatic spread of Scleroderris canker disease from the area quarantined in 1977. This suggests the quarantine action may have prevented the artificial long range dissemination of the disease.

The cost-benefit ratio of the regulatory program is favorable based on the: (1) preservation of markets for New York State plant products, and (2) prevention of dissemination and the potential loss of additional forest lands, Christmas trees and nursery stock.

THE TRANSFER OF SCLERODERRIS TECHNOLOGY IN CANADA

Henry L. Gross[1]

ABSTRACT

Technology transfer in Canada is presented as a communications process functioning among the various agencies managing the Canadian forest. Most of the forest land in Canada is provincially owned and provincial forestry agencies traditionally have played a key role in the dissemination of technology to forest managers. The federal mandate is to promote the development and protection of the forest and provide scientific and technical leadership. Disease and insect studies have always been a prominent part of the federal research program. Detection and assessment of pest problems are a part of the Canadian Forestry Service program and personnel of pest survey units distributed across the country provide a transfer link between researchers and forest managers.

The evolution of the current strategy for management of Scleroderris canker as caused by the North American race of *Gremmeniella abietina* is presented to illustrate how technology transfer has functioned in Canada. Incursions of the European race of *G. abietina* are currently treated by quarantine procedures. Technology transfer has made forest managers and the general public aware of the potential of this pest through publicity released to the news media, articles in various forestry publications, and a series of workshops held to discuss the potential of this pest and train foresters to recognize the problem.

INTRODUCTION

Technology transfer is one of the processes by which mankind evolves through time. The existence of useful knowledge and the quest for such knowledge stimulate the process. In a relatively free system of information transfer, as exists among pathologists, the rate of transfer is a function of the magnitude of the problem to be resolved. Resources applied and gains experienced are somewhat dependent upon how well technology is transferred. Scientists are rewarded for their endeavors when the knowledge they generate contributes to society. To contribute, however, the knowledge needs to be utilized, and this normally requires transfer to others.

[1] Disease Survey Officer, Department of the Environment, Canadian Forestry Service, P.O. Box 490, Sault Ste. Marie, Ontario, P6A 5M7.

Suppliers of knowledge frequently are recipients of knowledge as well as resource support for their endeavors and other rewards. The flow of knowledge rarely proceeds in only one direction. This is communication. To proceed efficiently, technology transfer requires a good atmosphere for communication as well as a demand or desire for the technology. At some point the user needs to become part of the evolution of technology. Occasionally, user interest has to be nurtured by the supplier, and sometimes the supplier has to be made aware of the needs of the user. A mutual commitment to communicate at all levels of use speeds the transfer process, and in effect, we are all users of knowledge. Support the system in terms of conference participation, client attention, personal contacts, publication of ideas etc., and the system will respond.

BACKGROUND

The flow of forest technology in Canada is related to the economic, geographic, and political nature of the nation. The total land area is 9.9 million km^2 of which 4.4 million km^2 is forested and 2.2 million km^2 is rated as commercially productive forest land (Bonner 1982). Forest products account for about 18% of all Canadian exports, and about 13% of the manufacturing work force is employed by the forest industry (Bonner 1982). The forest is important environmentally and esthetically as well as economically. Canadians relate well to the forest, and are concerned when pests threaten its existence.

The forest extends across the continent in a contiguous zone over 1000 km in width, and host tree species for *Gremmeniella abietina* (Lagerb.) Morelet are present throughout most of the forest. Hence, *Scleroderris* canker is extremely important, especially now that the forest is threatened by incursion of the European race of the fungus, which appears to be more virulent than the North American race, which has been present since the 1950s.

Politically, Canada consists of ten provinces and two large territories. Management and dispersement of public lands was assigned to the provinces at the time of Canada's confederation in 1867 by the British North America Act. Consequently, most (80%) of the productive forest land is provincially owned. Federal and private ownership components are 11 and 8% respectively, however, the ownership structure varies. Most of the federal lands are in the northern territories and private land ownership is greatest in the Atlantic provinces. In each province, administration of forested lands is vested in a natural resource oriented agency with departmental or ministerial status. Pest control and other jurisdictional authority for forests is assigned to these agencies, and most have research groups that are concerned primarily with silviculture, reforestation, marketing and tree improvement.

Forest industries own land and operate over large tracts of land licensed from the provinces. On licensed lands certain managerial responsibilities can be assigned to industry, but overall forest policy and management are provincial functions.

The federal forestry role is assigned to the Department of the Environment within which the Canadian Forestry Service (CFS) provides scientific and technical leadership in forestry, and promotes the economic development of the Canadian forest. Forestry expertise and assistance programs are available from the provincial forestry agencies and the CFS. The flow of forest technology to the forest user is mostly through the provincial agencies, probably because these agencies administer most of the forest land and are the usual forestry contact for forest users.

In the CFS, scientific and technical leadership is vested in the six research centres distributed regionally across Canada and two national research institutes. The overall program is coordinated by a senior managerial group, in our capital city of Ottawa, which also provides a governmental advisory service. The research centres each cooperate with one or more of the provincial forestry groups.

Research is also conducted at five accredited universities in Canada, and both the federal and provincial governments support this research.

Forest pest detection and surveillance activities are conducted by the Forest Insect and Disease Survey (FIDS) of the CFS. Each of the six research centres has a FIDS unit. FIDS also provides a pest identification and extension service to all forest users. Pest detection and surveillance could logically be a provincial jurisdiction, and some provincial forestry groups are now assuming a greater role in these activities. However, when forest pest surveillance was started in the 1930s, at the request of provincial and industrial foresters, most of the forest pest expertise in Canada was then in the Canada Department of Agriculture. Interagency cooperation has been excellent, with provincial agencies often providing resource support for FIDS by way of aircraft services, building facilities, and supplementary personal. The close association of FIDS with forest managers has resulted in a liaison and extension role for FIDS in pest technology. Pathologists and entomologists assigned to FIDS units often become involved in operational tests of the pest technology developed at the research centres, and pest appraisal efforts by FIDS frequently identify research needs and epidemiological information useful to pest research.

Annual and seasonal pest reports and meetings are part of pest surveillance activities in all provinces. The status of important pests such as G. abietina is discussed, and frequently committees are set up to deal with specific pest situations. For example, in all eastern provinces there are committees coordinating plans and actions with respect to the European race of G. abietina. Workshops and operational tests of pest managment techniques are frequently generated by these committees.

There are cooperative agreements in most provinces for conducting forest research and managing pests. Probably the most formal arrangement is the Canada Ontario Joint Forestry Research Committee (COJFRC). This group, whose members represent universities, industry and government, reviews and directs support to various research projects in Ontario. COJFRC also sponsors symposia and workshops where experts are assembled to consider forest technology. Pest problems are often featured at these assemblies.

Several international meetings have provided a medium for scientific exchange of information about forest pests. The Northeastern Forest Pest Council sponsors a winter pest review and summer field trip. The Central International Forest Insect and Disease Conference (CIFIDC) and the Northeastern Forest Pathology Workshop are annual assemblies, and a United States/Canada Bilateral Study Group on Scleroderris canker meets when appropriate and coordinates research in the two countries.

HISTORICAL

Information about Scleroderris canker was transferred in North America long before G. abietina was recognized as the causal agent. The common failure of red pine plantings was discussed and observed at conferences such as CIFIDC, and investigative teams in the United States and Canada representing a variety of scientific disciplines were aware that a common disease syndrome was present for most failures. Isolation and inoculation difficulties, and the facultative nature of G. abietina and other associated fungi, prevented easy identification of G. abietina as the causal agent.

Gremmeniella abietina was ultimately identified in 1965 as the cause of the red pine malady, to be known as Scleroderris canker following isolation in 1962 and pathogenicity tests by Jorgenson (Ohman 1966). The disease was reported as the cause of extensive pine mortality in the 1965 FIDS annual report (Sippell et al., 1965), and the surge of publications about Scleroderris canker that occurred in the North American literature for 1966 and 1967 by researchers such as Ohman, Punter, Skilling, Smerlis and Teich is testimonial to a good prevailing atmosphere for technology transfer. After the pathogen was implicated, pathologists were able to communicate about isolation, incubation and inoculation technique, and this information solved numerous difficulties experienced by researchers. Also, considerable information was available from the European literature and scientific community.

Descriptions of the Scleroderris canker disease were soon published (Ohman 1966, Skilling and Cordell 1966, Punter 1967), and there has since been a series of descriptive brochures (Anon 1972, Lavallee 1972, Gross and Dorworth 1978). The presence and role of G. abietina could then be established for each problem area. CFS pathologists D. Punter and E. Smerlis worked closely with the FIDS groups on methods for detecting and rating damage caused by the disease, and G. abietina was determined to be well established in Ontario (Sippell et al., 1966, Punter 1967) and Quebec (Martineau and Ouellette 1967, Smerlis 1967). The pattern of occurrence suggested that seedlings were being infested at certain tree nurseries and that this was contributing to spread of the fungus (Punter 1967, Dorworth 1970a). Disease control recommendations were formulated for the nurseries by C. Dorwoth (1970b) who assumed the lead role in Scleroderris canker research in Ontario in 1967 when D. Punter joined the staff of the University of Manitoba. By 1970 nursery inspections and controls prescribed by Dorworth had virtually eliminated the hazard associated with diseased nursery seedlings. Most of this was accomplished by shifting red pine seedling production to unaffected nurseries. In addition, windbreaks were converted to non-host species and fungicide applications were used to control the disease. Initially fungicides known to be effective in Europe were recommended for use in Canadian nurseries. Smerlies (1976, 1979 and 1980) screened fungicides for effectiveness in controlling infection by G. abietina of seedlings in the nursery and after planting in the field. Several materials provided effective disease control. Most nurseries in eastern Canada are currently applying fungicide sprays as a precaution to prevent infection and spread by the European race of G. abietina.

The North American race of G. abietina intensified and spread in a manner typical of an introduced pathogen (Dorworth 1971). When the magnitude of the disease hazard became evident to forest managers, the inoculum source aspect of the problem was shown to be important. The fungus was spreading to the natural forest (Martineau and Lavallee 1970, Dorworth 1971, 1972a) and at that time jack pine appeared almost as susceptible to infection as red pine (Dorworth 1972a, 1973). The potential for damage by Scleroderris canker appeared awesome. Currently, the potential of the European race of G. abietina causes similar apprehension in Canada, hence, the policy of quarantine and eradication of this fungus wherever it is detected. Pathological insight used in assessing these situations is technology, and eliciting an adequate response from administrators is a vital part of the communications process. Also, managers need to be apprised of a pathogen's potential so that they can make decisions with respect to harvest schedules, species to plant, and the necessity for pest control operations.

Several factors stimulated discussion and technology transfer in the early 1970s. The distribution of G. abietina had been confined largely to areas in which infected planting stock was used, but in 1970 young stands of natural jack pine were becoming infected in Ontario and Quebec at large distances from known sources of inoculum (Martineau and Lavallee 1970, Dorworth 1971, 1972a). G. abietina was detected in New Brunswick in 1971 and in Nova Scotia in 1972 (Forbes et al., 1971, 1972). Planting of red pine had been abandoned in many locations with long histories of planting failure. At other sites the infection hazard was increasing and planting failures were becoming common. Some forest managers were attempting to control the disease by sanitizing the trees surviving in partial failures, but control efforts were not completely satisfactory. Meanwhile, pathologists were becoming more knowledgeable about disease epidemiology (Dorworth 1972b, 1973, 1974a). Certain site characters, the lack of infection above a height of 1.5 m, and recognition that an inoculum hazard often existed on naturally occurring pines (mostly jack pine) near planting sites, seemed to indicate that disease control through sanitation was possible.

Discussions leading to management strategies for coping with the Scleroderris canker disease problem were held by responsible agencies. The Ontario experience was typical of the actions taken in other provinces. Forest managers acknowledged the folly of attempting to plant red pine where there was a disease infection hazard, but they felt the hazard zones needed to be defined by pest technologists. Managers also stressed the need for control technology, and felt obligated to use the best technology then available to control the disease. In response, the Ontario FIDS program for 1972 included rating of the infection hazard of G. abietina at all locations where the fungus was known to occur. Starting in 1973 a three-year moratorium was applied to planting of pine in the 51 high hazard areas (ranging in size from 200 ha to 4000 km^2 that were defined by the survey. Three of the high hazard zones were selected for investigation of the effectiveness and cost of disease control by sanitation, and technical guidance for the tests was provided by C. Dorworth and H. Gross. The work was performed by provincial crews who also compiled cost efficiency data. Test plantings showed that controls were effective at two sites, but the third site required more sanitation work than was anticipated to remove sources of inoculum for a distance of 1 km. The results at the third site demonstrated that sanitation for a distance of 100 m was unsatisfactory, as about one-third of the test seedlings became infected.

Currently, most pathologists and foresters feel that both red pine and jack pine can be managed as timber species within the range of the North American race of G. abietina in eastern Canada. Technology for controlling the disease is available for red pine management in a field manual prepared by Dorworth (1976). Studies by Dorworth (1977, 1982) show that jack pine is more resistant to the disease than red pine and that control is not necessary for jack pine on most sites. On harsh sites or sites with high inoculum densities the control measures recommended for red pine (Dorworth 1976) seem necessary for jack pine. Another option for these sites is the use of planting stock reared from disease resistant seed sources (Yeatman 1976).

Implementation of control strategies for red pine and jack pine management has required a promotional campaign. Discussions of control technology at pest reviews and with forest managers experiencing damage caused by the disease often leads to the presentation of workshops where detection, assessment, and control technology are demonstrated. Also, FIDS and research staff frequently work with foresters when disease management programs are initiated in a district. This personal contact and commitment seems to be a necessary part of technology transfer.

Interestingly, scientific curiosity about differences in the mode of action of the fungus G. abietina in Europe and in North America resulted in the technology necessary for effective definition of races and for the imposition of quarantine (Dorworth 1974b, Dorworth and Drywienczyk 1975, Dorworth et al., 1977). Serological affinities of the New York isolates were consistent with situations where G. abietina appeared to have more virulence and where the disease syndrome was more typical of what was reported for Europe.

At the 1975 CIFIDC conference Dorworth and I first heard of the occurrence of what now is regarded as the European race of G. abietina. In 1977 we visited the area in New York where extensive damage occurred in 1976 and in 1978 we attended a workshop with pathologists from the other Canadian research centres conducted by the USDA Forest Service and the New York Department of Environmental Conservation. By then the presence of two races of G. abietina was established and quarantines were in effect. Charles Dorworth assisted the Canadian Department of Agriculture in preparing the quarantine and also conducted several workshops demonstrating field identification, symptomatology and serological technique. A series of workshops to discuss the potential of the European race on the Canadian forest and to train foresters to detect evidence of G. abietina have been conducted in all of the eastern provinces, and detection surveys by the various FIDS groups are conducted each year.

American scientists are including Canadian tree species and seed sources in resistance screening, and we Canadians are extremely interested in the research being conducted on G. abietina in the United States.

SUMMARY

Application of Scleroderris canker technology has required a personal commitment and desire to communicate by scientists, pest monitoring and control personnel, and forest managers. Contact at pest reviews, workshops, symposia and various meetings provide an opportunity to promote disease control. The history of Scleroderris canker disease control technology shows how the information has been assimilated into forest management in Canada. In retrospect, the record of transfer has been good. Technology transfer is something than can always be improved, and this is our future challenge.

LITERATURE CITED

Anon. 1972. Learn to recognize Scleroderris canker of pines. Dep. Environ., Can. For. Serv., Fredericton, New Brunswick. Leaflet.

Bonnor, G.M. 1972. Canada's Forest Inventory. 1981. Dep. Environ., Can. For. Serv., Ottawa, Ont., 79 p.

Dorworth, C.E. 1970a. Scleroderris canker in Ontario forest nurseries. Dep. Environ., Canad. For. Serv., Sault Ste. Marie, Ont., Inf. Rep. O-X-148. 9p.

Dorworth, C.E. 1970b. Scleroderris lagerbergii Gremmen in the pine replant problem in Central Ontario. Dep. Environ., Can. For. Serv., Sault Ste. Marie, Ont., Inf. Rep. O-X-139. 12 p.

Dorworth, C.E. 1971. Diseases of conifers incited by Scleroderris lagerbergii Gremmen: A review and analysis. Dep. Fish. For., Can. For. Serv., Ottawa, Ont., Publ. No. 1289. 42 p.

Dorworth, C.E. 1972a. Scleroderris lagerbergii Gremmen in the boreal forest of Ontario. Dep. Environ., Canad. For. Serv., Sault Ste. Marie, Ont., Inf. Rep. O-X-146. 9 p.

Dorworth, C.E. 1972b. Epidemiology of Scleroderris lagerbergii in central Ontario. Can. J. Bot. 50:751-765.

Dorworth, C.E. 1973. Epiphytology of Scleroderris lagerbergii in a kettle frost pocket. Eur. J. For. Pathol. 3:232-242.

Dorworth, C.E. 1974a. Disease problems in intensively managed forests -- Scleroderris lagerbergii. Possibilities for selection and breeding for resistance. Eur. J. For. Pathol. 4:228-232.

Dorworth, C.E. 1974b. Comparison of soluble proteins of Ascocalyx abietis and Gremmeniella abietina by serology and electrophoresis. Can. J. Bot. 52: 919-922.

Dorworth, C.E. 1976. Reducing damage to red pine by Gremmeniella abietina in the Great Lakes -- St. Lawrence forest region of Ontario. Dep. Environ., Canad. For. Serv., Sault Ste. Marie, Ont., Inf. Rep. O-X-252. 22 p.

Dorworth, C.E. 1977. Relative susceptiblity of red pine and jack pine to Gremmeniella abietina. Dep. Fish. Environ., Can. For. Serv., Ottawa, Ont., Bi-Mon. Res. Notes 33:6.

Dorworth, C.E. and C.N. Davis. 1982. Current and predicted future impact of the North American race of Gremmeniella abietina on jack pine in Ontario. Dep. Environ., Canad., For. Serv., Sault Ste. Marie, Ont., Inf. Rep. O-X-342. 18 p.

Dorworth, C.E. and J. Krywienczyk. 1975. Comparison among isolates of Gremmeniella abietina by means of growth rate, conidia measurement and immunogenic reaction. Can. J. Bot. 53:2506-2525.

Dorworth, C.E., J. Krywienczyk, and D.D. Skilling. 1977. New York isolates of Gremmeniella abietina (Scleroderris lagerbergii) identical in immunogenic reaction to European isolates. Plant Dis. Rep. 61:887-890.

Forbes, R.S., G.R. Underwood and G.A. Van Sickle. 1971. Maritimes Region. p. 19-33 in Annual Report of the Forest Insect and Disease Survey. Dep. Environ., Can. For. Serv., Ottawa, Ont.

Forbes, R.S., G.R. Underwood and G.A. VanSickle. 1972. Martimes Region. p. 19-32 In: Annual Report of the Forest Insect and Disease Survey. Dep. Environ., Can. For. Serv., Ottawa, Ont.

Gross, H.L. and C.E. Dorworth. 1978. Gremmeniella (Scleroderris) disease of conifers. Dep. Environ., Canad. For. Serv., Sault Ste. Marie, Ont. Leaflet.

Lavallee, A. 1972. Le chancre Scleroderrien du pin. Dep. Environ., Can. For. Serv., Inf. Bull. 3, 8 p.

Martineau, R. and G.B. Ouellette. 1967. Quebec Region. p. 34-50 In: Annual Report of the Forest Insect and Disease Survey. Can. Dep. For. Rur. Dev., For. Br., Ottawa, Ont.

Martineau, R. and A. Lavallee. 1970. Quebec Region. p. 32-48 In: Annual Report of the Forest Insect and Disease Survey. Dep. Fish. For., Can. For. Serv., Ottawa, Ont.

Ohman, J.H. 1966. Scleroderris lagerbergii Gremmen: The cause of dieback and mortality of red and jack pines in Upper Michigan plantations. Plant Dis. Rep. 50:402-405.

Punter, D. 1967. Scleroderris lagerbergii Gremmen, a new threat to nurseries in northern Ontario. For. Chron. 43:161-164.

Sippell, W.L., B.W. Dance and A.H. Rose. 1965. Ontario Region p. 44-65 In: Annual Report of the Forest Insect and Disease Survey. Dep. For., Ottawa, Ont.

Sippell, W.L., B.W. Dance and A.H. Rose. 1966. Ontario Region. p. 51-74 In: Annual Report of the Forest Insect and Disease Survey. Dep. For. Rur. Dev. For. Br., Ottawa, Ont.

Skilling, D.D. and C.E. Cordell. 1966. Scleroderris canker on national forests in Upper Michigan and Northern Wisconsin. USDA For. Ser., Res. Rap. N.C. 3. 10 p.

Smerlis, E. 1967. The occurrence and pathogenicity of Scleroderris lagerbergii Gremmen in Quebec. Plant Dis. Rep. 51:584-585.

Smerlis, E. 1976. Evaluation of fungicides for control of Gremmeniella abietina. I. Laboratory and preliminary field assays. Dep. Environ., Can. For. Serv., Ste. Foy, Quebec Inf. Rep. LAU-X-23. 34 p.

Smerlis, E. 1976. Evaluation of fungicides for control of *Gremmeniella abietina*. II. Results of 1975, 1976, 1977 field assays. Dep. Environ., Can., For. Serv., Ste. Foy, Quebec, Inf. Rep. LAU-X-38. 44 p.

Smerlis, E. 1976. Evaluation of fungicides for control of *Gremmeniella abietina*. III. Results of 1978 field assays. Dep. Environ., Can. For. Serv. Ste. Foy, Quebec, Inf. Rep. LAU-X-46. 24 p.

Yeatman, C.W. 1976. Seed Origin -- First, last and always. Dep. Environ. Can. For. Serv. Chalk River, Ont., Inf. Rep. PS-X-64. 12 p.

TECHNOLOGY TRANSFER OF SCLERODERRIS RESEARCH RESULTS

Thomas H. Nicholls[1]

ABSTRACT

One way to minimize the impact of Scleroderris is to make sure that research results get to people who will implement them. To do this, users and their needs must be identified and research information must be packaged in a format that can be easily understood and used. Development of a technology transfer plan to accomplish this objective is discussed.

INTRODUCTION

One of the best ways to minimize the impact of Scleroderris canker is to make sure that research results get to the people who will apply them. Without application, the information developed by researchers is meaningless. Good information delivery systems are needed to distribute and implement research findings to users in a systematic, understandable way.

A major problem with most research is that a vast amount of knowledge is developed but not disseminated. In general, more is known about generating information than about transferring it to users and getting it applied. Thousands and sometimes millions of dollars are spent doing research, but often little money is spent on packaging the results in a way that will ensure the results will be put to use to solve problems. To avoid this situation, technology transfer (TT) planning and funding must be included in the research planning process. This paper focuses on the importance of identifying the audiences who will utilize the new knowledge developed by research. It also discusses the importance of developing an effective TT plan that will help researchers get their results applied.

FOREST SERVICE SCLERODERRIS RESEARCH

In 1978, the Forest Disease Research Project of the North Central Forest Experiment Station was given the USDA Service responsibility to study Scleroderris in the United States. To accelerate Scleroderris research, Cooperative Agreements totaling $350,000 were funded by the Forest Service with universities and with the New York Department of Environmental Conservation. In addition, a Forest Service team led by Darroll Skilling, Principal Plant Pathologist, has been conducting research on the life cycle of the pathogen, chemical controls, host range susceptiblity and genetic resistance. This team also

---

1 Principal Plant Pathologist, North Central Forest Experiment Station, USDA Forest Service, 1991 Folwell Avenue, St. Paul, MN 55108

helped develop serological techniques that can be used to distinguish the North American and European strains of the fungus. The money for this research was appropriated by Congress at the request of several groups that were or could be affected by Scleroderris, potentially one of the most important forest diseases in North America.

As a result of this research, a great deal of new knowledge has been acquired during the past five years. We now have a major responsibility to report our findings to the people we serve. This is important because we are accountable to these people. Research scientists need to document outputs and benefits of their research. We need to let users and administrators know that forest pathology research does payoff! Also, it may help us justify future research expenditures if we can document significant past accomplishments.

SCLERODERRIS TECHNOLOGY TRANSFER PLAN

To make sure that Scleroderris research results are used, the North Central Forest Experiment Station developed a TT plan. Its objective is to transfer new knowledge and technology needed to prevent or reduce damage from Scleroderris canker. This information will enable forest managers to establish conifer forests that can survive with minimal impact from Scleroderris. This information will also help prevent introduction of the disease into new areas. The TT plan has six parts. This Symposium and its published proceedings are one part of the plan. We invited all of the Forest Service cooperators to summarize their research results and to present and publish them in the Symposium proceedings. In addition, we invited other forest pathologists who have been active in Scleroderris work to report on their recent findings. The resulting Symposium publication will be a collection of the most current research papers on this disease. It will be an important reference tool, but one that will be used primarily by scientists.

Researchers have generally done a good job of communicating with each other, and they still need to do this, but they often have done a poor job of communicating with practitioners. Therefore, we recognize that the Symposium Proceedings should not be the only step in the Scleroderris TT process. We still have an obligation to package information in formats that can be used by Christmas tree growers, nursery managers, extension personnel, foresters, administrators, students, and the general public. One of the best ways the impact of Scleroderris can be minimized, or its spread slowed, is by ensuring that the "doers" are made aware of and receive the information they need in a usable format to solve or prevent Scleroderris problems.

To meet the needs of practitioners, the Forest Service Scleroderris TT plan calls for the development of a "How To" publication that will describe all known Scleroderris controls. This "How To" will be aimed at helping forest managers decide which control or prevention procedures are best for them. The success of the Forest Service "How To" type publication has been well documented (Nicholls and Prey 1982). For example, our Forest Pathology Project has produced 25 "How To" publications. They have been received well by many different audiences as indicated by the 468,000 copies printed and the 26 reprintings required to fill the demand. The "How To Identify Scleroderris Canker" publication alone has been reprinted twice with a total of 45,000 copies printed (Skilling and O'Brien 1979).

The Scleroderris TT plan also includes development of a slide-tape program, a poster, and workshops that can be used to help train people in identifying and controlling Scleroderris canker. To inform the general public interested in forestry, an article for AMERICAN FORESTS has been written. These kinds of TT planning efforts are helping the Forest Service accomplish the SERVICE part of its mission. I would like to share the Forest Service TT planning process with you in hopes that it will help you plan for the dissemination of your research results to appropriate user groups.

TECHNOLOGY TRANSFER PLANNING PROCESS

The main objective of a TT plan is to make sure that developed knowledge is transferred to potential users. The success of the TT process depends upon user acceptance of developed information. No matter how well research is packaged or advertised, if it doesn't meet user's needs, it won't be used. Therefore, it is important that communication be good between researcher and research client. This will minimize the difficulties of getting research findings into use. It will also ensure that research efforts are more relevant and will generate answers in a form usable for solving the problems.

The TT goal is to deliver useful research results to those that need it using media that will ensure that the information is easily understood and put into use in the shortest time possible. Many factors can help or hinder TT efforts (Nicholls and Prey 1982). Awareness of such factors can determine the success or failure of any TT effort.

H.G. Marx (1980) wrote a guide to help people develop a TT plan. I would like to briefly summarize its major points. A good TT plan need not be complicated. It should be short, simple, and direct. The Forest Service's TT process consists of the following eight steps: (1) identify the technology available and needing transfer; (2) identify the target user group(s) to whom the technology or information is to be transferred; (3) develop an objective and formal or informal plan of application; (4) package the knowledge or technology for easy understanding; (5) select the most effective media for transfer; (6) directly involve scientists and/or specialists with users, especially innovators and adoptors; (7) arrange for trouble shooting and feedback and; (8) evaluate the process and results.

A simple TT plan can be delevoped to include the following items as outlined in Marx's publication: (1) brief title; (2) message -- what is being transferred; (3) audience -- expected users; (4) objective -- expected accomplishments; (5) team -identify people and skills needed; (6) media -- select media that will best reach users; (7) budget -- what will it cost in dollars and time; (8) evaluation -- assessment of program; (9) assignments -- list assignments and time table for carrying out action and; (10) approval -- signed by appropriate officials. Once approved, it is easier to obtain funding and other support to carry out the plan. Even though there is more than one way to plan applications of technology, the procedures mentioned have been successfully used to implement forestry research results by the Forest Service.

CONCLUSION

Most of the emphasis on Scleroderris has been on producing research results. The challenge I give you is to place significant effort into identifying user groups and their needs and packaging, disseminating, and implementing usable research results. Only then will we have completed the research life cycle, served the needs of our clients, and shown that forest pathology research does payoff!

LITERATURE CITED

Marx, H.G. 1980. Guide to help develop a technology transfer plan. U.S.D.A., Forest Service. Washington, D.C. 26 p.

Nicholls, T.H. and A.J. Prey. 1982. Providing information: researchers to practitioners. In: Parks, B. (ed.) Urban and suburban trees: Pest problems, needs, prospects, and solutions. Symp. Proc. Michigan State University, East Lansing, MI. 156-162 p.

Skilling, D.D. and J.T. O'Brien. 1979. How to identify Scleroderris canker. U.S.D.A., Forest Service, North Cent. For. Exp. Stn., St. Paul, MN. 6 p.

OVERVIEW AND SUMMARY OF THE SCLERODERRIS CANKER SYMPOSIUM AND FUTURE RESEARCH NEEDS

Paul D. Manion and Darroll D. Skilling[1]

Scleroderris canker caused by the fungus Gremmeniella abietina has caused major damage to pine and spruce forests in Europe since the original report by Brunchorst in 1888. The disease seems to have been most damaging to Pinus nigra, P. sylvestris, P. cembra and other conifer stands especially when the trees were planted as exotics. Disease epidemics have been periodic throughout the European countries and epidemics appear to be primarily related to years with favorable climatic conditions for infection combined with the availability of highly susceptible hosts.

In many situations, the disease is best described as a dieback rather than a canker. In the dieback situation, conidia (Brunchorstia pinea) are the primary inoculum source. Current year's shoots are infected and killed throughout the tree crowns.

The North American experience with Scleroderris canker is much more recent. The first damage by Scleroderris was reported about 1951. The causal organism was not identified until the early 1960's. By that time, the fungus was well established in Ontario and the northern portion of the Lake States. The disease developed near forest nurseries, pine nursery stock became infected, and spread was very rapid in both Canada and the Lake States until about 1957. Once the role of infected nursery stock was determined, sanitation of hedgerows and complete fungicide spray programs in infected nurseries drastically reduced the spread of this disease. Because the fungus did not infect trees taller than 2 m, plantations of red and jack pine, that were able to remain free from this disease during their first 5 years, suffered very little mortality.

The historical and present status reports of the Symposium provide an up-to-date picture of this disease in the provinces of Canada (Dorworth 1983, Sterner 1983, Gross 1983, Lachance 1983, Magasi 1983), the United States (O'Brien 1983), and Europe (Gibbs 1983). The Canadian and United States situation shows the similarities of the disease in North America and the high degree of cooperation between the two countries that has developed to understand, contain, and minimize losses to this disease. The European historical and current experience presented by each of our European guests is an invaluable perspective that we in North America very much need and appreciate.

---

1 Professor, SUNY College of Environmental Science and Forestry, Syracuse, NY 13210 and Principal Plant Pathologist, North Central Forest Experiment Station, USDA Forest Service, St. Paul, MN 55108.

The development of a serious disease epidemic in the mid 1970's in New York focused attention on a new facet of the Scleroderris disease in North America. Death of large trees caused people to consider Scleroderris canker in a different perspective. Identification of a serologically different strain of the fungus, causing significant losses, initiated quarantine regulation and extensive research efforts to contain and better understand the problem. A situation similar to New York has recently developed in Newfoundland (Laidlaw 1983, Sterner 1983).

In 1977, Dorworth et al. characterized the strain of the fungus from New York as European using the strain identification procedure described two years earlier (Dorworth and Krywienczyk 1975). Now, we have numerous strain, race, or serotype designations that confuse both research specialists and anyone else interested in the Scleroderris problem. It, therefore, seems appropriate to provide some perspective on the _G. abietina_ race topic. A primary point of this topic is a lack of agreement on what to call the subdivisions of the _G. abietina_ fungus. The terms race, strains, serotype, pathotype, variety, and group are used interchangeably.

The initial separation of three races, (North American, European and Asian) by Dorworth and Krywienczyk (1975) was done on the basis of serological and morphological characteristics of the cultures. Dorworth et al. (1977) characterized the New York strain as identical serologically to the isolates from Finland. Wendler et al. (1980) refocused attention on the unique characteristic of the northeastern _G. abietina_ by describing a Vermont intermediate on the basis of a more refined serological procedure than had been previously utilized. Other "intermediate" types are mentioned by Magasi (1983) from New Brunswick and O'Brien (1983) from Maine. Setliff and Manion (1980), examined morphological growth characteristics of numerous conidial and ascospore isolates of _G. abietina_ from New York in comparison with standard cultures from the Lake States, Ontario and Vermont. They also had the cultures serotyped by the USDA Forest Service Lab at St. Paul, MN and the University of Vermont. Neither laboratory knew the origin of the specific isolates they were testing until after they had made a judgment. A clear difference of opinion became evident. Isolates from New York typed as European by the Forest Service lab were typed as Intermediate by the Vermont Lab. Our cultural observations likewise typed the New York isolates derived from conidia as similar to the Vermont intermediate and different from most of the Lake States, Ontario and European isolates. It is impossible to clearly determine if the above mentioned "intermediates" and the intermediate (Skilling 1981) or "hybrid" used by Furnier et al. (1983) from New York are identical.

Interestingly enough, ascospore isolates are extremely variable and can have cultural characteristics identical to standard North American, European, or intermediate isolates. Furnier et al. (1983) found that individual ascospore derived cultures, from the same apothecium, varied serologically. Both European and North American serotypes were found in the same apothecium. Furnier et al. (1983) described apothecium from New York, but Setliff and Manion (unpublished) found a similar situation in an isolate from Wisconsin.

A number of authors in this Symposium proceedings showed additional evidence for further subdivision of _G. abietina_ isolates. Gibbs (1983), Capretti (1983), Donaubauer (1983) and others, all make reference to more than one race in Europe. These subdivisions are based on numbers of septations in the conidia and relative abundance of the perfect stage. Morelet (1980) characterized two varieties, Brunchorstia pinea var. tipica and B. pinea var. cembrae alpine strains of the imperfect stage of the fungus. Yokota (1983) describes at least two strains of G. abietina from Japan on the basis of host range. Clearly there is no single European or Asian race of the fungus.

In North America, Benhamou et al. (1983) separated the isolates into six groups on the basis of electrophoretic protein band patterns. All but one New York isolate and numerous pine isolates from Quebec, grouped with the European standards. A North American standard group was also identified. Additional groups show the complexity of the strain designation. One pine isolate from Quebec serotyped North American, two pine isolates typed European, another non-serotyped pine isolate from Quebec, and four pine isolates from New Brunswick were different enough to be set in a group separate from North American, or European. Ayer and Hiratsuka (1983) using high pressure liquid chromatography compared metabolites of various cultures. The New Brunswick strain they used was previously serologically typed European but in their tests it produced a metabolite spectrum very similar to the standard North American type culture from Ontario. The spectra of these two Canadian cultures was different from a New York isolate and a Swedish isolate.

These examples are provided to illustrate the imperfections and inconsistencies in the present race or strain concepts. If one further attempts to assign greater or lesser virulence qualities to these poorly defined races, the credibility of the science is stretched too far. Clearly, there is a serious need to properly define the race or strain structure for G. abietina and to develop a system that appropriately fits the complex collection of isolates that occur throughout the worldwide range of the pathogen.

The simplistic approach of ascribing the mid 1970's Scleroderris epidemic in New York to a "new virulent European strain" should remain as a hypothesis but other hypotheses should be carefully considered and likewise tested.

The experiences in Europe suggest over and over again that environmental and host factors play prominant roles in epidemic development. Carvin (1980) did a preliminary evaluation of weather records for the 1970's in New York in relation to disease development. The early part of the decade can be characterized as having cooler than normal summers while the later part of the decade had more normal or warmer summers. The European situation would suggest that these early 1970's conditions were ideal for development of an epidemic problem. A detailed analysis of weather records in relation to disease development in New York and Newfoundland would surely be appropriate in order to predict possible future problems.

The European experience with Scleroderris canker demonstrates another general characteristic of this disease. Kurkela (1983), Capretti (1983), Donaubauer (1983) and Gibbs (1983) all point out the relationship between off-site climatically stressed trees and epidemics. Trees from non-local seed sources are often environmentally stressed and more prone to develop Brunchorstia dieback. Locally adapted seed sources and natural populations may be affected by Scleroderris canker but the dieback phase does not develop. The proximity of native and introduced tree population affect each other. The native population may be the source for initial inoculum on the introduced population. The epidemic in the introduced population may likewise spill over to the native population. Canadian experience with jack pine shows differences in disease severity with different seed sources although the climatic adaptability of the seed sources does not seem to be as apparent as with the European situation.

No one has seriously considered the possible role of non-local seed sources and stressed tree populations in the New York dieback problem on red and Scots pine. Plantings in New York of both species are from non-local seed sources. It would seem appropriate to properly assess the effects of seed sources on the disease problem in New York. The present studies of Skilling and Riemenschneider (1983) show that most seed sources of red and Scots pine are killed by high inoculum loads of the fungus. This shows that there are no highly resistant lines. It does not demonstrate if there are differences in the various seed sources which can affect an epidemic of the pathogen. What we really need to know is if there are seed sources of red pines that can be grown in the Scleroderris infested regions that may become cankered but do not develop into an epidemic dieback problem. Even in the absence of extensive experimental evidence, forest managers contemplating large-scale plantings of pines should be very conscious of the importance of using local seed sources.

The Symposium brings together considerable detailed information on the biology of G. abietina. Furnier et al. (1983) demonstrated a heterogenic incompatibility system with no major barriers for nuclear or cytoplasmic exchange. This system provides no barrier to inbreeding thereby allowing isolated populations to develop into distictive races. The fungus with no barrier to cytoplasmic or nuclear exchange is provided with an ability to adjust to varying environmental and host conditions even in the absence of sexual cycle. This type of information demonstrates the potential for future problems from this pathogen through movement of infected material with differing genetic potential. The quarantines on movement of Scleroderris infected material are clearly an appropriate stop-gap measure. The genetics of the fungus also suggest that future epidemics might also develop from genetic rearrangement with existing populations. As such, there is no reason to expect that epidemics in various parts of the world will be caused by the same race of the pathogen.

Our imperfect understanding of inoculum dispersal and its role in epidemics is exemplified in the papers of Bergdahl (1983), Bergdahl and Ward (1983) and Luley and Manion (1983). Inoculum moves and contributes to infection differently in different plantations. In some situations, it is possible to reduce local inoculum through pruning infected branches. The impact of region-wide inoculum and the concentrating of inoculum through specific weather patterns is poorly understood. These papers just begin to scratch the surface of this topic.

The details of the infection process and environmental factors influencing invasion are brought out in the two papers by Blenis, Patton, and Spear (1983). Their work building on the previous work of Siepmann (1976) and Lang and Schutt (1974) show that initial infection through stomata on bracts subtending needles causes a sparse mycelial colonization of the bract tissue, which crosses the periderm and invades cortical tissues after some poorly defined environmental conditions and period. The specific details on stimulatory and inhibitory environmental condition affecting invasion need further investigation to be able to properly design inoculation experiments and also to better understand the role of environmental conditions on disease epidemic development.

The possible role of acid rain, SO_2 air pollution, insects, and bacteria in Scleroderris canker development were investigated by Barkland et al. (1983), Bragg and Manion (1983), Laurence et al. (1983), Lanier et al. (1983) and Knudsen and Hudler (1983). Leakage of ions due to acid rain and associated effects on spore germination suggest a hypothesis involving on interaction of acid rain and Scleroderris canker (Barkland et al 1983). Direct acid rain treatments, bark pH measurements, and spore germination tests suggest no significant impact of acid rain or a slight suppression of Scleroderris canker by acid rain (Bragg and Manion 1983). Scleroderris canker interaction with SO_2 pollution showed a suppression of the disease in relation to increasing concentration of pollutant (Laurence et al. 1983). Although no specific insects could be found that likely contribute to Scleroderris canker incidence, the possible role of spittlebugs in producing wounds was suggested. Treatment with the systemic insecticide Furadan significantly reduced Scleroderris canker infection suggesting that insects in some way affect disease development (Lanier et al. 1983). Experiments with epiphytic bacteria that show effects on G. abietina under controlled conditions were difficult to duplicate in the outdoor environment (Knudsen and Hudler 1983). These types of studies may uncover a route to direct disease control or a better understanding of what happens when the disease gets out of control. They only partially examine the many other possible interacting factors of this complex disease.

The quantification of disease increase over time shows the wildfire explosive potential of the disease (Manion and Schneider 1983) and the up and down nature to the disease (Carvin and Manion 1983). This type of data provides a quantitative means for testing experimental treatments as well as monitoring of factors affecting disease development on local and regional scales.

The impact of Scleroderris infection on growth and yield were topics addressed earlier by Dorworth (1979), Teich (1972) and again by Kurkela (1983). An understanding of the disease progress in relation to management objectives generates a decision making process for resource managers. The question is not one of totally eliminating the disease but rather one of quantifying the disease impact in relation to quantified growth and yield projections to provide a sound basis for management decisions.

Skilling and Riemenschneider (1983) enumerated a long list of hosts for the New York source of the pathogen through high inoculum dose testing of seedlings. These studies show the differing effects of environment between test sites which complicate the process of rating hosts for susceptibility. The information is useful in looking at worse case situations but totally misses the real life situation. Consequently comparison of seed sources of red and Scots pine show minimal differences. A more realistic inoculum level is needed to properly assess these tests and also the tests to determine appropriate fungicidal control procedures for nurseries (Skilling et al. 1983).

Stephan et al. (1983) Yeatman (1983) and Martinsson (1983) working with clonal or progeny test plantation, demonstrated genetic difference to disease development. These studies show the advantage of genetic resistance in reducing disease losses. They also point out the interaction of genetics and site factors. Many additional progeny and clonal tests are clearly needed for other species and locations to identify proper matches between planting sites and tree species. Additional advantages will be gained from these progeny tests on that other diseases and insect problems can also be properly evaluated and minimized by proper selection.

Stephan et al. (1983) utilized a clonal test plantation to look for chemical markers related to resistance. This type of approach, if successful, will greatly speed up the identification of the more disease tolerant plants.

The identification of diseased plantations and quantification of disease losses in the forest is a difficult, time-consuming task. Regulatory agencies attempt to put realistic boundaries on quarantined areas and keep up with expansion of the diseases beyond the quarantine boundaries (Laidlaw 1983, Barrett 1983). Hopkins and Abrahamson (1983) developed an aerial photography survey technique to facilitate the assessment process. The integration of aerial photographic procedure with disease assessment needs should be developed.

Gross (1983) and Nicholls (1983) emphasize the need and procedures for transferring of technical information such as has been presented at the Symposium to the users. This Symposium proceedings is just one of the many avenues for dissemination of this information. It is hoped that these proceedings will provide an up-to-date state-of-the-art document on Scleroderris canker and also provide some perspective of where we go from here.

LITERATURE CITED

Ayer, W.A. and U. Hiratsuka, 1983. Metabolites of Gremmiella abietina.

Barklund, P., G. Axelsson and T. Unestam. 1983. Gremmeniella abietina in Norway spruce, latent infection, sudden outbreaks, acid rain, predisposition. p. 111-113.

Barrett, R.O. 1983. The use of a quarantine to control Scleroderris canker in New York. p. 243-247.

Benhamou, N., G.B. Ouelette, A. Asselin and E. Maicas. 1983. The use of polyacrylamide gell electrophoresis for rapid differentiation of Gremmiella abietina isolates. p. 68-76.

Bergdahl, D.R. 1983. Dispersal of conidia of Gremmeniella abietina related to weather. p. 77-81.

Bergdahl, D.R. and T.M. Ward. 1983. Pruning as a silvicultural tool in the management of Pinus resinosa infected with Gremmeniella abietina. p. 166-176.

Blenis, P.V., R.F. Patton and R.N. Spear. 1983. Effect of environmental factors on the post-infection behavior of Gremmeniella abietina. p. 104-110.

Bragg, R.J. and P.D. Manion. 1983. Evaluation of possible effects of acid rain on Scleroderris canker of red pine in New Yorkk. p. 130-141.

Brunchorst, J. 1888. Uber eine neue verheerende Krankheit der Schwarzfohre (Pinus austriaca Hoss). Bergens Museums Aarsberetning, VI 1887:1-16.

Capretti, P. 1983. The movement of Gremmeniella abietina from the Alps to the Appenines. p. 153-157.

Carven, R.P. 1980. Factors affecting intensification of Scleroderris canker in young red pine plantations. M.S. Thesis, State University of New York, College of Environmental Science and Forestry, Syracuse, N.Y. 43 p.

Carven R.P. and P.D. Manion. 1983. Five-year survey of Scleroderris canker in young red pine plantations. p. 142-147.

Donaubauer, E. 1983. Experiences with Scleroderris canker on Pinus cembra L. in afforestations of high altitude. p. 158-161.

Dorworth, C.E. History of Scleroderris canker in Canada. Scleroderris canker of Conifers, Proceedings. p. 5-7.

Dorworth, C.E. 1979. Stand reduction of red pine by Gremmeniella abietina. Can. J. For. Res. 9:36-322.

Dorwroth, C.E. and J. Krywienczyk. 1975. Comparison among isolates of Gremmeniella abietina by means of growth rate, conidia measurement and immunogenic reaction. Can. J. Bot. 53:2506-2525.

Dorworth, C.E., J. Krywienczyk and D.D. Skilling. 1977. New York isolates of Gremmeniella abietina (Scleroderris lagerbergii) identical in immunogenic reaction to European isolates. Plant Dis. Rep. 61:887-890.

Furnier, G.R., N.A. Anderson and D. Skilling. 1983. Genetic variation in Gremmeniella abietina. p. 59-67.

Gibbs, J.N. 1983. Brunchorstia dieback in Europe. p. 32-41.

Gross, H.L. 1983. Scleroderris canker in the province of Ontario, Canada. p. 11-15.

Gross, H.L. 1983. The transfer of Scleroderris technology in Canada. p. 248-256.

Hopkins, P.F. and L.P. Abrahamson. 1983. Scleroderris canker survey using aerial photography. p. 226-235.

Knudsen, G.R. and G.W. Hudler, 1983. Interactions between epiphytic bacteria and conidia of Gremmeniella abietina. p. 217-225.

Kurkela, T. 1983. Factors affecting the development of disease epidemics by Gremmeniella abietina. p. 148-152.

Kurkela, T. 1983. Growth of trees affected by Gremmeniella abietina. p. 177-180.

Lachance, D. 1983. Scleroderris canker in the province of Quebec, Canada. p. 16-20.

Luley, C.J. and P.D. Manion. 1983. Inoculum potential of Gremmeniella abietina in New York. p. 82-95.

Lang, K.J. and P. Schutt. 1974. Anatomische Untersuchungen zur Infectionsbiologie von Scloderris lagerbergii Gr. [Brunchorstia pinea (Karst.) von Hohn.] Eur. J. For. Pathol. 4:166-164.

Lanier, G.N., L.P. Abrahamson and H. Schoenek. 1983. The role of insects in the dynamics of Scleroderris canker in New York. p. 114-121.

Laurence, J.A., K.L. Reynolds, D.C. MacLean, Jr., G.W. Hudler, and L.S. Dochinger. 1983. Effects of sulfur dioxide on infection of red pine by Gremmeniella abietina. p. 122-129.

Magasi, L.P. 1983. Scleroderris canker in the Maritine Provinces of Canada. p. 21-24.

Manion, P.D. and B. Schneider. 1983. Disease increase rates for Scleroderris canker in Lewis County, New York. p. 162-165.

Martinsson, O. 1983. Resistance of lodgepole pine to Scleroderris canker in northern Sweden. p. 207-211.

Morelet, M. 1980. La maladie a Brunchorstia 1. Position systematique et nomenclature du pathogene. Eur. J. For. Pathol. 10:268-277.

Nicholls, T.H. 1983. Technology transfer of Scleroderris research results. p. 257-260.

O'Brien, J.T. 1983. Historical and current Scleroderris situations in the United States. p. 26-31.

Patton, R.F., R.N. Spear, and P.V. Blenis. 1983. The infection process in pines by Gremmeniella abietina. p. 96-103.

Setliff, E.C. and P.D. Manion. 1980. Forest pathology and acid rain. In Effects of Acid precipitation on a forest ecosystem. New York State Energy Research and Development Authority, Albany, N.Y.

Siepmann, R. 1976. Ein Beitrag zur infektionsbiologie des durch Scleroderris lagerbergii verursachten Schwarzkieferntriebsterbens. Eur. J. For. Pathol. 6:103-109.

Skilling, D.D. 1981. Development of an intermediate strain of Gremmeniella abietina in New York. Phytopathology 71:255.

Skilling, D.D. and D.E. Riemenschneider. 1983. Screening conifers for resistance to Gremmeniella abietina. p. 189-196.

Stephen, B.R., R. Scholz, and U.P. Singh. 1983. Physiological and biochemical factors in Austrian pine clones with different susceptibility to Gremmeniella abietina. p. 181-188.

Sterner, T.E. 1983. Scleroderris canker, Gremmeniella abietina (Lagerb.) Morelet, in Canada -- An overview. p. 8-10.

Teich, A.H. 1972. Damage caused by the fungus Scloerderris in plantations and nurseries in Europe and North America. Eur. F. For. Pathol. 2:11-15.

Wendler, P.B., A.R. Gotlieb, and D.R. Bergdahl. 1980. A newly discovered serotype of *Gremmeniella abietina*. Can. J. Bot. 58:1923-1928.

Yeatman, C.W. 1983. Response in Canada of jack pine provenances to *Gremmeniella abietina*. p. 197-206.

Yokota, S. 1983. Pathogenicity and host range of races of *Gremmeniella abietina* in Hokkaido. p. 47-53.

INDEX

GPSR Compliance

The European Union's (EU) General Product Safety Regulation (GPSR) is a set of rules that requires consumer products to be safe and our obligations to ensure this.

If you have any concerns about our products, you can contact us on ProductSafety@springernature.com

In case Publisher is established outside the EU, the EU authorized representative is:

Springer Nature Customer Service Center GmbH
Europaplatz 3
69115 Heidelberg, Germany

Batch number: 10761697

Printed by Printforce, the Netherlands